Proceedings of the 3rd International EFC Workshop on Microbial Corrosion

EARLIER VOLUMES IN THIS SERIES

Number 1   **Corrosion in the Nuclear Industry**
Prepared by the Working Party on Nuclear Corrosion

Number 2   **Practical Corrosion Principles: a manual of corrosion experiments**
Prepared by the Working Party on Corrosion Education

Number 3   **General Guidelines for Corrosion Testing of Materials for Marine Applications**
Prepared by the Working Party on Marine Corrosion

Number 4   **Guidelines on Electrochemical Corrosion Measurements**
Prepared by the Working Party on Physico-Chemical Methods of Corrosion Testing

Number 5   **Illustrated Case Histories of Marine Corrosion**
Prepared by the Working Party on Marine Corrosion

Number 6   **Corrosion Education Manual**
Prepared by the Working Party on Corrosion Education

Number 7   **Corrosion Problems Related to Nuclear Waste Disposal**
Prepared by the Working Party on Nuclear Corrosion

Number 8   **Microbial Corrosion**
Prepared by the Working Party on Microbial Corrosion

Number 9   **Microbiological Degradation of Materials — and Methods of Protection**
Prepared by the Working Party on Microbial Corrosion

Number 10 **Marine Corrosion of Stainless Steels: Chlorination and Microbial Effects**
Prepared by the Working Party on Marine Corrosion

Number 11 **Corrosion Inhibitors**
Prepared by the Working Party on Inhibitors

Number 12 **Modifications of Passive Films**
Prepared by the Working Party on Surface Science and Mechanisms of Corrosion and Protection

Number 13 **Predicting CO_2 Corrosion in the Oil and Gas Industry**
Prepared by the Working Party on Oil and Gas

Number 14 **Guidelines for Methods of Testing and Research in High Temperature Corrosion**
Prepared by the Working Party on Corrosion by Hot Gases and Combustion Products

European Federation of Corrosion Publications
NUMBER 15

Microbial Corrosion

Proceedings of the 3rd International EFC Workshop

Edited by
A. K. Tiller AND *C. A. C. Sequeira*

Published for the European Federation of Corrosion by The Institute of Materials

THE INSTITUTE OF MATERIALS
1995

Book Number 591
Published in 1995 by The Institute of Materials
1 Carlton House Terrace, London SW1Y 5DB

© 1995 The Institute of Materials

All rights reserved

British Library Cataloguing in Publication Data
Available on application

ISBN 0-901716-62-6

Neither the EFC nor The Institute of Materials
is responsible for any views expressed
which are the sole responsibility of the authors

Design and production by
PicA Publishing Services,
Drayton, Nr Abingdon, Oxon

Made and printed in Great Britain

Contents

Series Introduction ix

Preface xii

Part 1 **An Interdisciplinary Approach to Microbially Influenced** 1
Corrosion of Copper

1 An Interdisciplinary Approach for Microbially Influenced Corrosion 3
   of Copper
   A. H. L. Chamberlain, W. R. Fischer, U. Hinze, H. H. Paradies,
   C. A. C. Sequeira, H. Siedlarek, M. Thies, D. Wagner and J. N. Wardell

2 Physical Behaviour of Biopolymers as Artificial Models for 17
   Biofilms in Biodeterioration of Copper. Solution and Surface
   Properties of Biopolymers
   M. Thies, U. Hinze and H. H. Paradies

3 Bacteria Associated with MIC of Copper: Characterisation 49
   and Extracellular Polymer Production
   J. N. Wardell and A. H. L. Chamberlain

4 Membrane Properties of Biopolymeric Substances 64
   C. A. C. Sequeira, A. C. P. R. P. Carrasco, D. Wagner,
   M. Tietz and W. R. Fischer

5 Corrosion Behaviour of Biopolymer Modified Copper Electrodes 85
   D. Wagner, H. Siedlarek, W. R. Fischer, J. N. Wardell
   and A. H. L. Chamberlain

Part 2 **Microbial Corrosion: Mechanisms and General Studies** 105

6 Contribution of Microbiological Phenomena in the Localised 107
   Corrosion of Stainless Steels
   F. Colin, M. J. Jourdain, G. D'Ambrosio, A. Pourbaix and D. Noel

7 Attachment of *Desulfovibrio vulgaris* to Steels: Influence of Alloying 119
   Elements
   D. Feron

8	Electrochemical and Surface Analytical Evaluation of Marine Copper Corrosion B. J. Little, P. A. Wagner, K. R. Hart, R. I. Ray, D. M. Lavoie, W. E. O'Grady and P. P. Trzaskoma	135
9	Microbial Degradation of Fibre Reinforced Polymer Composites P. A. Wagner, R I. Ray, B. J. Little and W. C. Tucker	143
10	The Modelling of Microbial Soil Corrosion on Iron Oxides and Hydroxides T. S. Gendler, A. A. Novakova and L. E. Ilyina	152
11	The Influence of Corrosion Experiments on Microorganisms and Biofilms in the Sessile Phase D. Wagner, J. T. Walker, W. R. Fischer and C. W. Keevil	158
12	The influence of Metal Ions on the Activity of Hydrogenase in Sulphate Reducing Bacteria C. W. S. Cheung and I. B. Beech	169
13	A Kinetic Model for Bactericidal Action in Biofilms P. M. Gaylarde and C. C. Gaylarde	181
14	Biocorrosion of Mild Steel by Sulphate Reducing Bacteria I. T. E. Fonseca, A. R. Lino and V. L. Rainha	188
15	Effect of Marine Biofilms on High Performance Stainless Steels Exposed in European Coastal Waters J.-P. Audouard, C. Compere, N. J. E. Dowling, D. Feron, D. Festy, A. Mollica, T. Rogne, V. Scotto, U. Steinsmo, K. Taxen and D. Thierry	198
16	Detection and Characterisation of Biofilms in Natural Seawater by Analysing Oxygen Diffusion under Controlled Hydrodynamic Conditions A. Ambari, B. Tribollet, C. Compere, D. Festy and E. L'Hostis	211
17	Determination of Biofilm on Stainless Steel in Seawater in Relation to the Season by Analysing the Mass Transport of Oxygen B. Tribollet, C. Compere, F. Darrieux and D. Festy	223
18	The Importance of Bacterially Generated Hydrogen Permeation Through Metals J. Benson and R. G. J. Edyvean	233

Part 3   Microbial Corrosion: Case Studies 241

19 Influence of Metal–Biofilm Interface pH on Aluminium Brass 243
Corrosion in Seawater
P. Cristiani, F. Mazza and G. Rocchini

20 Effect of Seasonal Changes in Water Quality on Biofouling and 261
Corrosion in Fresh Water Systems
R. P. George, P. Muraleedharan, J. B. Gnanamoorthy, T. S. Rao
and K. V. K. Nair

21 Corrosion Behaviour of a Carbon Steel Valve in a Microbial 276
Environment
J. C. Danko, C. D. Lundin, N. J. E. Dowling and W. Hester

22 Sulphide-producing, not Sulphate-reducing Anaerobic Bacteria 293
Presumptively Involved in Bacterial Corrosion
M. Magot, L. Carreau, J.-L. Cayol, B. Ollivier and J.-L. Crolet

23 Biofilm Monitoring and On-line Control: 20 Month Experience 301
in Seawater
G. Salvago, G. Fumagalli, P. Cristiani and G. Rocchini

24 Challenges to the Prediction and Monitoring of Microbially 314
Influenced Corrosion in the Oil Industry
T. S. Whitham

25 Biofilm Development on Stainless Steels in a Potable Water System 322
L. Hanjangsit, I. B. Beech, R. G. J. Edyvean and C. Hammond

26 Bacteria and Corrosion in Potable Water Mains 328
I. B. Beech, R. G. J. Edyvean, C. W. S. Cheung
and A. Turner

27 Study of Corrosion Layer Products from an Archaeological 338
Iron Nail
A. A. Novakova, T. S. Gendler and N. D. Manyurova

Part 4   Microbial Corrosion: Prevention and Control 347

28 Electrochemical Sensor of Water and of the Amount of Microbial 349
Proliferation in Fuel Tanks
B. M. Rosales

29 Biocorrosion in Groundwater Engineering Systems 354
P. Howsam, A. K. Tiller and B. Tyrrell

30 Laboratory Evaluation of the Effectiveness of Cathodic Protection 367
in the Presence of Sulphate Reducing Bacteria
K. Kasahara, K. Okamura and F. Kajiyama

31 Studies on the Response of Iron Oxidising and Slime Forming 375
Bacteria to Chlorination in a Laboratory Model Cooling Tower
K. K. Satpathy, T. S. Rao, V. P. Venugopalan, K. V. K. Nair
and P. K. Mathur

32 Monitoring and Inspecting Biofouled Surfaces 384
S. W. Borenstein and G. J. Licina

33 Corrosion Behaviour of Steel in Coal Mining Water in the Presence 398
of *Thiobacillus thiooxidans* and *Thiobacillus ferrooxidans*
S. M. Beloglazov and A. N. Choroshavin

Index 405

European Federation of Corrosion Publications
Series Introduction

The EFC, incorporated in Belgium, was founded in 1955 with the purpose of promoting European co-operation in the fields of research into corrosion and corrosion prevention.

Membership is based upon participation by corrosion societies and committees in technical Working Parties. Member societies appoint delegates to Working Parties, whose membership is expanded by personal corresponding membership.

The activities of the Working Parties cover corrosion topics associated with inhibition, education, reinforcement in concrete, microbial effects, hot gases and combustion products, environment sensitive fracture, marine environments, surface science, physico–chemical methods of measurement, the nuclear industry, computer based information systems, corrosion in the oil and gas industry, and coatings. Working Parties on other topics are established as required.

The Working Parties function in various ways, e.g. by preparing reports, organising symposia, conducting intensive courses and producing instructional material, including films. The activities of the Working Parties are co-ordinated, through a Science and Technology Advisory Committee, by the Scientific Secretary.

The administration of the EFC is handled by three Secretariats: DECHEMA e. V. in Germany, the Société de Chimie Industrielle in France, and The Institute of Materials in the United Kingdom. These three Secretariats meet at the Board of Administrators of the EFC. There is an annual General Assembly at which delegates from all member societies meet to determine and approve EFC policy. News of EFC activities, forthcoming conferences, courses etc. is published in a range of accredited corrosion and certain other journals throughout Europe. More detailed descriptions of activities are given in a Newsletter prepared by the Scientific Secretary.

The output of the EFC takes various forms. Papers on particular topics, for example, reviews or results of experimental work, may be published in scientific and technical journals in one or more countries in Europe. Conference proceedings are often published by the organisation responsible for the conference.

In 1987 the, then, Institute of Metals was appointed as the official EFC publisher. Although the arrangement is non-exclusive and other routes for publication are still available, it is expected that the Working Parties of the EFC will use The Institute of Materials for publication of reports, proceedings etc. wherever possible.

The name of The Institute of Metals was changed to The Institute of Materials with effect from 1 January 1992.

A. D. Mercer
EFC Scientific Secretary,
The Institute of Materials, London, UK

EFC Secretariats are located at:

Dr J A Catterall
European Federation of Corrosion, The Institute of Materials, 1 Carlton House Terrace, London, SW1Y 5DB, UK

Mr R Mas
Fédération Européene de la Corrosion, Société de Chimie Industrielle, 28 rue Saint-Dominique, F-75007 Paris, FRANCE

Professor Dr G Kreysa
Europäische Föderation Korrosion, DECHEMA e. V., Theodor-Heuss-Allee 25, D-60486, Frankfurt, GERMANY

Proceedings of the Third European Federation of Corrosion Workshop on Microbial Corrosion held in Estoril, Portugal, 13–16 March 1994, and co-sponsored by JNICT (Junta Nacional de Investigação Científica e Tecnológica).

Preface

This publication contains the papers from the third European Federation of Corrosion workshop on microbial corrosion organised by the Instituto Superior Técnico, Lisbon, Portugal in collaboration with the EFC working party on microbial corrosion. The purpose of the workshop was to summarise the actual European experience in this field while also drawing on the recent experience from other countries. It represents the current concerns of the problems associated with corrosion induced by the activity of a wide range of microorganisms and, as such, encompasses a spectrum of interest from aerobic and anaerobic situations to the control and prevention of the problem or the assessment of new on-line monitoring techniques.

Microbially influenced corrosion (MIC) is by definition, corrosion associated with the action of microorganisms present in a system. MIC is, therefore, an interdisciplinary subject that embraces the fields of materials science, chemistry, microbiology and biochemistry. Since the first reports of MIC at the end of the 18th century in the UK much has heen done to understand the role of microorganisms in the corrosion of materials. However, the mechanisms by which microorganisms enhance corrosion of many materials still remains unclear.

This workshop was broadly divided into four main topic areas: (i) an interdisciplinary approach to MIC, (ii) mechanisms and general studies in the laboratory and in the field, (iii) case studies, and (iv) prevention and control.

Experience with a variety of environments and materials is a feature of the reports in this volume. Thus, data are reported from potable water, river water, seawater and soils; the materials considered include stainless and carbon steels, copper, aluminium brass, fibre reinforced composites, etc.

The best example of the interdisciplinary approach which is required to understand the problem is given in the first section of the volume which is concerned with the MIC of copper in potable water systems. The model which has been developed to explain the mechanism by which copper is corroded includes the role of the biofilm and its permselectivity. During the workshop new theories were proposed for the anaerohic corrosion of iron and stainless steel by the sulphate-reducing bacteria involving the role of the thiosulphate ion.

It is also interesting to note that several new experimental techniques have been developed and used to characterise the nature of the biofilm which forms on the metal surface. Atomic Force microscopy and mass transport measurements using a spinning disc electrode have heen particularly successful in this respect.

New biosensors have been developed for the control and prevention of MIC and these should become commercially available in the near future. However, the use of biocide and cathodic protection still remain the standard approaches even if their overall efficiency may occasionally be questioned.

The case histories that are reported illustrate the wide and diverse range of materials that are susceptible to the problem of MIC including some of the newer materials such as plastic/metal composite matrices.

From the data presented during the workshop it was obvious that the mechanism of MIC remains a major concern and that there is still a considerable amount of R and D to be undertaken. Nevertheless, this kind of workshop has continued to improve the general awareness of the subject to industrialists and to academia.

Improved communications between microbiologists, corrosion scientists and engineers will eventually result in a better comprehension of the mechanisms of microbial corrosion.

D. Thierry
Chairman of EFC working party on microbial corrosion

C. A. C. Sequeira
Instituto Superior Técnico, Lisboa

A. K. Tiller
Lithgow Associates/University of Portsmouth, UK

(Members of the organising committee)

This volume was edited by A. K. Tiller and C. A. C. Sequeira

Proceedings of the 1st and 2nd EFC Workshops on MIC were published as follows:

1 Microbial Corrosion (Proceedings of the 1st International EFC Workshop)(Eds C. A. C. Sequeira and A. K.Tiller), 1988, Published Elsevier, New York, 1988.

2 Microbial Corrosion (Proceedings of the 2nd EFC Workshop, 1991), European Feleration of Corrosion Publication No. 8, (Eds C. A. C. Sequeira and A. K. Tiller), 1992, The Institute of Materials, London.

Part 1

An Interdisciplinary Approach to Microbially Influenced Corrosion of Copper

Part 1

1

An Interdisciplinary Approach for Microbially Influenced Corrosion of Copper

A. H. L. CHAMBERLAIN*, W. R. FISCHER, U. HINZE, H. H. PARADIES,
C. A. C. SEQUEIRA†, H. SIEDLAREK, M. THIES, D. WAGNER
and J. N. WARDELL*

Märkische Fachhochschule, Laboratories of Corrosion Protection, Biotechnology and Physical Chemistry,
Post Box 20 61, D-58590 Iserlohn, Germany
*School of Biological Sciences, University of Surrey, Guildford, Surrey, GU2 5XH, UK
†Instituto Superior Tecnico, Technical University of Lisbon, Av. Rovisco Pais,
PT-1096 Lisbon Codex, Portugal

ABSTRACT

An interdisciplinary approach for studying the microbially influenced corrosion of copper is presented which takes into consideration the metal site, the interphase between copper, the exopolymeric materials produced by certain microorganisms, and the bulk water containing the relevant cations and anions. The electrochemical results are explained on the basis of diffusion processes and multiple Donnan equilibria under the assumption of ion selectivities of mainly anionic exopolymers.

1. Introduction

Copper as a metal is used widely because of its good resistance to corrosion combined with mechanical workability, excellent electrical and thermal conductivity, and ease in soldering and brazing. Vast quantities, amounting to several million kilometres of copper tube are in service for potable water distribution systems in countries throughout the world [1].

The exposure of foreign surfaces to a biologically active environment usually results in their modification. This may be simply adsorption of organics or, if the surface is not excessively toxic, the development of a complex biofilm. In order to characterise these steps for copper, it is essential to discriminate between the primary adsorption and secondary developmental events, which lead to the modification and even ultimately, destruction of the material's surface. Several types of pitting corrosion have been reported but the level of failures is remarkably low. Reiber [2] indicated that copper corrosion is mainly general attack, with pitting only occurring under limited and rare conditions. A range of welldefined types of pitting corrosion of copper were reviewed by Mattsson [3], but none of these was considered to be associated with microbial activity. However, since then two forms of pitting have been reported which do appear to have microbial origins. One of these, termed 'pepper-pot pitting' was reported from large institutional buildings in S.W. Scotland, UK [4] whilst the other, frequently termed Type $1^1/_2$ pitting, has been observed in tubes from Saudi Arabia, Germany and England [5, 6].

Most of the observations published to date have concentrated upon the presence of copious biofilms, the detection of microorganisms within the films by electron microscopy techniques and the biochemical activities of the bacteria *in vivo* and *in vitro* [7, 8]. At the start of the microbial corrosion process the initial adsorption phase may include soluble and particulate biopolymers, such as the microbial extracellular polymers, chiral metabolic products of high and low molecular weight and secreted glycoproteins. All of these could be involved in linking organisms to the surface and/or producing a complex layer, which may exhibit membrane-like properties. Because both the bulk and the biofilm phases are fully hydrated there appears to be a continuum, however the physico–chemical properties of the surface-associated polymers may actually produce two (or more ?) contiguous, but functionally discrete, compartments.

Various types of pitting corrosion have been examined and chloride ions are believed to be a critical feature for the initiation of pitting attack [9, 10]. However, pit morphology was not always a satisfactory indicator of pitting type and the corrosion products overlying the pits depended entirely upon the chemical composition of the water and did not determine an important phase of the pitting process [10]. Therefore, any feature covering the cuprous oxide layer of the copper metal surface and preventing chloride ion migration towards the copper metal could possibly control the onset of pitting. An important question is how this may operate at the atomic or molecular level, particularly at the onset. Another parameter which must be considered in the deterioration of copper and its dissolution into free ions or colloidal forms is the presence of micro-organisms which can both promote and accelerate corrosion of metals and in a few cases even act as inhibitors.

An interdisciplinary approach is now described taking the biological macromolecules into consideration to explain electrochemical results relating to diffusion processes, proton products and multiple Donnan equilibria and is based upon the possession of certain ion selectivities by the mainly anionic biopolymers.

This contribution introduces the subject and with the four following papers considers different aspects of this interdisciplinary approach.

2. Surface and Near Surface Chemistry of Copper

Most of the phenomena of the deterioration of copper surfaces, including the influential role of microorganisms, exclude physical processes, despite the large number of observations, and experiments reported to delineate a consistent picture of events. Notwithstanding the availability of modern surface analytical tools, most of the contributions on the deterioration of metal surfaces do not consider:

- the surface structure (physical and chemical) at the molecular level (including specific surface energies);

- the chemical composition of the extracellular polymers, including sequence and three-dimensional structure (folding) despite the length of time since they were shown to play a crucial role in microbially influenced corrosion (MIC); this may have been due to purification difficulties, low solubility or a lack of sequencing techniques for these kinds of polysaccharides;

- adhesion and orientation processes of the macromolecules at the surface, including the interface between metal surface and polymers, particularly polysaccharides;

- phase phenomena, gelled vs ungelled polymers, including membrane-like assemblies, since release of solubilised corrosion products from the gelled cubic phase is considerably slower than from the ungelled material revealing a controlled release mechanism;

- diffusion and interactions in gels and solutions at the metal–polymer interface *and* polymer-bulk solution interface, particularly for charged systems;

- capillary rise of solutions on the metal surface when grafted with long (charged) polymer chains, which can act as mediators of adhesion;

- development of corrosion products, e.g. Cu_2O, $CuCl$, CuO, $Cu_xO_y(OH)_z$ etc., particularly of Cu_2O due to particle increase under the influence of the macromolecular nature of the charged/neutral film;

- diffusion potentials and Donnan equilibria at the interfaces between solid (metal) and film, as well as film/bulk solution which is pH dependent, including micellar catalysis due to ultra-fast proton transfer reactions;

- the establishment of a minimum thickness of the film layer in order to achieve measurable effects on Cu–metal corrosion, which is accessible through standing X-ray waves or ellipsometry;

- membrane-like behaviour inducing characteristic events, e.g:

(i) anodic stimulation,

(ii) maintenance of low interfacial pH,

(iii) alteration of localised dissolved oxygen concentrations to serve entrapped consortia of bacteria, or enzymatic systems (development of HO_2^- due to the presence of mono-oxygenases, superoxide dismutase, cyclo-oxygenases etc.)

(iv) action as diffusion barriers of the charged species, including Donnan effects on the distribution of small ions across the barrier of the 'membrane'-like exopolymers.

These points demonstrate the complexity of the mineralization (corrosion) of metal surfaces (any solid surface) in the presence of microorganisms whether dead or alive since some of the enzymatic systems still function even when immobilised in a macromolecular matrix.

In this paper we restrict ourselves to the cause of the anodic dissolution of Cu surfaces in the presence of model biopolymers carrying negative charges when deprotonated, e.g. alginic acid, various forms of xanthan including de-acetylated and de-pyruvylated forms, though these particular biopolymers have never been observed in any deterioration processes in nature! Figure 1 is an hypothetical model which encompasses most of the considerations outlined above and the data derived from our investigations since 1988 [11]. The original model starts with the phase boundaries between bulk copper including $Cu_xO_y(OH)_z$ and a small solution layer and the biopolymer (of thickness $d = 200$ nm) (Side 2). Side 1 is indicated as the phase boundary between the biopolymer and the electrolyte (pH 7–9) at the beginning of the corrosion process. The reaction layers formed at the onset of the corrosion process are also included in Fig. 1.

3. Establishment of the Model by Cyclic Voltammetry

Cyclic voltammetric experiments have been performed in aerated sodium chloride solution. Figure 2 shows the voltammogram obtained with a bare copper electrode. The curve starts in the hydrogen evolution region, followed by a first plateau corresponding to oxygen reduction [12]. At more positive potentials, a second plateau occurs with negative total current densities.

In this potential range of 40 to 160 mV_H a thin tarnish layer is formed on the copper surface, but its formation is superimposed by an oxygen reduction reaction under proton consumption. A small oxidation peak is found at 270 mV_H (peak b) and a reduction peak is situated at 100 mV_H (peak c). These peaks correspond to the formation and reduction of copper(I) chloride. The threshold potential is situated at 300 mV_H, as indicated by a strong increase of the anodic current densities (maximum α). The scan was reversed at a potential of 310 mV_H to avoid the formation of copper(II)-corrosion products. The reduction peak at -420 mV_H (peak a) corresponds to the reduction of a typical reaction layer consisting of copper(I) oxide or copper(I) hydroxide formed in electrolytes containing chloride ions [13].

Figure 3 (p.8) depicts the voltammogram of a copper electrode coated with 0.3 % (w/w) xanthan/0.1 % (w/w) agarose/0.05% (w/w) alginate in aerated sodium chloride solution [14]. The following differences have been evaluated compared to the results obtained with a bare copper electrode (see Fig. 2).

- The two peaks described in Fig. 2 (peaks b, c) do not occur. So chloride ions cannot pass this coating and cannot form copper(I) chloride underneath the

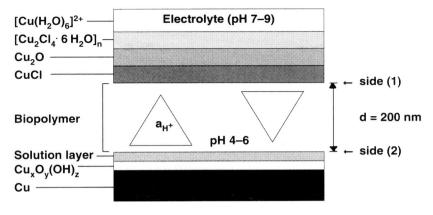

Fig. 1 Working model for microbially influenced corrosion of copper.

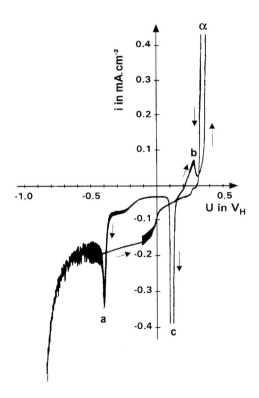

Fig. 2 Potentiodynamic current density potential curve of copper in sodium chloride; [Cl$^-$] = 200 mg L^{-1}; pH 6–6.5; aeration; scan rate: 0.01 mV s^{-1}; start potential: –950 mV$_H$; reverse potential: +360 mV$_H$.

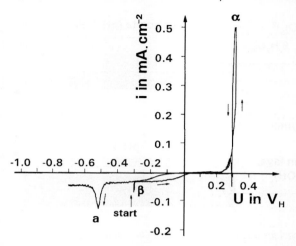

Fig. 3 *Potentiodynamic current density potential curve of copper covered with 0.3% (w/w) xanthan/0.1% (w/w) agarose/0.05% (w/w) alginate in 1 mM sodium chloride; pH 6–6.5; aeration; scan rate: 0.1 mV s^{-1}; start potential: –300 mV$_H$; reverse current density: 0.1 mA cm^{-2}.*

coating within the time scale of the experiment. The occurrence of these two peaks is not affected by the different concentrations of chloride ions in the electrolytes.

The integral charge of the peak corresponding to the reduction of copper(I) oxide or copper(I) hydroxide at – 520 mV$_H$ (peak a) is much smaller and formed through the reaction with water under proton production at positive potentials. This confirms the assumption that the pH underneath the coating is lower than in the bulk electrolyte.

- The total current densities corresponding to the oxygen reduction reaction in the first scan are considerably smaller on the coated electrode indicating that the coating is acting as a diffusion barrier.

- No notable inhibition of the anodic partial reaction (maximum α) can be observed.

These results are in excellent agreement with the model proposed in Fig. 1 [14]. The biopolymeric substance is cation selective, thus permeable for copper ions and hydrated protons, whilst chloride ions cannot diffuse through the coating. This discloses that at the phase boundary Cu$_x$O$_y$(OH)$_z$/biopolymer only corrosion reactions involving water can occur. Due to the high water content of the coating, sufficient water is available at the phase boundary Cu$_x$O$_y$(OH)$_z$/biopolymer, leading to the formation of hydrated or unhydrated cuprous oxide or cuprous hydroxide. However, the acidic nature of the electrolyte at this phase boundary limits considerably the amount of oxide formed underneath the coating.

In chloride ion containing electrolytes the same reaction products are obtained for a coated electrode as for a bare copper electrode. However, the precipitated copper(I) oxide formed via hydrolysis of the underlying copper(I) chloride is now situated on top of the biopolymeric coating. Therefore, a reduction of these layers in the reverse scan of a voltammogram is not possible.

4. Structure of the Model Matrix, Membrane Mimetic Behaviour and Donnan Equilibria

Linear polysaccharides, e.g. alginates and depyrulyvated xanthan, can be used as model membranes mimicking the biofilm matrix produced by microorganisms, in the form of extracellular polymers.

These polymers are primarily polysaccharides but may also incorporate protein moieties and even nucleic acids from lysed cells. Allison [15] has very recently considered the nature and role of the exopolysaccharides in biofilms. The presence of the polysaccharides can be demonstrated quite easily on copper surfaces using the modified histochemical staining procedures developed by Chamberlain et al. [16]. These also allow detection of acidic components such as uronic acids which are believed to be important in copper binding [17]. Thus, the two alcian stains, alcian blue and alcian yellow used at pH 1.0 and 2.5 respectively, can be used to discriminate between sulphated and carboxylated polysaccharides.

The exopolysaccharides may be produced as a result of attachment of the bacteria to a surface, or alternatively they may be secreted by planktonic populations. However, this second type of polymer can still coat surfaces with which it makes contact. Polymers derived from the planktonic phase are more likely to form relatively continuous films over the surface, particularly if the flow regimes are low and non-turbulent. Convective movements will aid overall coverage of a pipe interior.

Growth and recruitment at a surface will lead to the establishment of a biofilm which is usually discontinuous and almost always heterogeneous.

The quantities of polymers produced will not only vary from species to species but will also be dependent upon the local environment and the nutritional conditions. In the case of *Ps. aeruginosa* a low molecular weight polymer is produced in response to solid surfaces [5], whilst growth as a biofilm organism results in transcriptional activation of the *algD* gene yielding alginate [18].

In addition, a specific protein has been isolated from *Ps. aeruginosa* of apparent molecular weight of 92 000, which seems to be involved in the production of the specific exopolymer depending on the nature of the surface. However, alginates consist of mannuronate and guluronate residues arranged in a nonregular, clockwise pattern along the linear chain [19]. An important property of this biological macromolecule is its ability to form a gel by reaction with monovalent and divalent ions, e.g. H_3O^+, Ca^{2+}, Cu^{2+}. These gels, which resemble a solid in retaining their shape and resisting shear stress, consist mainly of H_2O with only 0.5–1.0 wt% alginate. Moreover, due to the carboxylate groups, this acidic exopolymer (and other similar exopolymers) do show Donnan phenomena associated with proton uptake in a pH dependent manner (unpublished results), which can be described as

$$\frac{\mu^o_{w_2} - \mu^o_{w_1}}{q} + \frac{RT}{nF} \ln\left(\frac{w_1}{w_2}\right) = \psi \tag{1}$$

with w_1 and w_2=weight fractions respectively of the gelatinous exopolymer and the orientated film adsorbed at the copper surface as assessed by the availability of free protons (w_1 and w_2 refer to saturated conditions). $RT/nF = +0.155$ V (0.155 V at pH 7.0); μ^o = the standard chemical potentials and ψ the Donnan potential. This simplified expression can be solved for (i) $w_1 \rightarrow$ pH, and (ii) N-site density indicating the optimal number of available sites for H$^+$, or Na$^+$, respectively, not considering the equilibrium binding constant for H$^+$ (Na$^+$). So the Donnan potential ψ for this linear structure elongated as a rod, calculates to 0.155 V, since the additional Cl$^-$ accompanying the extra Na$^+$ (Cu$^+$) to the solution has to be orientated towards the more positive phase by reducing the potential. Furthermore, generally the addition of NaCl increases the ionic strength by $\frac{1}{2}\Sigma c_i z_i^2$, and the conductivity, so rendering it more difficult to support a difference potential only. Surface potentials of the alginate as a model acting like a cation selective membrane have been determined also by the use of a dye (2-naphthol-3,6-disulphonic acid di-sodium salt) according to:

$$pK_a^{obs} = pK_a^o - F\psi / 2.303\, RT \tag{2}$$

The measured pK_a^{obs} is a function of ($\alpha/1-\alpha$) and therfore pK_a^o, which is the intrinsic apparent equilibrium constant for the alginate can therefore be obtained by plotting log ($\alpha/1-\alpha$) vs pH, where α is the degree of ionisation of the alginic acid as obtained from the absorption spectra of the probe, yielding a straight line. The slope of this line should be unity if the surface potential is not varying with pH. The log ($\alpha/1-\alpha$) = 0 intercept gives the pK_a^{obs}. A pK_a^{obs} = 4.12 for the alginate yields a potential ψ = 0.145 V at pH 7.0. The deviation of the Donnan potential of RT/nF = 0.155 V as determined from electrochemical measurements is reasonably close to the ψ = 0.145 V as determined by the use of the dye thus indicating the complementary nature of the two experimental methods.

Our model system is comprised of the alginate (Alg–H $\rightleftharpoons$ Alg$^-$ + H$^+$) having the charge z on one side (2) as indicated in Fig. 1 when functioning as a semi-permeable membrane with freely equilibrating, ionisable salt MeX (side (1)). The salt dissociates according to MeX $\rightleftharpoons$ Me$^+$ + X$^-$; and the chemical potentials at equilibrium are:

$$\mu_{MX} = \mu_{M^+} + \mu_{X^-} \tag{3}$$

$$\mu_{MX} = \mu^0_{MX} + RT \ln(MX) = \mu^0_{M^+} + \mu^0_{X^-} + RT \ln[(M^+)(X^-)] \tag{4}$$

At equilibrium, $\mu_{MX}^{(1)} = \mu_{MX}^{(2)}$.

Assuming that the standard chemical potentials

$\mu^0_{M^+}$ and $\mu^0_{X^-}$ are the same on both sides of the membrane, eqn (3) holds:

$$\left(X^-\right)^{(1)}/\left(X^-\right)^{(2)} = \left(Me^+\right)^{(2)}/\left(Me^+\right)^{(1)} = r_D \quad (5)$$

with r_D the Donnan ratio. Since a charged macromolecule, e.g.

$$\text{Alg} - \text{H} \rightleftharpoons \text{Alg}^- + \text{H}^+,$$

is present r_D deviates significantly from unity:

$$(Me^+)^{(1)} = (X^-)^{(1)}$$

$$(Me^+)^{(2)} - z\,(Alg^-) - (X^-)^{(2)} = 0 \quad (7)$$

Substituting (7) into eqn (5), we obtain

$$r_D\left(Me^+\right)^{(1)} - z\left(Alg^-\right) - \left(X^-\right)^{(1)}/r_D = 0 \quad (8)$$

Assuming, the salt (S) is completely dissociated, according to eqn (6), then (S) can be taken as the total salt concentration on side 1 yielding the final result as

$$r_D^2 - r_D^2\left(Alg^-\right)/S - 1 = 0 \quad (9)$$

so that

$$r_D = z\left(Alg^-\right)/S \quad (9a)$$

Equation (9) reveals that $r_D \neq 1$, so Na$^+$(H$^+$) and Cl$^-$ do not distribute themselves equally across the EPS-membrane. Furthermore, this unequal distribution occurs even though the system is at equilibrium, assuming no binding of H$^+$ (Na$^+$) to alginate. Considering the magnitude of the Donnan effect, the important parameter is z(alginate)/2(S) which is the ratio of the concentration of charge from the alginate (biopolymer) to the total concentration of ionic species [(Na$^+$, H$^+$) + (Cl$^-$) = 2(S)] on side 2, which is towards the metal surface side. In the case where z(alginate)/2(S) $\cong$ 10, r_D = 20, so there is a 20 fold excess of [(Na$^+$, H$^+$)] on side 1, and a 20 fold excess of (Cl$^-$) on side 1 (towards the bulk volume side) of the membrane (Fig. 1). This situation can be achieved at rather low concentrations e.g. 10^{-5}–10^{-6} M, resulting in a net charge on the alginate of –10, and a monovalent salt, e.g. NaCl concentration, (S), of 45–50 µM on side 1, and vice versa. Considering the polarisation curve shown in Fig. 3, a satisfactory explanation with respect to r_D and concentrations of NaCl (50µM vs 1 mM in the experiment) cannot be offered entirely in terms of membrane equilibria. This needs to be further investigated. The Donnan effect clearly deserves serious consideration in this particular situation where the binding of an ion to a macromolecule is studied by means of equilibrium dialysis across a semipermeable membrane. This effect is completely neglected by Jang *et al.* [20], in their contribution studying binding of Cu^{2+} to alginate. Not recognising these effects can lead to serious errors in calculating the amount of bound ions from the total amount of the particular ion present on each side of the artificial membrane. Furthermore, taking [H^+], and [OH^-] or [Cl^-] concentrations into account the Donnan ratio r_D can be expressed as:

$$pH^{(2)} - pH^{(1)} = \log r_D \qquad (10)$$

by fixing the hydrogen and chloride (or hydroxyl) ion concentrations on each side of the membrane as well as the ratios of alginic acid = alginate + H^+, Na^+; and Cl^-, respectively.

Taking the value for $r_D = 20$ as mentioned above (and determined experimentally), the difference in pH is greater than one pH-unit. Assuming membrane-mimetic activities of this alginic acid/alginate system, we can calculate the membrane polarisation or potential according to

$$\Delta \Phi = \Phi^{(1)} - \Phi^{(2)} = \Phi_m + (R \cdot T/F) \ln r_D \qquad (11)$$

with $\Delta \Phi = \Phi^{(1)} - \Phi^{(2)}$ the potential difference between side 1 and 2, and F is the Faraday constant (96 487 Coulomb $mole^{-1}$)[21]. Equation (11) is only valid under the assumption that Φ_m is a junction potential analogous to a liquid–liquid junction potential encountered when a saturated salt bridge is placed between two solutions. In the present case, Φ_m is the membrane potential, showing (i) that it is equal and opposite to that generated by ion gradients; (ii) that the membrane itself is polarised as a result of ion gradients, and (iii) for $r_D \cong 20$ the membrane potential $\Phi_m = 75$ mV, and for $r_D \cong 1.05$ the value $\Phi_m = 0.27$ mV.

However, it should be mentioned, that a liquid junction is due to unequal mobilities of cations and anions. The membrane potential is due to unequal activities of cations and anions inside and outside the membrane at equilibrium. The unequal concentrations of ions also give rise to a concentration cell whose potential is equal and opposite in sign to the membrane potential. This problem needs to be further elucidated.

Figure 4 shows the measured potential difference for alginate in 5×10^{-4} M/5×10^{-5} M sodium chloride solutions as a function of time. A maximum potential difference value of 108 mV was obtained shortly after the start of the experiment. This potential difference was separated into a concentration contribution of 55 mV and a membrane contribution of 53 mV. This membrane potential $\Phi_m = 53$ mV indicates a Donnan ratio of $r_D \cong 14$. Experimentally, a continuous decrease of the potential difference was noticed with increasing time. It could be seen during the measurements that this substance dissolves slowly but continuously into the solution, and this might be the cause for the observed decrease in the potential difference values.

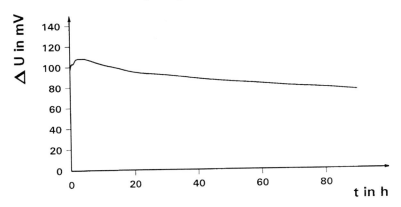

Fig. 4 *Potential difference of 0.5% (w/w) alginate in 5×10^{-4} M/5×10^{-5} M sodium chloride solutions as a function of time.*

The membrane potential studied for alginic acid/alginate as a membrane mimetic system might be significant in magnitude, particularly at high salt concentrations, e.g. through Cu^{2+} ions. The high salt concentration can suppress the Donnan effect to a point where the membrane potential is eliminated. However, due to the cation selectivity of the alginate and the accelerated crossing of the copper ions through the membrane the Donnan potential is not at steady state equilibrium.

Geesey et al. [21] showed that exopolysaccharide (from a river isolate) bound free cupric ions were approximately six times more immobilised at pH 5.0 than at pH 7.0. This correlated with the earlier observation by Mittelman and Geesey [17], that copper binding resulted in a release of H^+ from the exopolymer, hence a synergistic effect was obtained. The authors further suggested that the establishment of adjacent colonies of bacteria secreting polymers of differing copper-binding affinity would lead to differential regions of dissolved copper ions, i.e. a copper concentration cell; those areas with the lower concentrations would act as the anodic regions. Subsequent dissolution of metal from the surface followed by differential binding would initiate a general attack but it is important to note that this process in itself can only proceed until all the exopolymer binding sites are filled. The process may become of extended duration if biofilm development is continued and metabolic activities result in more acidic conditions within the polymer bulk. However, it is essential to stress that a similar surface dissolution can be shown with a wide range of acidic polysaccharides, including agar. This simple removal of metal from the surface will only constitute pitting corrosion if it is a self-sustaining localised attack. In the case of copper the pitting mechanism normally depends on the localised activity of chloride ions [14].

The presence of biofilm exopolymer on the copper surface could have a central role to play in pitting corrosion if the migration of Cl^- ions to the metal surface was prevented. Such permselective membranes are well known as ion exchange membranes and comprise a matrix with water filled pores which carry fixed charges along their length.

At low electrolyte concentrations the pores are only filled with the counter-ions. This means that in the case of the bacterial exopolysaccharide layers, the presence of negative charges associated with the uronic acid and other substituents such as the ketal-linked pyruvyl groups could create a polymer through which only cations could pass. Thus, copper ions released at the metal/biofilm phase boundary could migrate through the negatively charged pores to the biofilm/bulk electrolyte phase boundary. The retention of anions at this phase boundary would create an electrochemical potential gradient drawing the copper cations away from the metal surface. However, by preventing the passage of Cl^- ions to the metal surface it would preclude the establishment of the pitting reactions described by Lucey [9] and Shalaby et al. [10].

Moreover, the diffusion of an ion of finite size, e.g. $(H_3O)^+$, $[Cu(H_2O)_6]^{2+}$, Cu^{2+}-bidentates, in such membrane systems is influenced not only by electrostatic interactions but also by obstruction defects. These effects are accessible through measurements of time dependent diffusion coefficients according to eqn (6)

$$\frac{D(t)}{D_o} = \frac{D}{D_o} + \left(1 - \frac{D}{D_o}\right) e^{2\tau/\tau^{\frac{1}{2}}} \qquad (12)$$

where D is the long-time self-diffusion coefficient, and τ the characteristic time for the relaxtion from start-time to long-time behaviour. To estimate the obstruction effects, D/D_o vs the concentration of Cu^{2+}, we also estimated the diffusion of an uncharged molecule, glycerol, in this system. This molecule is of comparable size to $[Cu(H_2O)_6]^{2+}$, as determined from the diffusion coefficients obtained in aqueous solutions. Furthermore, the diffusion coefficient, particularly D/D_o is very sensitive to the conformation of the polymer system and the polymer volume fraction which is involved in the diffusion process, e.g. H^+, $[Cu(H_2O)_6]^{2+}$, Cl^-, and the bulk water between the polymer chains. Furthermore, for a charged molecule, the diffusion will be slower in a helical macromolecule system due to a decreased concentration in the bulk by accumulation of mobile ions close to the polyions. Increasing the size of the charged molecule in a mixture with small ions results in an increase of the diffusion of the larger ion due to the preferential accumulation of the smaller ions close to the polyion. Even for larger diffusing ions, e.g. Cu^{2+}–pyruvated complexes, nanoparticles of Cu embedded in interstitial spaces of the alginate — including the fractal behaviour of the $Cu_xO_y(OH)_z$ complex (resulting in the long term in Cu_2O formation) — a subsequent decrease in diffusion (20–30 fold depending on the size of the diffusing molecule), will also be expected due to obstruction.

A naturally occurring membrane system comprising an uncharged molecule e.g. glycolipid A or mono-olein as biocompatible lipids with the synthetic alginate membrane system, exhibits two continuous critical phases for water concentrations above 20–30 wt%, i.e. as also found in naturally occurring membranes.

These critical phases are seen as continuous since a convoluted lipid bilayer divides them into two discrete regions which are simultaneously continuous. However, these cubic phases reveal some unique features:

(i) a large interfacial area between the two separate regions;

(ii) uniform nanopores where size can be perfectly controlled;

(iii) a biocompatible and stable backbone comprising lipid (or glycolipids) and alginate (charged, strongly hydrated), and

(iv) the capability of incorporating active enzymes *or* transporting H^+, $[Cu(H_2O)_6]^{2+}$, to the outside towards the bulk solution. An interesting property of this model membrane is that it can exist with an excess of water and cations at various concentrations in both the gelled and ungelled state.

5. Conclusion

The heterogeneous chemical composition of the exopolymeric substance and the unknown folding of these biological macromolecules makes it difficult at present to model the electrochemical results. However, using model polysaccharides, e.g. alginate or xanthan, we were able to show some of the events which can cause copper corrosion. The same results have been obtained using exopolymeric substances produced from microorganisms isolated from perforated copper pipes.
The results do seem to indicate a beneficial effect due to expelling chloride ions from the film at low chloride concentration and to restriction by the exopolymeric film of diffusion to and from the surface. However, in practice the biofilm is unequally distributed over the entire surface revealing interruptions in the biofilm. In the areas of the these interruptions the formations of local anodes leading in some cases to the perforation of a copper tube has been observed.Notwithstanding at the molecular level further variables, e.g. the sequence comprised of bacteria, biomembrane (EPS) properties and composition and morphology of the copper surface have to be regarded we were able to show that the biofilm can play a crucial role in this process. The detailed results of this interdisciplinary approach are described in the following sequence of papers.

6. Acknowledgements

The financial support of this work within the BRITE/EURAM Project No. 4088 (Contract-No. BREU-CT91-452) "New Types of Corrosion Impairing the Reliability of Copper in Potable Water Caused by Micro-organisms" is gratefully acknowledged.

References

1. J. C. Nuttall, in *Corrosion and Related Aspects of Materials for Potable Water Supplies* (Eds P. McIntyre and A. D. Mercer), The Institute of Materials, London, UK, 1993, 65–83.

2. S. H. Reiber, *JAWWA*, 1989, 114–122.
3. E. Mattsson, *Brit. Corros. J.*, 1980, **15**, 6.
4. C. W. Keevil, J. T. Walker, J. McEvoy and J. S. Colbourne, in *Biocorrosion* (Ed. C. Gaylarde), Biodeterioration Society, Kew (1988), 99–117.
5. P. Angell and A. H. L. Chamberlain, *Int. Biodeterior.* 1991, **27**, 135–143.
6. H. S Campbell, A. H. L. Chamberlain and P. J. Angell, in *Corrosion and Related Aspects of Materials for Potable Water Supplies* (Eds P. McIntyre and A. D. Mercer), The Institute of Materials, London, UK, 1993, 222–231.
7. D. H. Pope, D. J. Duquette, A. H. Johannes and P. C. Wagner, *Mat. Perform.*, 1984, **24**, 14–18.
8. A. H. L. Chamberlain and P. Angell, in *Microbially Influenced Corrosion and Biodeterioration* (Eds N. J. Dowling, M. W. Mittelman and J. C. Danko), The MIC Consortium: Knoxville, TN, 1991, 3.65–3.72.
9. V. F. Lucey, *Symp. on Corrosion of Copper and Alloys in Building*, Japan Copper Development Association, Tokyo, 1982, 1.
10. H. M. Shalaby, F. M. Al-Kharafi and V. K. Gouda, *Corrosion*, 1986, **45**, 536–547.
11. D. Wagner, W. R. Fischer and H. H. Paradies, 23. GDCh-Hauptversammlung (Gesellschaft Deutscher Chemiker), München, 1991, 233.
12. W. Fischer and W. Siedlarek, *Werkstoffe und Korros.*, 1979, **30**, 695–699.
13. W. R. Fischer and B. Füßinger, in *Proc. 12th Scand. Corros. Congr. & EUROCORR '92*, Dipoli (Finland) 1 (1992), 769.
14. H. Siedlarek, D. Wagner, W. R. Fischer, and H. H. Paradies, *Corros. Sci.*, 1994, **36** (10), 1751–1763.
15. D. G. Allison, *Microbiology Europe*, 1993, 16–19.
16. A. H. Chamberlain, P. Angell and H. S. Campbell, *Brit. Corros. J.*, 1988, **23**, 197.
17. M. Mittelman and G. G. Geesey, *Appl. Environm. Microbiol.*, 1985, **49**, 846.
18. B. D. Hoyle, L. J. Williams and J. W. Costerton, *Interf. Immunol.*, 1993, **61**, 777.
19. R. Kahn, *Pure Appl. Chem.*, 1975, **42**(3), 371.
20. L. K. Jang, N. Harpt, D. Grasmick, L. N. Vuong and G. G. Geesey, *J. Phys. Chem.*, 1990, **94**, 482.
21. G. G. Geesey, M. Mittelman, T. Iwaoka and P. R. Griffiths, *Mat. Perform.*, 1986, 37–40.

2

Physical Behaviour of Biopolymers as Artificial Models for Biofilms in Biodeterioration of Copper.

Solution and Surface Properties of Biopolymers

M. THIES, U. HINZE and H. H. PARADIES

Märkische Fachhochschule Biotechnology and Physical Chemistry, D-58590 Iserlohn, P.O. Box 2061, Germany

1. Introduction

When a foreign surface is exposed to a biological environment, there exists a natural tendency to destroy the foreign object or, failing that, to 'wall it off', to cover or encapsulate the object. The biochemical components which are involved in this process are proteins and cells which are capable of coating the foreign surface with biopolymers (or cells). The composition and organisation of this layer will influence the subsequent cellular and surface events. Therefore, it is essential to characterise and reproduce the surface of any materials, e.g. metallic, non-metallic or biomaterials.

The biopolymers ('extracellular polymers', or 'exopolymers') produced and secreted by microorganisms into aqueous solutions are chemically composed of polysaccharides, glycolipids, and glycoproteins having short sequences of α-amino acid residues which are spaced into discrete skeletons to oligosaccharide chains that are thought to be linked to metal deterioration processes, causing aquatic pollution with metal ions, and high economic losses [1–3]. These polymeric materials of organic origin produced by certain micro-organisms reveal to some extent surface tension activities, akin to those found for cationic, non-ionic or anionic surfactants, particularly when grown in media consisting of hydrocarbons as an organic carbon source vs bicarbonate as an inorganic carbon source.

The microbial exopolysaccharides can be divided into two groups: monopolysaccharides, e.g. cellulose, scleroglucan, homosialic acid, dextrans, pollulan, elsinan and sialic acids, and heteropolysaccharides which are composed of repeating units of varying sizes ranging from disaccharides to octasaccharides, e.g. xanthan, succinoglycan, XM6, emulsan, and gellan (for review see Ref. 4). Following the chemical nomenclature for surfactant molecules with regard to the nature of the polar groups, i.e. cationic, anionic, and non-ionic, most of the exopolymers belong to the groups of anionic or nonionic detergents, respectively, not implying that all exopolymers are surface active due to their polar or apolar grouping. These biological macromolecules appear as amphiphilic materials: the hydrophobic part is built of long chain fatty acids, hydroxy fatty acids of the type of α-alkyl-β-hydroxy fatty acids. The hydrophilic part within the macromolecule can be e.g. de-acylated amino residues, the polyhydric nature of the sugar backbone, a phosphate, a carboxylic acid or a sulphate.

Furthermore, a sugar molecule within the frame of its macromolecular structure, e.g. oligo- or polysaccharide, can be considered also to be hydrophobic depending on packing considerations in solution including intramolecular hydrogen bonding and intermolecular hydrogen bonding between adjacent chains or regimes of the coiled macromolecule and solvent molecules, particularly water molecules [5].

Some of the structures of the exopolymers are well characterised, e.g. alginic acid and xanthan, and some are not, particularly those materials which are biotransformed by microorganisms. However, their surface activities have not been studied as extensively as for natural occurring lipids or 'soaps' [6]. This is especially true for biosurfactants, mixtures of biosurfactants with other extracellular and intracellular polymeric materials, including peptides, nucleotides and metabolites, as e.g. pyruvate, (R)-lactate, (R)- or (S)-β-hydroxy butyric acids [7].

Since the surface activities of these materials are intimately related to their physical structure, e.g. aggregational behaviour, by lowering the surface tension, adhesion, spreading and wettability, these surface activities have been studied in detail in order to obtain physical parameters, and meaningful data at the interface liquid/air, thus subsequently at the solution/copper interface.

In light of the still unknown mechanism of MIC we can discriminate two main steps at least on the microbiological site, which are (i) initial binding of the microorganisms to surfaces, and (ii) accumulation (or aggregation) of microorganisms at surfaces accompanied by a production of exopolymeric substances, often considered as a 'biofilm' including the consortia of microorganisms residing within these exopolymeric materials. Furthermore, recent experiments do indicate in case of *Xanthomonas comprestis* of the atypically salt regulation strain, and *Pseudomonas aer.* 44Tl, the existence of a certain protein having a molecular weight of 96 000 and a submit stoichiometry of the α, β type, to be responsible for the excretion of exopolymers. This has also been shown by *in vitro* experiments in the presence of copper rings, supporting the occurrence of a specific protein of molecular weight 30 000 according to SDS-polyacryl-gel electrophoresis (Hinze, Thies, Paradies, unpublished results). Expression of the protein activities as measured by production of exopolymeric substances is dependent on the nature of the surface (hydrophob vs hydrophil) and the copper metal.

Physically, the microbially influenced corrosion process (MIC) is involved in at least four different sets of interfaces:

The clean Cu-surface (i), as prepared and studied in ultra high vacuum (UHV), can be characterised by physical sensitive techniques, e.g. LEED, SIMS, XPS, EXAFS and X-ray reflectivity [8]. The Cu-surface (ii) exposed or covered with water (liquid or vapour) can be studied, i.e. by X-ray or neutron reflectivity and/or ellipsometric methods [9, 10], whereas surface reactions (iii) and (iv) can be followed also by ellipsometry, XPS, electron microscopic methods and X-ray reflectivity in combination with contact angle measurements[11].

X-ray reflectivity and neutron reflectivity techniques, including those applying

Scheme 1	
Surface (Me)	Modified Surface, Interface
(i) Cu:	UHV
(ii) Cu+(H$_2$O)$_{bulk}$:	Cu-O-H$_2$O-HOH
(iii) Cu-O-H$_2$O + EPS:	Cu-O-EPS(H$_2$O)$_{STR}$*
	Cu-O-EPS + H$_2$O
(iv) Cu-O-EPS(H$_2$O)$_{STR}$: ↗ ↘	Cu-O-EPS(H$_2$O)$_{STR}$ -EPS(H$_2$O)$_{STR}$

*(H$_2$O)$_{STR}$ means structural water as opposed to bulk water

standing waves using synchrotron radiation are particularly well applicable due to tunable wavelengths, high intensity and polarisation of the radiation including fluorescence [12]. These specific techniques can also be applied to study wettability, critical micelle concentrations (CMC) at interfaces to solid surfaces [13, 14]. Since wettability includes a number of variables, e.g. orientation of the substratum, surface roughness, low or high energy surfaces (hydrophilic vs hydrophobic), and water chemistry, wettability can be explained in terms of surface signals which can be extended to microorganisms also, encompassing long range and short range interactions as well as dispersion forces.

The wettability of surfaces which is dealt with in this contribution with relevance to MIC is determined in terms of the contact angle with liquids, commonly designated as θ. The work of adhesion can also be determined according to eqn (1):

$$W^A = \gamma_L(1 + \cos\theta) \quad (1)$$

with γ_L the surface tension of the liquid (solution) and θ the contact angle, respectively.

The quantity W^A represents the reversible work which is required to separate unit area of the liquid–solid interface so as to form unit areas of solid and liquid surfaces. This implies that the contact angle formed on a solid by a particular solution can be measured and analysed if the surface energies of the solution (liquid) including that of the solid surface are known.

A solid–liquid contact angle is the angle formed at the three-phase contact of a drop of liquid with a surface [15, 16]. An advancing contact angle (θ_a) is the static, kinetically, stationary angle formed after a drop has advanced across a surface. A receding contact angle (θ_r) is the angle formed after a drop has receded across the surface, e.g. after liquid has been removed from the drop.

Young's equation (2) describes the different interfacial energies as:

$$\cos\theta = \frac{\gamma_{SV} - \gamma_{SL}}{\gamma_{LV}} \qquad (2)$$

where θ = contact angle, γ = interfacial free energy, where the interface is specified by subscripts: S = solid, L = liquid and V = vapour. The term $\cos\theta$ is proportional to the difference in interfacial energies ($\gamma_{SV} - \gamma_{SL}$). Since contact angle measurements are intimately related to the interfacial energies, measurements of the wettability of metal surface, e.g. Cu, Cu_xO_y, $Cu_xO_y(OH)_z$, including Cu–H_2O interfaces with adsorbed exopolymers (EPS) which create a complete new surface area in contact with H_2O (Scheme 1), can provide qualitative information about identity, distribution, and orientation of interfacial functionality, especially with EPS at the interface.

Hystereses in wetting experiments, e.g. $\theta_a - \theta_r$, or correspondingly $\Delta(\cos\theta) = \cos\theta_a - \cos\theta_r$, on modified Cu surfaces were high (40–95°) in the presence of H_2O, metabolites e.g. pyruvate, (R)- or (S)- lactate, (R)-(S)-B-hydroxo butyrate, acetate, propionate, acetoacetate in all cases, indicating that the system is not at equilibrium at all. However, when the Cu-surface was modified with EPS, hystereses were low (35–40°), the system is apparently in equilibrium, as was found in the presence of 1 µM H_2O_2.

Causes of hysteresis also include heterogeneity and roughness of the surface as well as reorganisation of the functional groups within the interfacial region. The chemical heterogeneity at a Cu-surface can be related to contact angles which are averaged over the interfacial functionality [17], and can be described according to Cassie (eqn 3):

$$\cos\theta_{obs} = \Sigma\, A_i \cos\theta_i \qquad (3)$$

which can be correlated to the roughness factor ν (eqn 4):

$$\cos\theta_{obs} = \nu \cos\theta_i \qquad (4)$$

The alternative approach uses the fact that the Cu-surface is a composite of Cu, CuO, CuO.H_2O, CuO · H_2O · EPS (Scheme 1), i.e, consisting of small particles of various kinds. So considering two kinds of particles, e.g. $CuO_x(OH)_y$ and $CuO_x(OH)_y$ · EPS, occupying a fraction of the surfaces f_1 and f_2, it follows:

$$\gamma_{LV}\cos\theta_c = f_1(\gamma_{SV} - \gamma_{SL}) + f_2(\gamma_{S2V} - \gamma_{S2L}) \qquad (5)$$

$$\cos\theta_c = f_1\cos\theta_1 + f_2\cos\theta_2 \qquad (6)$$

The Zisman parameter, θ_c or $\cos\theta_c$ [18], which is an empirical parameter, can also be used to characterise energetically the surface: a low value indicates a relatively hydrophobic nonwettable material with low surface energy, while a high value indicates a hydrophilic easily wetted high-surface energy solid [19]. Furthermore, the critical surface tension of wetting, γ_c, and hence θ_a, depends also on the crystallinity of the Cu-surface, and — when covered with EPS — on the chemical nature of the polysaccharide, ionic strength, ageing and temperature.

This contribution is restricted to the surface tension properties of xanthan, alginic acid, agarose, and the two glycolipids A and B which have been identified as components within the biofilm formation during fermentation in the presence of metallic copper and as model compounds for MIC. In addition, attempts have been made to determine the folding of the chains of xanthan adhering to a copper surface in contact with water by X-ray reflectivity measurements. Since the conformation of the xanthan has been determined by static light scattering measurements in solution, the appropriate shape parameters, e.g. radius of gyration, second virial coefficient etc. are known, and the obtained X-ray reflectivity spectra can be compared easily with those measured by inelastic light scattering and small-angle X-ray scattering results in aqueous solutions.

2. Materials and Methods

Static and quasi-elastic light scattering measurements were performed in an ALVLSE 3000 goniometer (Langen, Germany) as described in detail in [20].

Version surface tension measurements were performed using a Kruss Tensiometer K12 (version 2.0) at 20°C, using the Du Noüy (Pt–Ir) ring method applying the appropriate correction factors [21]. The surface tension values were converted to surface pressure (π) by using the relationship:

$$\pi = \gamma_{H_2O} - \gamma \tag{7}$$

The ring, plate, and all glassware were cleaned in freshly prepared chromic acid, and then rinsed with double-distilled water. The ring was subsequently flowed. The instrument was calibrated with pure water. Applying the ring method the surface tension is determined from the maximum force excepted on the ring without detachment of the meniscus. Applying correction factors for obtaining final values the relative accuracy is better than 0.5%; the sole source of error is the assumption that wetting of the ring or plate is complete.

Contact angle measurements and absorption measurements were performed with the same instrument (Version 1.09c, 1990) applying the static method. Data for advancing and receding angles were recorded.

X-ray reflectivity measurements were performed on a small-angle X-ray goniometer using Cu-radiation (λ = 0.1541 nm). The incident beam was collimated and monochromatised by a graphite crystal and various slits. The residual divergence amounted to 0.01°. The intensity of the incident beam and the background levels are

3×10^6 and 10.0 counts per second, respectively. The X-ray reflectivity curve was recorded in steps of 0.01 between 0.1° and 2.25°.

The theory of total external reflection is based on Fresnel's formulaes of classical electrodynamics, and has been applied to X-ray optics by Parratt [22], which has also been applied here.

Multilayers of xanthan films were prepared by the conventional Langmuir Blodgett dipping technique. The subphase consisted of a solution of H_2O or 0.01 M TRIS-HCl, pH 7.0, containing 0.1 M NaCl. As substrates Cu-wafers have been used. Monolayer transfer was performed at a constant surface pressure of ~ 20 mN m^{-2}, and at a dipping speed of 10 mm/min for xanthan, 5 mm/min for alginic acid which is pH dependent, and 10 mm/min for agarose.

Ellipsometric measurements were performed with a Rudolph & Sons (Type 43603200E) instrument, using light from a NEC-Laser, 50 mW He–Ne of wavelength of 63.28 nm. The ellipsometric data were analysed by a computer program developed by McCrackin [23].

Exopolymers were produced by fermentation in the presence of Cu-metal rings: the fermentation broth consisted of 10 g glucose, 2.5 g KH_2PO_4, 2.5 g albumine, 10 g urea, 0.8 g peptone in 2.5 L of double distilled water, the pH was maintained at 6.0 and automatically regulated. Inoculation was performed with commercially available micro-organisms (American Tissue Culture, CA, USA) from soil and water, respectively, comprising 12 strains of microorganism (EPA approved, Washington, D.C.) among them *Pseudomonas aeruginosa* 44T1, *Xanthomonas comprestis* (ATCC 3/923), *Xanthomonas comprestis*, atypically salt-regulative, (ATCC # 13951), producing alkali-labile deacylated polysaccharides; *Bacillus subtilis vulgatus subsq. hydrolyticus* (ATCC 6984), and *Torulopsis* spp (ATCC 34356). In addition the presence of other aerobic-facultatively anaerobic bacterial genera are present as *Acinetobacter*, *Agrobacterium* (~5%), *Alcaligenes* (22%), *Nocardia Arthrobacter* (~20%), *Cellulomonas* (~2%), *Mycobacterium* (~1%) and *Flavobacterium* (~2%) [24]. The temperature was maintained at 19.5°C through temperature controlled thermostated baths over 2–10 days. The broth was aerated with pO_2 and pCO_2 controlled, and stirred at 600 rpm [25]

The cells were harvested, and centrifuged at 1000 rpm (4°C, Beckman Centrifuge Model J-2-21, Rotor 21) washed with 0.1 M salt, and stored at – 20°C, or lyophilised after adsorbing to silica. The turbid supernatant was concentrated by flow dialysis, analysed for EPS (Morgan–Elson reaction), and glycolipids A and B were extracted by acetonitrile, analysed by TLC, GC and FAB-MS (fast atomic bombardment mass spectroscopy, positive ions, matrix thioformamide–glycerol) [26]. The chemical analysis yielded the following composition: glycolipids A and B; xanthan, acetylated, deacetylated, pyruvated and de-pyruvated, pulluhn, dextran and 3-poly-3-hydroxy butyrate in an amount of 10% (w/w), and approximately 5% (w/w) sulphated organic material, bound to the carbohydrate. Protein determination was performed as described in [27], and found to be of the order of 5% (w/w).

3. Results
3.1. Surface Tension Measurements

Figures 1 and 2 show the FAB (MS) results for purified glycolipids A and B; Figs. 3 and 4 their chemical structures, which are in accord with those found in the literature [28]. However, we are more interested in the surface active properties, e.g. surface tension, so we measured the dependency of the surface tension vs concentration of both glycolipids in aqueous solutions over various pH (Figs. 5–8). Figures 9 and 10 show the change of the surface tension vs concentration of the organic extract of the fermentation broth at different pH since we observed during fermentation a considerable decrease of the surface tension in a pH dependent manner and flocculation of fermenting materials in the presence of copper rings. Apparently the presence of Cu-metal induces an increase in the production of glycolipids A and B accompanied by a decrease of the surface tension. Apart from the low CMC of the order of $4–5 \times 10^{-8}$ mol L^{-1} (20°C) for these two glycolipids, the drop in surface tension from 72 to 20 or even to 5 mN m^{-1} is quite surprising. The Gibbs free energy of aggregation ($\Delta G°$) has been calculated to 51 kJ mol^{-1} for glycolipid A and 45 kJ mol^{-1} for glycolipid B. Similar values were obtained for $\Delta G°_{ad}$, the Gibbs free energy of adsorption at the air/water interface.

Since surface and/or interfacial reduction depends on the replacement of solvent molecules at the interface by e.g. glycolipid molecules, the efficiency of the glycolipids A, B, and for xanthan (Figs. 11, 12) in reducing surface tension should reflect the concentration of these biosurfactants at the interface relative to that in the bulk liquid phase, hence influencing the interface interactions with the solid Cu surface. A reasonable measure for the efficiency with which a biosurfactant of this structure

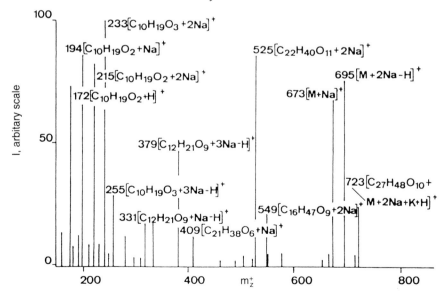

Fig. 1 Fast atomic bombardment–mass spectroscopy (FAB–MS).

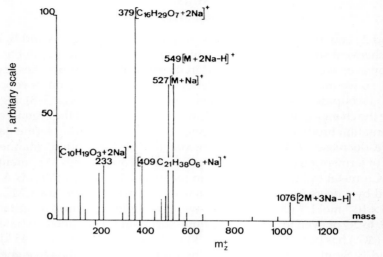

Fig. 2 FAB-MS spectrum of glycolipid B.

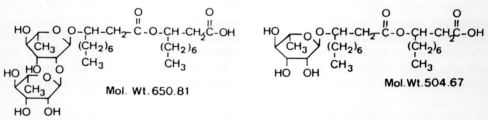

Fig. 3 Molecular structure of glycolipid A.

Fig. 4 Molecular structure of glycolipid B.

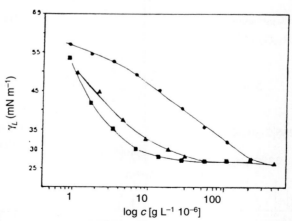

Fig. 5 Surface tension (γ) vs concentration of the organic extract from fermentation in the presence of Cu rings at different pH-values; ▲ pH 7.0, ■ pH 4.5, ● pH 7.9.;unpurified glycolipids (~80% pure).

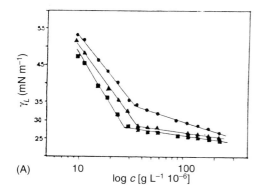

(A)

Fig. 6 Surface tension (γ) vs concentration of pure glycolipid A at different pH values. Symbols are the same as shown in Fig. 5.

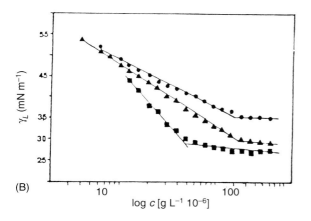

(B)

Fig. 7 Surface tension (γ) vs concentration of the organic extract as described in Fig. 5, for glycolipid B, contaminated with native xanthan.

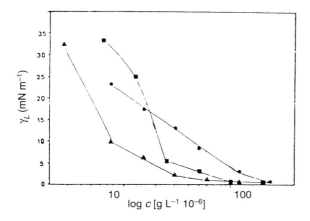

Fig. 8 Surface tension (γ) vs concentration of pure glycolipid B at different pH values. Symbols are the same as shown in Fig. 5.

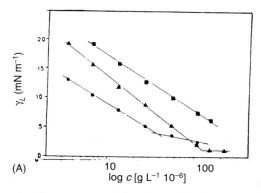

Fig. 9 Interfacial tension $\gamma L_{1,2}$ vs concentration of pure glycolipid A at different pH-values in the presence of n-hexane as a sub-phase. Symbols are the same as shown in Fig. 5.

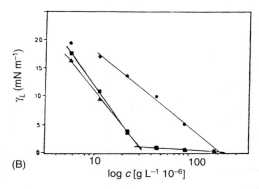

Fig. 10 Interfacial tension $\gamma L_{1,2}$ vs concentration of pure glycolipid B at different pH-values in the presence of n-hexane as a ninterphase. Symbols are the same as shown in Fig. 5.

can perform this function would therefore be the ratio of the concentration of biosurfactant at the surface to that in the bulk phase at equilibrium, e.g. $[c_1^{surf}/c_1]$, where both concentrations are in mol L^{-1}. The surface concentration of biosurfactant $[c_1^{surf}]$, in mol L^{-1}, is related to its surface excess concentration Γ_1, in mol cm^{-2} by the relation

$$[c_1^{surf}] = (1000\Gamma_1/d) + c_1 \qquad (8)$$

where d equals the thickness of the interfacial region, in cm. Γ_1 was found to be in the range of $\sim 5 \times 10^{-10}$ mol cm^{-2}, while $d = 50 \times 10^{-8}$ cm or even less and c_1 is of the order 10^{-6} M or less. Thus $[c_1^{surf}] = 1000\Gamma_1/d$ without significant error, and relates to

$$[c_1^{surf}]/c_1 = 1000\Gamma_1/d \cdot C \qquad (9)$$

Since the surface tension has been reduced on the average by 20 mN m^{-1}, the

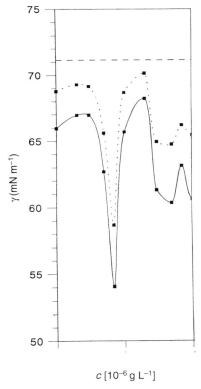
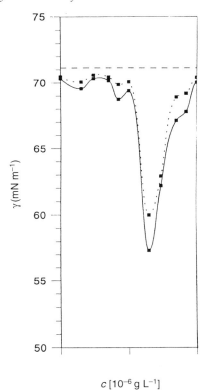

Fig. 11 Interfacial tension γ vs concentration of native xanthan in water (20°C).

Fig. 12 Interfacial tension γ vs concentration of native xanthan in 0.01 M TRIS-HCl, pH 7.5, containing 0.01 M NaCl(20°C).

value of Γ_2 is close to its maximum value Γ_m, so most of the glycolipids are lying slightly tilted to the interface (tilt angle ~10–15°). Assuming that the thickness of the interfacial region d (5.0–5.5nm) is determined by the height of the glycolipids normal to the interface, then d is inversely proportional to the minimum surface area per adsorbed molecule a_m^{surf}. So a larger value of a_m^{surf} indicates a smaller angle of the biosurfactants with respect to the interface, and a smaller value of a_m^{surf} indicates an orientation of the glycolipid more perpendicular to the interface since $a_m^{surf} = (K/\Gamma_1)$ or $(1/d)$. The quantity Γ_1/d can be considered to be constant, and $[c_1^{surf}]/c_1 = (K_1C_1)_{\pi\,=\,20}$, where K and K_1 are constants, which are temperature and pH dependent parameters. This indicates that the bulk concentration of the glycolipids studied are necessary to produce a 20 dyn cm^{-1} reduction as seen in Figs 5–8, which is not only a measure of the affinity of adsorption at the liquid–vapour interface but also a measure of the efficiency of surface tension reduction by the glycolipids, hence influencing the work of adhesion considerably according to eqn (1). It is amazing that such small concentrations of glycolipids as well as xanthan are able to reduce the surface tension from 70 dyn cm^{-1} (H$_2$O) to 20 dyn cm^{-1} and less.

Similar reasoning with regard to surface tension reduction can be performed for xanthan in aqueous solutions although the absolute amount of this particular component is much less than for the glycolipids. Furthermore, the orientation of this particular macromolecule is much more complicated due to the presence of salt, thermal instability of this system with time due to drastic conformational changes in solution, aging and flocculation (Fig. 12).

3.2. Solution Studies of EPS in Water

Static and dynamic light scattering measurements of native xanthan, de-acetylated xanthan, de-pyruvated xanthan at 20°C in the concentration ranges of (0.2–0.4) $\times 10^{-3}$ g mL^{-1} were performed. This also included measurements from materials obtained through fermentation without any further purification by means of turbidity measurements [29]. The static light scattering data obtained from common Zimm-plots are listed in Table 1, and the one obtained for native xanthan is shown in Fig. 13 in 0.15 M NaCl (20°C).

Since the angular dependency in the Zimm plot at large $q = (4\pi/\lambda) \sin \theta/2$ reveals a strong deviation from a straight line toward lower values, we plotted the light scattering data as qR_θ/Kc vs $qR_{g,\,app}$ where $R_{g,\,app}$ is the apparent radius of gyration at the concentration c_2 (Fig. 14) [30]. At large q values we observed a plateau of constant value, which is characteristic of rigid rods having a magnitude of $\pi \cdot M_L = qR_\theta/(Kc)$, where M_L is the mass of a rod section per nanometer of length (Table 1).

A surprising result of the static light scattering experiments is the high maximum at $q = 1.69$ for freshly prepared solutions of native xanthan, and materials from the supernatant obtained from fermentation before and after extraction of the glycolipids A and B (Fig. 13). Although it is not quite unexpected that this system is polydisperse, $M_W/M_n = 2.1$–2.3, the height of the maximum for freshly prepared solutions of xanthan differs considerably from those so far known in the literature [31]. In addition the position of the height of the maximum is different for fermented material, revealing a change in the number of Kuhn segments per chain where the length of the Kuhn segment is a characteristic parameter of the chain stiffness [32]. Without going into further details of evaluation and validation of the data for xanthan and their derivatives obtained by static and inelastic light scattering, it can be said that native xanthan and de-acetylated xanthan behave like a double-stranded chain, but de-pyruvated xanthan reveals a linear mass density M_L. This is an indication of approximately 1.6 (1.5) strands on the average. The Kuhn segment length has increased from 250 nm for native xanthan to 320 nm for de-pyruvated xanthan.

Surprisingly, the value of 320 nm is very close to that observed for the material obtained from fermentation in the presence of Cu rings. One explanation of these changes can be offered by an increase of ordered structure of the de-pyruvated xanthan which becomes more stabilised. This is also consistent with experiments from measurements of optical activities using linear polarised light at various temperatures of the different fractions of EPS obtained by fermentation.

The increase in chain rigidity for the essentially single-stranded de-pyruvated xanthan, particularly for the specimens obtained from fermentation in the presence

Table 1. Light scattering results of native xanthan (X_n), pyruvate-free xanthan (X_{py}) and acetyl-free xanthan (X_{ac}) in dilute aqueous solutions (20 °C)

	X_n	X_{py}	X_{ac}
$M_w \times 10^{-6}$, Da/nm	2.80	1.35	1.85
$<R_g^2>_2^{1/2}$, exp, nm	2.91×10^2	2.39×10^2	2.09×10^2
$<R_g^2>_z^{1/2}$, exp, nm	2.90×10^2	2.41×10^2	2.10×10^2
$B_2 \times 10^3$ (mL mol) · g^{2-}	5.14	5.20	6.25
$(M/L)_{exp}$, Da/nm	182 ± 70	12.3 ± 1.5	160.0 ± 40
$(M/L)_{calc}$, Da/nm	910*	845*	850*
	$(102.2)^\dagger$	$(964.0)^\dagger$	$(92.5)^\dagger$

*computed with $l_o = 1.08$ nm; † calculated with $l_o = 0.98$ nm.

The Na+ -form of xanthan has been studied; the pyruvate residues were cleaved from xanthan by boiling native xanthan in the presence of 0.1 M NaCl and 2×10^3 M formic acid at 70°C under reflux for 1 h. The O-acetyl groups were removed in the presence of 0.1 M NaCl and sodium ethoxylate (1.5×10^3) at 20°C for 2 h, followed by neutralisation and dialysis.

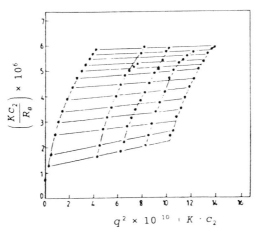

Fig. 13 Zimm plot of the light scattering measurements from a pyruvate –free preparation of xanthan in 0.05 M NaCl (20°C); $c = 0.5 \times 10^{-3}, 0.9 \times 10^{-3}$ and 1.2×10^{-4} g. mL^{-1}.

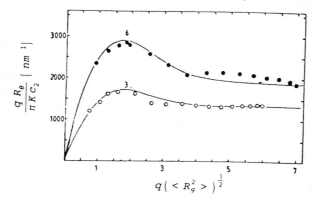

Fig. 14 A plot of $qR_\theta/pK \cdot C$ (Holtzer-plot) vs $q <Rg^2>_z^{1/2}$ at $c = 0$ for native xanthan (•) and de-pyruvated xanthan (o) at 20 °C.

of Cu-rings, may be an indication for stabilisation of the structure by winding the side chain around the backbone. If so, the terminal hydroxyl group at C4 is situated behind the acetyl group of the first sugar by forming a strong interaction.

The repeat unit length of $l_0 = 1.08$ nm was taken as a basis length for calculation of M_L which is close to that a stretched cellulose backbone (Table 1). As a result, the xanthans belong to the most rigid polymers in nature having an average thickness of $d_{th} \cong 2.0–2.2$ nm, and a segment length of $l_K \cong 240$ nm. Similar reasoning is possible for experiments with alginic acid, however, the Cu^{2+} and Na^{2+}, Me^{2+} dependency of shrinking by the polymannuronic acid also has to be taken into account [33].

3.3. Solution Behaviour of Alginic Acid in the Presence and Absence of $Cu^{\pm 0}$ Colloids

Dynamic light scattering measurements of solutions of alginic acid (0.1 % w/w) in the absence and presence of $Cu^{\pm o}$ colloids which are prepared as described in [34] were performed for determining the thickness of adsorption of the organic material onto $Cu^{\pm o}$ particles.

Figure 15 shows the adsorption isotherms for alginic acid on $Cu^{\pm o}$ colloids, whereas Fig. 16 reveals the dependence of the hydrodynamic thickness, R_H, on ionic strength for alginic acid and $Cu^{\pm o}$ colloids, respectively, as determined through photon correlation spectroscopy.

Since the hydrodynamic thickness can be (to a first approximation) related to the change of the effective diffusion coefficient, a measure of the hydrodynamic changes, as measured through the radius of gyration for comparison purposes only, can be calculated. Here R_g is the radius of gyration of the particular alginic acid sample in bulk sodium chloride or water solution. The only case where R_H values comparable to 2 R_g are found is for the low surface charge density of alginic acid layers which cover the Cu-surface, for which high Γ_2 values are found at pH 5.0–6.0. These observations give evidence for segments of alginic acid occurring as extended tails on the low charge surface due to interactions of the protonated carboxylic acid residues of

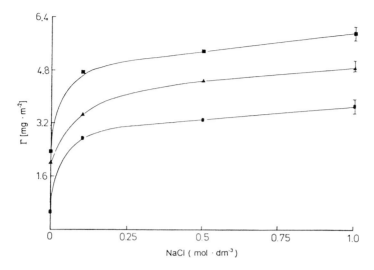

Fig. 15 Adsorption isotherms of adsorbed alginic acid to $Cu^{\pm 0}$ at an equilibrium concentration of 200 ppm at 20°C.   deprotonated alginic acid, ■ alginic acid, protonated at pH 4.5.

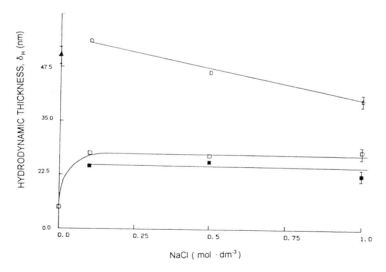

Fig. 16 Variation of RH with ionic strength for alginic acid (20°C):   protonated form, ■ deprotonated form.

alginic acid with CuO–H$_2$O. However the electrostatic attraction between the segments and the surface is much weaker. Indirect evidence for this attitude comes from the fact that at moderate ionic strengths (0.01 M) photon correlation spectroscopic measurements could not be made, because of strong precipitation and/or flocculation of the Cu-colloids particles in the presence of alginic acids. This is also consistent with the view of long tails and bridging flocculation.

Preliminary segment density profiles of bound native xanthan on Cu-surfaces were obtained from light scattering measurements at low ionic strength, $\mu = 0.01$ M. The δ (z) profile for the native xanthan reflect the generally flat conformation with folding rapidly to a value of 2.0 nm. However, there is strong evidence for tails from the shape of the density profiles: δ_H for the xanthan system is of the order of 15–18 nm, indicating there must be some tails of the xanthan exposed to the solution site. Furthermore, the shapes at low z-values suggest that there can be some loop or re-coil development as the ionic strength is increased to $\mu = 0.05$ M. This seems not to be unreasonable as the intersegmental length (Table 1) and intersegmental repulsions that occur in re-coils would be suppressed at higher ionic strengths. The higher adsorbed amounts with decreasing ionic strength is consistent with this picture. The situation is reversed in case of alginic acids, which is also pH dependent.

In the absence of added salt or in H_2O at pH 4.0 the adsorbed amount is greatly increased which is consistent with film thickness measurements by ellipsometric techniques of Cu-gels covered with alginic acid layers. For the charged surface there is an initial increase in H_2O on increasing the pH and increasing the ionic strength. This would also tend to favour the formation of an extended layer, but is offset at low pH and high ionic strength by the coil concentration due to the reduction in the intrachain repulsion. The first — effect and its pH dependency — is likely to be more important when the adsorbed layer is initially rather flat, i.e. at rather low surface charge. Above a certain salt concentration, 0.01 M, the values for all the surfaces reveal a decrease as the two effects balance one another. Similar results were found by conductivity measurements of the supernatant during later steps of the adsorption revealing increasing amounts of polyelectrolytes remaining in solution having a flexible conformation rather than a coiled one.

3.4. Ellipsometry, CuO-Film Thickness

In order to estimate the size of the oxide layer on the Cu-surface, the film formation of this oxide layer has been followed with time by ellipsometry. Figures 17 and 18 show the variation of the film thickness vs time for deaerated and aerated solutions, respectively, at pH 5.5. It is interesting to consider the mechanism of multilayer film formation. The average film thickness is approximately 20 nm for deaerated solutions, and 40 nm for aerated solutions. The film thickness on Cu-surfaces in contact with H_2O over a period of time of 4–6 h is of the order of 18–20 nm, reaching a thickness of 10 nm within 1 h.

3.5. X-Ray Reflectivity

In order to understand the surface structure of the Cu substratum including oxide layers of limited depth and of varying density, X-ray reflectivity measurements were performed. Assuming, that oxidation is complete and uniform to a certain depth, and that there is no loose packing in the oxide or in the underlying pure copper, a

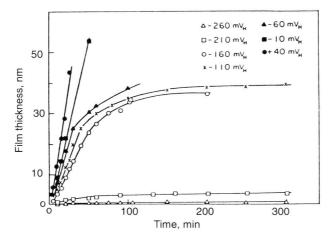

Fig. 17 *Growth behaviour of the Cu-oxide film formed at potentials more negative than the critical passive potential in pH 5.5 solutions as determined by ellipsometry.*

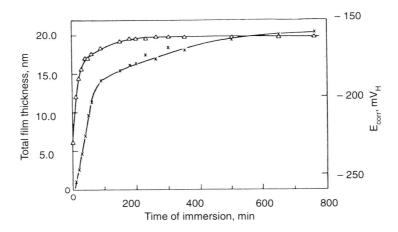

Fig. 18 *Growth behaviour of the Cu-oxide film produced in aerated solutions at pH 5.5 ($\triangle = -210\,mV$), and the change of corrosion potential as a function of time (open circuit potential measurements).*

three-media curve for this model can be proposed, and is represented in Fig. 19 for each of several oxide thicknesses.

Shape analysis of the curve of X-ray intensity vs glancing angle in the region of total reflection provides a technique for studying structural properties of the copper surface or any metal surface about 1 to several tens of nm deep. A spongy aggregate of copper crystallites was found with the average crystallite size increasing, and the intercrystalline space decreasing with depth into the surface. The lattice constant

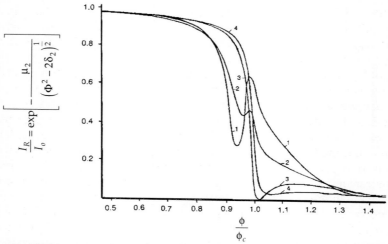

Fig. 19 *X-Ray reflectivity measurements of Cu-surface in the presence of oxide layers and xanthan (1) Cu, Cu_2O, Cu and xanthan d = 39 nm; (2) Cu, Cu_2O, d_2 = 20 nm; (3) Cu, Cu_2O with d = 15 nm and (4) Cu, clean, experimental; $\delta_2 = {}^1\!/_2\, \phi_c^2$, where ϕ_c is the critical angle of reflection.*

determined Cu_2O(0.426 nm) is larger than that of copper (0.361 nm), and so the copper crystallite in the surface swells significantly as it changes to Cu_2O. This is possibly one reason for fluffing of the oxide or oxide–bicarbonate. At some critical depth, the intercrystalline space fills up with oxide first. This effect operates as a 'seal' since continued fluffing would require a rather violent eruption of the larger mass of material above the seal. The copper crystallites below the seal remain loose packed with a thin coverage of oxide or hydrocarbonate. The electron density, calculated from the experimental data and compared with a theoretical model, may vary with depth somewhat, and with the broadening of the seal for loose-packed pure copper and for partially oxidised Cu. Oxidation below the seal continues (but much more slowly) at a rate which is determined, perhaps, by the mobility of the cation. In this situation the enrichment of organic origin is considerably enhanced. The effect of the 'reflection trap' in the surface can result from the degree of close-packing of crystallites which gradually increase in size with depth. The 'crystallite' size may be first as small as individual molecules, i.e. a liquid like surface with essentially no intercrystalline space. Also, in competition with oxidation, especially in the presence of a biofilm, or of a liquid like surface affecting the density maximum, limited-depth intercrystalline adsorption of oils, biofilms, metabolic products, e.g. pyruvic acid or debris of bacteria can occur. Furthermore, electrons which move from inside the copper (or oxide) surface towards the oxygen, carbon dioxide or groups of the biofilm in a process similar to chemisorption, will result in an excess of electrons at the surface (*n*-type material). This would contribute to the electron density maximum and minimum configuration, however the relative magnitude of this contribution is only small in the presence of oxides, although significant in the presence of a biofilm. It can be assumed that in the early stages of film deposition the copper crystallites grow vertically, and 'bridge over' so as to enclose holes at the substrate.

Then, as the film thickness increases, such vertical aggregation with subsequent lateral bridging will continue to occur, so that holes that are much smaller than those adjacent to the substrate will be enclosed and the more deeply embedded holes will tend to collapse — except at the substrate.

The curve of Fig. 17 contains interference maxima and minima in the region $\phi/\phi_c > 1$ which are not included for clarity, but the second maximum in the 18.0 nm curve appears just beyond $\phi = 10^{-2}$ rad. The curve of thickness gives a good agreement with the experiment, and shows also the reduced intensity, but not the rigid shape in the region $\phi/\phi_c > 1$, which is in accord with findings reported by Parratt [22]. Since we can assume that the (evaporated) copper surface is somewhat spongy, and that interior oxidation is probably present, and that the oxide on the surface is not expected to penetrate more than about 2–3 nm, we can assume the presence of a spongy aggregate of Cu crystallites. This results in an increase of the crystalline space, so the intercrystalline space decreases with depth into the larger surface [22]. Since the lattice constant of Cu_2O is 0.426 nm and considerable larger than that of pure copper of 0.361 nm (ignoring the different lattices), the Cu-crystallites in the surface will swell considerably as they change to Cu_2O.

At some specified depth and in the presence of water introduced by the hydrated biofilm, the intercrystalline space will be filled up with oxide. According to Parratt [22] the mechanical forces involved in fluffing and loosening of the Cu-surface can, at this depth, effect a 'seal' rather than continue fluffing. If such a 'seal' is being formed, further free access of oxygen *below* this seal is prevented and any further oxidation is not possible. This implies, that the Cu-crystallites below the 'seal' will remain loose packed, although with a thin coverage of oxide, probably amorphous. On continuing exposure to oxygen, the 'seal' extends upward towards the Cu-surface, hence becoming thicker. However, on exposing the oxidised Cu-surface to a solution of EPS in water (5 ppm xanthan, or 4–6 ppm glycolipid, or 5 ppm alginic acid) a large eruption of this 'seal' can occur resulting in the occurrence of a large mass of Cu-material above the seal (Fig. 19).

Our measurements were carried out under *in vitro* conditions. These conditions do not permit reasonable separation and control of the physical variables as well as chemical variables, such as complex formation, or flocculation of Cu^+ or Cu^{2+} materials with the EPS which is, believed to be involved in the deterioration of the Cu-surface. It is impossible to examine Fig. 17 in terms of a realistic combination of the three surface features (Scheme 1) involved, i.e. uniform porosity, an oxide layer of limited thickness of varying density (amorphous vs crystalline) and the huge reflection trap identified in the presence of EPS. More theoretical and model calculations are needed to explain the X-ray reflectivity data obtained for this particular problem, e.g. by combining neutron reflectivity measurements (determining the influence and conformation of the EPS) with X-ray reflectivity experiments, and molecular dynamic simulations. This is particularly important in light of the recent findings of the fractal nature of flocculated colloid copper hydroxide sols prepared in the presence of EPS 1351. In addition, recent studies have shown that self-assembled anisotropic microstructures, including those from lipids, could be metallised in the presence of inorganic oxides or unmineralised with metal carbonates and/or hydroxides by chemical precipitation [36, 37]. This is also consistent with the X-ray reflectivity experiments, since chemical precipitation was found not to be highly specific towards organic materials. The tailoring of exopolymeric materials in the presence of copper at the surface or beneath the surface ('seal') apparently yielded large exopolymeric microstructures of high anisometric configuration as shown phenomenologically in Fig. 19. These structures can also serve as nucleation sites for copper–exopolymer composites of anisometric morphology.

3.4. Contact Angle Measurements

Changes or chemical modification of the surface, e.g. metallic or oxidised surface as well as surfaces of the biopolymers, alter their free energy of interaction with contacting solids or liquids while leaving the bulk physical properties largely unchanged. The ability to control interfacial interactions by chemical modification of a surface is important for adhesion, biocompatability [38], static discharge [39] which all depend on wetting or on hydrophilicity [40]. Studying wettability through measuring contact angles enables us to evaluate the relations between the structures of the surface (Scheme 1) and of surface-modified exopolymers. This is also true for the fundamental mechanisms governing bacterial adhesion to solids, which apply a thermodynamic approach to adhesion.

Applying the semi-empirical approach of Fowkes [41], contact angles θ as measured with pure Lifshitz–van der Waals liquids (LW) yield γ_s^{LW} of the solid, once γ_L^{LW} of the liquid is known according to

$$\cos\theta = -1 + \left(\frac{2\gamma_s^{LW}}{\gamma_L}\right)^{\frac{1}{2}} / \gamma_L \qquad (10)$$

where γ_s^{LW} denotes an effective surface tension due to dispersion or van der Waals components of the interfacial tension for the solid, γ_L is the surface tension of the Lifshitz–van der Waals energy for the liquid. For liquids where $\gamma_L^{LW} = \gamma_L$, eqn (10) infers that Zisman's critical surface tension corresponds to γ_s^{LW}. Moreover, if both the solid and the liquid are polar, the following equations are important:

$$\gamma_L(\theta \leq 0) \leq \left(\gamma_s^{LW} \ \gamma_c^{cw}\right)^{\frac{1}{2}} + \left(\gamma_s^+ \gamma_L^-\right)^{\frac{1}{2}} + \left(\gamma_s^\ominus \cdot \gamma_L^\oplus\right)^{\frac{1}{2}} \qquad (11)$$

$$\gamma_s = \gamma_s^{LW} + 2\left(\gamma_s^+ \cdot \gamma_s^-\right)^{\frac{1}{2}} \qquad (12)$$

Physically, eqns (11) and (12) indicate that for the condition, θ = 0, for polar liquid l on a solid s, the relevant surface parameters can be achieved in a number of different ways [42, 43]. Equations (11) and (12) can be applied for monopolar solids that are Lewis bases, and for monopolar solids which are Lewis acids. An independent criterion for identifying solids and interfaces that are monopolar bases or acids, respectively, are measurements of contact angles with a monopolar liquid that has the same polarity yielding the same apparent γ_s^{LW} as does an apolar liquid, yielding the values listed on Table 2, whereas Table 3 lists the measured contact angles relevant for solids adhering to copper surfaces. Using dimethylsulphoxide (DMSO), and the apolar liquids diiodomethane or α-bromonaphthaleine as test liquids with known γ_L, γ_s^{LW}, γ_L^+ and γ_L^- values, the following biopolymers appear to be γ^- monopoles: cellulose acetate, hydrolysed gelatin, serum albumine. Bipolar solids are found to be α-D-glucose, cellulose, and pyruvate. On the basis of structure, sucrose and agarose as well as cellulose acetate are effectively γ^θ monopoles, and alginic acid and native xanthan and derivatives are, as expected to be bipolar, since the protons of the, hydroxyl groups should be electron acceptors. One explanation for the high γ^θ value is that there are fewer numbers of hydroxyls pointing outward to the adjacent phase. In these structures all the hydroxyls of the surface carbohydrate rings are Lewis-neutralised by hydrogen bonding to the Lewis base oxygen atoms of neighbouring hydroxyls.

However, when a stronger Lewis base is present in the other phase or in H_2O, it is expected that these particular hydroxyls are so bonded to oxygen atoms that these oxygens are being taken away by the external Lewis base, and are thus turning from 'inward to the bulk phase' to complete 'outward orientation'. According to Table 1 there are grave distinctions between deacetylated, depyruvated xanthans vs native xanthan of freshly prepared solutions and aged ones! There are also severe differences with films composed of alginic acid.

Another explanation is surface hydration because adsorbed water molecules can be bound tightly to surfaces so that they are not easily desorbed, and the Lewis acid character of underlying groups may not be manifested. Proteins or proteinaceous material are also effectively γ^θ monopoles. This is consistent with AT-FTIR spectroscopy and surface enhanced Raman scattering experiments because of the orientation of the NH_2 groups toward Lewis base groups in the bulk as described above for the case of hydroxyls [44]. Interestingly, the γ_s^{LW} and γ^θ-values can be related theoretically for dispersion interactions, and to the Hamaker constants [45] which is particularly of interest at low ionic strength (≤ 0.01 M), since the Debye length λ_{DH} varies with

$$\lambda_{DH} = \frac{3.04}{(salt\ concentration\ [M/L])^{\frac{1}{2}}} \mathring{A}$$

for 1:1 electrolytes. The free energy of interaction between two planar surfaces as revealed by the copper surface, and the orientation of the exopolymers on the metal surface per unit area is

$$W(R) = -\frac{A_H}{12\ \pi\ D^2} \qquad (13)$$

where A_H is the Hamaker constant, which for interactions between pure hydrocarbons or liquid bilayers (e.g. glycolipids) across solutions aqueous is approximately $(4-8) \times 10^{-14}$ erg. A_H depends strongly on the nature of the particles and the medium, and this normally in the range $10^{-12}\gamma < A_H < 10^{-19}\gamma$. For small values of distances x, the attractive parts of the Lifshitz–van der Waals forces decreases as x^{-1} whereas for $x \gg 1\ \gamma_L$ it decays as x^{-6}. So the repulsive term ($V_1^R(x)$) of the total potential between copper surface and exopolymer dominates for large x and $(V_1^R(x))$ for small values of x if the factors are of comparable magnitude. Adjustment of the ionic strength which is very low in the studied case alters the values of y according to

$$\beta V_{11}^R(x) = \frac{\gamma \exp(-bx)}{x} \quad (a_p/\lambda_{DH} \ll 1) \tag{14}$$

with $\beta = 1/k_B \cdot T$, $b = 2a_1/\lambda_{DH} = d_1/R_{DH}$, $x = r/d$ is the reduced distance of separation, and $\gamma = (\beta \epsilon\, a_1 \Psi_0^2 /) \exp(b)$ within the thin film atmosphere limit applicable to our measurements, the following expression is valid

$$\beta V_{11}^R(x) > \gamma \ln[1 + \exp(-bx)] \quad a_p/\lambda_{D11} \gg 1 \tag{15}$$

a_p/λ_{DH} equals the ratio of the macro-ion to the screening length, λ_{DH}.

It has long been known that the surfaces of synthetic and biological polymers carrying polar groups as well as functional groups are capable of reorientation, when in contact with polar condensed phases, to expose the polar functionality.

Alginic acid, glycolipids A and B, including aged xanthan provide a remarkable example of this behaviour: the sensitivity of wetting to conformational changes within the θ interphase revealing a surprisingly large change with pH in the wettability by water (Fig. 20). At low pH (4.0–6.5), the advancing contact angle of water is approximately 80°, i.e. the surface is more hydrophobic than 'unfunctionalised Cu/CuO-surface'. At pH above 6.0, the advancing contact angle is approximately 40°. The differences in the contact angles of water at low pH between Cu/CuO and the pH-dependence of the exopolymers cannot be interpreted solely in terms of hydrophobicity, since the surface of Cu/CuO is rougher than that of the biological polymers, particularly in light of the results listed in Table 2 obtained by calibrated monopolar liquids. This is also valid for the reverse behaviour of the contact angles for the glycolipids.

It can be inferred that the large decrease in the contact angle, e.g. a change of approx 30°, of water on the Cu/CuO surface as the pH of the contacting ring is varied from 4.0 to 8.0 is due to conformational changes of the 1,4-linked β-D-mannuric acid residues protonated or deprotonated, respectively (see Figs. 21, 22).

In addition, calculation of the interfacial energies between 1,4-linked β-D-mannuric acid and the α-L-anomeric isomer resulted in differences of – 10 kJ mol^{-1} applying the Fowkes theory [41]. Moreover, the hysteresis curve of advancing and receding contact angles for both pure substances including their pH is different from each other. Since alginic acid is a linear macromolecule with varying amounts of α-(1→4)L-glycosyluronic acid residues, of which the relative proportions vary with the botanical source and state of maturation, it is safe to state that this conformational change as observed for alginic acid is basicly dependent on the conformation of the polysaccharide in a pH-dependent manner. We suggest that when in contact with a basic pH, the carboxylate ions are exposed into solution (Figs. 21, 22) by rotating the pyranose ring around the other linkage through ~ 78°. This allows solvation, binding of counter ions and swelling, including the formation of multilayers. When the macromolecule occurs in contact with acidic water (pH 4.0), the protonated carboxylic

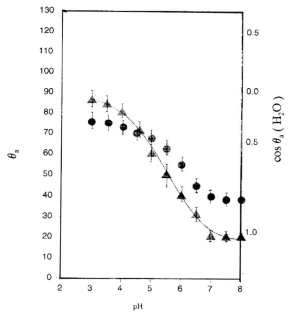

Fig. 20 Variation of the advancing contact angle θ_a of water as a function of pH on alginic acid on Cu-surfaces at 20°C (▲ alginic acid, • xanthan).

acid residues lost its functionality with its Cu/CuO surface, less through a transition state and oriented toward the bulk structured water, whereas the keto groups of the carboxyl groups are also hydrogen bonded. One question still remains: why should alginic acid or similar biological macromolecules have a conformation that minimises energetically favourable polar interactions, and maximises hydrogen bonding between the carboxylic groups by entropic ordering and the contacting water at the interface between Cu/CuO and the exopolymer? One explanation can be offered by fitting the contact angle results listed in Tables 2 and 3, as well as supporting Schemes 2 and 3 that strong hydrogen bonding (hydroxyl–H_2O, carboxyl-carboxylate interactions of the type hydroxyl–carboxylate, and carboxyl–carboxyl interactions having cyclic symmetry tending to the highly organised) between the functional groups present at the surface of Cu/CuO make it feasible for the macromolecular component to present a relatively low-energy surface (Fig. 23).

It is believed that an acid surface, in the absence of contacting water, forms a dimer structure. A schematic representation of this structure is reasonably well presented by:

$$(R_n^-\ CO_2H)_2 + n(H_2O)_{abs} \leftrightarrows (R_n \bullet CO_2H)_2 \bullet H_2O)_x \leftrightarrows 2(RCO_2H) \bullet (H_2O)_y$$

This structure is clearly perturbed by contacting water, at natural pHs as shown in Fig. 23.

The multilayer formation of the macromolecules onto the Cu/CuO-surfaces can appear in different ways as shown in Schemes 2 and 3.

Table 2. Surface parameters for relevant surfaces obtained from contact angles obtained by applying diiomethane, α-bromonapthalene, and glycerol and water

Solid	γ^{LW}(mJ m^{-2})	γ^+(mJ m^{-2})	γ^-(mJ m^{-2})
Xanthan, fresh	45	1.5	60.0
Xanthan, aged	47	2.0	55.0
Xanthan, deacetylated	40	0.14	25.0
Xanthan, de-pyruvated	37.5	0.35	21.5
Alginic acid, pH 7.0	38.5	0.21	27.0
Agarose	40	0.12	26.0
Glycolipid A,	39	0.4	22.5
Glycolipid B,	40	0.4	22.4
Celluloseacetate,	37.5	0.4	22.9
Serum Albumine	40	1.0	1.5
Hydrolysed Gelation	50	1.2	25.0
Gelatin	35	0.5	19.5
Broth Extract	41.5	2.1	22.5
Alginic Acid,m pH 7.8	38.9	0.35	24.0
Alginic Acid, pH 6.0	335.2	0	22.4
Alginic Acid, pH 4.8	34.3	0	25.0
Cast Film, Alginic Acid, dry	39–45	(0)	22.4

Scheme 2

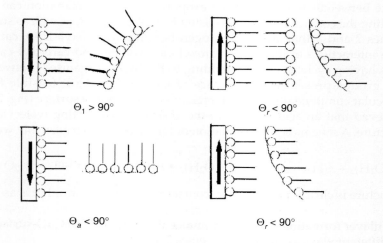

$\Theta_1 > 90°$ $\Theta_r < 90°$

$\Theta_a < 90°$ $\Theta_r < 90°$

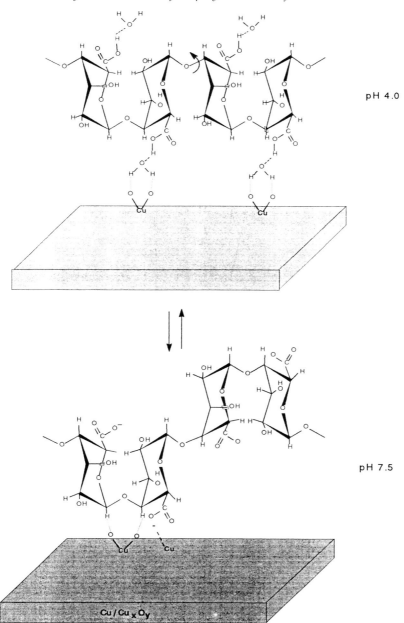

Fig. 21 *Model of the conformational changes of the β-anomeric form of alginic acid with pH.*

The most common and thermodynamically stable multilayer structures are of the Y-type in which layers are deposited in a head-to-tail, tail-to-tail fashion, particularly for alginic acids.

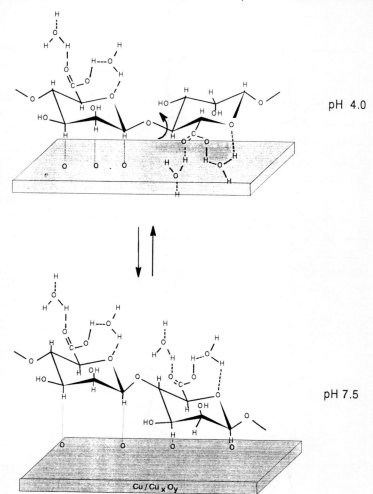

Fig. 22 *Model of the conformational changes of the α-L-polymannuric acid form with pH as measured by contact angles.*

Scheme 3

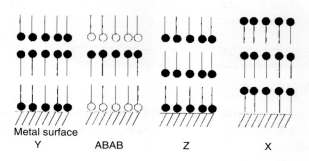

Table 3. Selected contact angle data obtained experimentally by the static method (see materials and methods)

Liquid γ (mN^{-1} m)		Solid Degrees	θ (20°C)	Remarks
H_2O	(72)	Cu	115	clean, UHV
H_2O	(72)	Cu	95	exposed to air
H_2O	(72)	Cu	45	exposed H_2O
H_2O	(72)	Carbon	80	evaporated onto Cu
H_2O	(72)	Fatty Acids	82-85	deposited on Cu
H_2O	(72)	Polyethylene	96	neat, MW 100 000
pH 7.5,	(60)	Xanthan	7.0–12.5	deposited on Cu
pH 7.0,	(60)	Alginic Acid	7.5–15.0	deposited on Cu
pH 7.0,	(58)	Agarose	17–21	deposited on Cu
CH_2I_2	(68)	Carbon, Paraffin	61	evaporated on Ci
Benzene	(30)	Paraffin	1	evaporated on Cu
n-Decane	(23)	Carbon	120	neat, on Cu/CuO
H_2O	(72)	CuO	35	

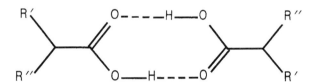

Fig. 23 Symmetry behaviour (C2) of the carboxyl–carboxylate interactions.

Polar Y-type films of the general ABAB... arrangement can be prepared by alternate deposition of A, i.e. alginic acid, and B, i.e. xanthan, with special two-compartment trough Z-type multilayers composed of molecules of the same kind in a head-to-tail arrangement, and are formed when depositing the solid substrate (Cu) to the outer side, so layering occurs only during the upstroke of the dipping cycle. Z-type structures should be obtained by decompression also, or removal of the monolayer from the air/water interface prior to the downstroke. From these examples it would be very interesting to make generalisations or predictions as to which of these amphiphilic compounds would tend to deposit in a polar structure. Therefore, it is of utmost interest in designing polar LB-films, to understand the deposition behaviour on a molecular level including the proposed membranous behaviour of the biofilm of the Cu-electrode, or any other relevant substratum.

Furthermore, the deposition behaviour is governed by the shape of the water meniscus during deposition which, in turn, is controlled by the wettability through the advancing and receding contact angles of the previously deposited layer. The hydrophilic character of the exposed surface in the various types of films increases in the order X < Y < Z. If the surface is strongly hydrophobic both advancing and receding contact angle are > 90 (Schemes 2 and 3).

In this case (hydrophobic surface) deposition occurs only during dipping of the substrate, but not on withdrawal (X-type). If the surface is less hydrophobic then the advancing contact angle is > 90° and the receding contact angle is < 90°, deposition will occur on both dipping and withdrawal (Y-type). Z-type multilayers are formed when deposition occurs only during withdrawal. Such behaviour requires both advancing and receding contact angles to be < 90° as shown in Scheme 2. Designing polar Z-type films, i.e. alginic acid or xanthan, requires a means to increase the hydrophilicity of the surface but without use of bipolar amphiphiles, i.e. xanthan, that can induce best conformations or mixtures of two different orientations at the air/water interface.

Modification of the solid surface ($Cu^{\pm o}$) influences the energetic characteristics and therefore the bacterial adhesion which is in accordance with thermodynamic predictions. This can be shown conclusively when the spreading pressure is accounted for by γ_B, which is the total surface free energy of the bacteria. The values determined are π_c; the spreading pressure of 100 mJ m^{-2}(± 8.7); γ_B^d 90.7 ± 5.9 mJ m^{-2} and γ_B^P = 9.7 ± 2.9 mJ m^{-2} (Hinze, Thies and Paradies, unpublished results). The adhesion of the bacteria is therefore driven by the balance between γ_S^P and γ_S^d, and not by the total γ_s or wettability of the metallic copper surface. So, polar interactions can reduce bacterial adhesion to a copper surface considerably!

4. Discussion

Our contact angle measurements and preliminary X-ray reflectivity results with regard to MIC for copper, are aimed at the elucidation of the influence of the surface state of the Cu-surface with respect to confined water molecules when dispersed in a solution containing EPS. This situation can be called water layering and can be calculated theoretically by a quantum mechanical approach and a statistical approach [42]. The theoretical results ultimately will be able to explain the wettability of the metal and metal-oxide surface as well as the influence of confined water molecules at the interphase of Cu/CuO and the exopolymers (Scheme 1). The interfacial energy of the Cu–H$_2$O-exopolymer system will predict its wettability, and swelling behaviour. The energy of the water-neutral Cu-surface (interface) originates from two contributions: water–water and water–Cu/CuO-surface interactions. In case of a clean non-oxidised Cu-surface the interfacial energy in the presence of water is weak, resulting in a low wettability of the Cu-surface, whereby the cohesion energy of liquid water is –80 kJ mol^{-1} and twice as large as the cohesion energy of bulk liquid water of –40 kJ mol^{-1}. However, the work of adhesion (eqn (1)) is considerably lowered by the reduced surface tension, γ_L of the water due to the presence of exopolymers.

Calculations to determine the heat released during a swelling process of the CuO-surface, taking into account the presence of 80–85% oxide as determined from contact angle measurements (eqn (1)) resulted in a nearly constant value of 0.8–1.0 J m^{-2}. This energy originates from the solvation of the Cu/CuO-surface, and is apparently independent of the presence of exopolymers. In addition this energy is much smaller than the energy required to cleave a Cu$_2$O crystal into two fragments, namely Cu (metal surface) and Cu$_2$O, explaining why Cu$_2$O particles (crystallites) immersed in H$_2$O do not swell spontaneously.

However, the total energy of the charged interface (CuO ~ H$_2$O + EPS, Scheme 1) is attractive, and of the order of –100 mJ mol^{-1}, and tends towards an EPS, interfacial energy per H$_2$O molecule close to –300 kJ mol^{-1}, e.g. one order of magnitude larger than the cohesion energy of liquid water, and four times larger than the interfacial energy calculated for Cu. However, the energies of the 'neutral' Cu-surface, and the oxidised surface cannot really directly be compared since the reference states are not the same, only the energy variations between the different states can be compared.

Hydration forces of the charge interface (Cu–O–H$_2$O + EPS) have to be considered also because the degree of hydration can be modulated through the presence of the charged or protonised exopolymeric molecules. Hydration forces of a charged interface can be defined to a first approximation as deviations of the diffuse layer theory [46]. Within the weak overlap approximation, the repulsive pressure is given by eqn (16)

$$\pi = 64 k_B \cdot T \rho^\circ \gamma^2 \exp(-\lambda_{DH} D) \qquad (16)$$

with γ as a constant related to the surface potential of the particle, φ°, and the mean salt concentration of the suspension and λ_{DH} the Debye screening constant defined as

$$\lambda_{DH}^2 = (2\epsilon^2 \varphi^\circ) / \epsilon_r \cdot \epsilon_o \cdot k_B \beta \cdot T$$

for 1:1 salt.

At large separation ($\lambda_{DH} \cdot D \gg 1$), which is the case for MIC, the swelling pressure of charged lamellar phases of detergents decreases exponentially [47], but at small distances the pressure deviates from the exponential law [48]. So, in the case of the surface active exopolymers studied here, it is tempting to assign the difference to repulsive hydration forces, but the weak overlap approximation (eqn (16)) is unfortunately only valid at large separation. But a full treatment of the diffuse layer without this approximation would describe perfectly the swelling behaviour of these charged interfaces [49].

For large separation, the pressure for the exopolymer system can be described by the diffuse layer theory, particularly at moderate salt concentrations, but it shows large oscillations for separations smaller than 2.0 nm. The periodicity of these oscillations corresponds to the sugar backbone with their hydrophilic hydroxyl groups and partially charged carboxyl groups. A partial dehydration of the interlamellar counterions, cannot explain the observed hydration forces since 2.0 nm exceeds the diameter of water and hydrated counterions. But considering the structure of the model exopolymers, it can be inferred from X-ray reflectivity measurements and contact angle experiments that near a solid particle the liquid is organised as successive layers with a complete modification of the structure of the bulk liquid. This implies that layering is short-ranged: more and more disorders appear as one moves further from the structurelising surface. The layering disappears when the solid–liquid interaction becomes comparable to the liquid–liquid or to the thermal energies.

5. Conclusion

To our knowledge the reported experiments reveal for the first time the energetics and the structural involvement of the EPS on copper surfaces. In addition, the presence of the EPS, e.g. alginate or the various forms of xanthan including the peculiar intrinsic surface tension of the glycolipids studied, provide unimpeachable proof of the molecular event leading to the destruction of the copper surface, hence dissolution of metallic copper , into the aqueous environment by organisms.

6. Acknowledgement

This work was supported by a grant from the BRITE-EURAM Project (# 4088). A copy of the Gaussian 92 program was obtained from the Quantum-mechanical Group of the University of Pittsburgh, PA, USA. The authors are grateful to Drs. W. R. Fischer, D. Wagner and A. H. L. Chamberlain for critical and extensive discussions on MIC.

References

1. R. F. Hadly, (Ed.), in *The Corrosion Handbook*, The Electrochemical Society, Inc., New York, N.Y., (1948), 466–481. Published by John Wiley & Sons, Inc., NY.
2. K. C. Marshal, *Interfaces in Microbial Ecology*, Harvard University Press, Cambridge, MA, 1976.
3. A. H. L. Chamberlain and B. J. Garner, *Biofouling*, 1988, **1**, 79–96.
4. I. W. Sutherland and M. I. Tait, in *Encyclopedia of Microbiology*, Vol. 1, (Eds M. Alexander, D. A. Hopwood, A. I. Laskin and B. H. Iglewski) pp. 339–350, Academic Press, (1992), New York.
5. A. D. French, and J. W. Brady, in *Computer Modelling of Carbohydrate Molecules*, pp. 1–19; ACS-Symposium Series 430, Washington DC, (1990).
6. Ch. Wandrey, *J. Amer. Oil. Chem. Soc.,* 1991, **68**, 124.
7. S. F. Clancy, D. A. Tanner, M. Thies and H. H. Paradies, ACS-Symp, San Fransisco, USA, Extended Abstract, Dir. of *Environm. Chem.* ,1992, **321**, 907–908.
8. M. A. Van Hove, W. H. Weinberg and C.-M. Chan, in *Low Energy Series in Surface Science 6*, Springer-Verlag, 1986, (Eds G. Ertl and R. Gower), pp. 13–46.
9. J. M. Walls, in *Methods of Surface Analysis*, pp. 1–19, (Ed. J. M. Walls), 1988, Cambridge University Press, Cambridge, UK.
10. D. G. Armour, in *Methods of Surface Analysis*, pp. 263–298, (Ed. J. M. Walls), 1988, Cambridge University Press, Cambridge, UK.
11. A. B. Christie, in *Methods of Surface Analysis*, pp. 127–168, (Ed. J. M. Walls), 198, Cambridge University Press, Cambridge, UK.

12. L. Kunz, in *Topics in Current Physics — Synchrotron Radiation, Techniques & Applications*, 1979, Springer Verlag, Berlin, pp. 1–24.
13. E. A. Sinister, E. M. Lee, R. K. Thomas and I. Penfold, *J. Phys. Chem.*, 1992, **96**, 1373.
14. E. A. Sinister, R. K. Thomas, I. Penfold, R. Aveyard, B. P. Brinks, P. Cooper, P. D. I. Fletcher, I. R. Lee and A. Sukolowski, *J. Phys. Chem.*, 1992, **96**, 1383.
15. R. J. Good, in *Surface and Colloid Science*, (Eds R. J. Good and R. R. Stromberg), 1979, Vol. 11, p. 1, Plenum Press, New York.
16. P. G. de Gennes, *Rev. Mod. Phys.*, 1985, **57**, 827.
17. A. B. D. Cassie, *Discuss. Faraday Soc.*, 1948, **3**, 11; A. B. D. Cassie and S. Baxter, *Trans. Faraday Soc.*, 1944, **40**, 546.
18. W. A. Zisman, *Adv. Chem. Ser.*, 1964, **43**, 1, and references therein.
19. R. J. Good and C. J. van Oss, in *Modern Approaches to Wettability — Theory and Applications* (Eds M. E. Schrader and G. L. Loeb), 1992, pp. 1–25, Plenum Press, New York.
20. H. H. Paradies, U. Hinze and M. Thies, *Berichte d. Bunsen Ges. Physikal. Chem.*, in press.
21. Kruess-Handbook, Hamburg, 1993.
22. L. G. Parratt, *Phys. Rev.*, 1954, **95**, 359.
23. F. L. McCrackin, A Fortran Program for Analysis of Ellipsometly Measurements, US-Government Printing Office, 1969.
24. S. F. Clancy, D. A. Tanner, M. Thies and H. H. Paradies, *J. Amer. Oil Chemist.*, in press.
25. S. F. Clancy, D. A. Tanner, M. Thies and H. H. Paradies, *J. Amer. Oil. Chemist.*, in press.
26. O. Dahl, K. H. Ziedrick, G. I. Marek and H. H. Paradies, *J. Pharm. Sci.*, 1989, **78**, 598.
27. H. H. Paradies, *J. Biol. Chem.*, 1979, **254**, 7495.
28. J. L. Pasra, J. Guivea, M. A. Mauresa, M. Robert, M. E. Mercade, F. Comelles and M. P. Bosch, *J. Amer. Oil Chemist.*, 1989, **66**, 141.
29. R. D. Camerini-Otero, R. M. Franklin and L. A. Day, *Biochemistry*, 1974, **13**, 3763.
30. A. Holtzer, *J. Polymer. Sci.*, 1955, 17, 432; E. F. Casassa and H. Eisenberg, *Adv. Protein Chem.*, 1964, **30**, 287.
31. G. Paradossi and D. A. Brant, *Macromolecules*, 1982, **15**, 874.
32. T. Coviello, O. K. Kajiwara, W. Burchard, M. Dentini and V. Crescenzi, *Macromolecules*, 1986, **19**, 2826.
33. There is a strong cooperative binding of Cu^{2+} ions on alginic acid, which is pH and ionic strength dependent. In addition, in the presence of Cu^{2+} at pH 7.0 considerable biomineralisation occurs yielding distinct micro-assemblies.
34. H. H. Paradies, M. Thies and U. Hinze, *Phys. Rev. Lett.*, submitted for publication.
35. H. H. Paradies, M. Thies and M. Hinze, submitted to *Ber. Bunsen Gesell. Physikal. Chemie*.
36. D. D. Archibald and S. Mann, *Nature*, 1993, **364**, 430.
37. S. Mann, *Nature*, 1993, **365**, 499.
38. F. J. Holy, in *Modern Approaches to Wettability —Theory and Applications* (Eds M. E. Schrader & G. L. Loeb), 1992, Plenum Press, New York, NY, pp. 213–246.
39. H. K. Christenson, in *Modern Approaches to Wettability — Theory and Applications* (Eds M. E. Schrader & G. L. Loeb), 1992, Plenum Press, New York, NY, pp. 29–52.
40. G. S. Ferguson and G. M. Whitesides, in *Modern Approaches to Wettability — Theory and Applications* (Eds M. E. Schrader & G. L. Loeb), 1992, Plenum Press, New York, NY, pp. 143–178.
41. F. M. Fowkes, *J. Phys. Chem.*, 1962, **66**, 682; F. M. Fowkes, Adv. Chem. Ser. No. 43, Amer. Chem. Soc, Washington, D.C., 1964; F. M. Fowkes and M. A. Mostafa, *I.E.C. Prod. Res. Der.*, 1978, **17**, 3.
42. C. J. van Oss, , M. K. Chandbury and R. J. Good, *Adv. Colloid. Interface Sci.*, 1987, **28**, 35–64.
43. C. J. van Oss, M. K. Chandburry and R. J. Good, *Chem. Rev.*, 1988, **88**, 927–991.

44. H. H. Paradies, D. Wagner, I. Hänßel and W. R. Fischer, submitted to *Ber. Bunsen Gesell. Physikal. Chem.*
45. H. C. Hamaker, *Physica (Amsterdam)*, 1937, **4**, 1058.
46. V. A. Parsegian, R. P. Rand and N. L. Fuller, *J. Phys. Chem.*, 1991, **95**, 4777.
47. M. Dubois, Th. Zemts, L. Belloni, A. Delrille, P. Levitz and R. Selton, *J. Chem. Phys.*, 1992, **96**, 2278.
48. H. K. Christenson and R. G. Harn, *Chem. Phys.*, 1983, **98**, 45.
49. I. Snook and W. van Megen, *J. Chem. Phys.*, 1979, **70**, 3099.

3

Bacteria Associated with MIC of Copper: Characterisation and Extracellular Polymer Production

J. N. WARDELL and A. H. L. CHAMBERLAIN

School of Biological Sciences, University of Surrey, Guildford, GU2 5XH, UK

ABSTRACT

Thirty-eight strains of heterotrophic bacteria have been isolated by conventional plating, and continuous-culture enrichment techniques from standing water and scrapings of corrosion products and biofilm from pipework exhibiting Type $1^1/_2$ pitting of copper. Isolated strains were characterised by standard microbiological methods and, where appropriate, by the API system to assign tentative identities.

The isolates were examined on an enriched solidified medium for their ability to produce extracellular polymeric material. Selected strains were grown subsequently in continuous-culture to produce extracellular polymer which was harvested and partially purified and characterised.

1. Introduction

It has been realised for many years that, although generally resistant to corrosion, copper tube will succumb to corrosive attack. Usually this general corrosion is associated with very soft, poorly buffered waters showing shifts to alkaline or acidic pH. However, a more specific type of corrosion, 'pitting' corrosion, has also been recognised [1], divided into two types (Type 1 and Type 2) depending on the temperature of operation and the corrosion products formed [2].

Within the last twenty years a further three types of pitting have been identified; Type 3, decribed by von Franque [3]; 'pepper-pot' pitting, mainly confined to the South West of Scotland where the water is soft, acidic and has a high humic content [4]; and third, Type $1^1/_2$ pitting, so called because it exhibits characteristics of both Type 1 and Type 2 [5, 6]. These latter types of pitting have been thoroughly investigated and in the case of Type $1^1/_2$ a microbial component has been identified [6]. A more satisfactory label for the Type $1^1/_2$ is 'hemispherical MIC pitting'.

Studies have indicated that the role of microorganisms in this type of corrosion is through the development of a biofilm comprising a polysaccharide matrix within which the microorganisms develop as consortia. A detailed investigation into the role of specific bacterial strains, isolated from a German hospital, carried out by Angell [7] produced evidence which suggests that specific types of microorganisms — which have a propensity to grow as a biofilm in oligotrophic environments and to produce extracellular polysaccharides — may be responsible for initiating the pitting process.

This paper describes an investigation designed to determine whether specific types of microorganism are indeed associated with biofilms and water contained within copper pipe exhibiting Type $1^1/_2$ pitting and the conditions under which they produce extracellular polysaccharide. To achieve this, strains of naturally occurring bacteria were isolated from samples of water and scrapings of corrosion products/biofilm from corroding copper pipe by direct plating and chemostat enrichment techniques. The resulting isolates, together with three strains previously isolated from a hospital in Germany were characterised and examined for their ability to produce exopolymer.

Selected strains were subsequently cultured in continuous-flow (chemostat) culture. Exopolymer was extracted and partially purified for further analysis and for use in corrosion studies.

2. Materials and Methods
2.1. Direct Plating

Samples of standing water from copper pipework known to be at risk from Type $1^1/_2$ pitting from a hospital in SW England were treated as follows. Decimal dilutions were prepared in $^1/_4$ strength sterile Ringer's solution (Oxoid) and the bacterial population enumerated by spread plates and the Miles and Misra technique using $^1/_4$ strength nutrient agar (Oxoid) [8]. Plates were incubated at 30°C for 7 days and then for a further 14 days at ambient room temperature at which time the total populations were determined and a differential count performed on the spread plates to determine the percentage population of the predominant colony types present. Representative colonies were then selected and subcultured for further study.

2.2. Continuous-culture (Chemostat) Enrichment

Culture vessel: LH500 series, working volume 1 L, fitted with a direct-drive, pitched-blade impellor (LH Fermentation Ltd.). Temperature and stirring speed were maintained at 30°C and 250 rpm respectively, while dissolved oxygen and pH were monitored but not controlled using commercially available electrodes (Uniprobe Ltd.) connected to the LH500 series instrumentation.

To encourage selection of copper tolerant organisms, and perhaps provide a niche for the attachment of periphytic forms, a length of copper wire (1 m weighing 20.7 g) was incorporated into the culture vessel; this was achieved by coiling the wire into a helix of three turns so that it fitted close to the walls, rising approximately $^1/_3$ of the height of the glass culture vessel. The wire was degreased with acetone before insertion into the lower third of the vessel prior to autoclaving. Keeping the copper submerged in reverse osmosis (RO) water during sterilisation of the vessel prevented the formation of copper (II) oxide.

Inoculum: The inoculum was 350 mL of water, collected aseptically, together with scrapings of corrosion products from within corroding pipework from a hospital in

SW England. The water and corrosion products were shaken together briefly and then used as the inoculum.

The culture vessel was filled to 1 L with medium, a time zero sample taken for viable counts and the medium flow established to give the required dilution rate.

Sampling: Samples were withdrawn from the culture vessel on a daily basis, serial decimal dilutions prepared and plated out onto $1/4$ strength Tryptone Soya Agar (TSA4) (Oxoid) and $1/4$ strength Nutrient Agar (NA4) (Oxoid) to determine the total viable count and, by differential counts, the % populations of the predominant organisms. In addition, samples of culture supernatant were collected and frozen for future analysis.

Experimental regimes: The culture was maintained for a total 2411 h during which five experimental regimes were examined. In each case the culture was allowed to equilibrate for four volume changes ($4 \times 1/D$) before initiating the next regime. The dilution rate (D) was maintained at 0.02 h^{-1} throughout.

Experimental regimes were:

1. Synthetic pitting water medium (STTW) + head-space gas exchange only.

2. STTW + forced aeration.

3. STTW + head-space gas exchange only.

4. STTW anaerobic (O_2-free N_2 bleed through sparger).

5. STTW + 100 ppm Cu^{2+} + aeration.

2.2.1. Characterisation of isolates

Each of the organisms isolated was characterised employing the following tests: colony morphology, Gram's stain, catalase, oxidase, oxidative/fermentative growth and arginine dihydrolase activity (Stewart's A-G medium) [9]. The API 20NE system (Bio Merieux) was used to provide a biochemical profile and subsequently to assign a provisional identification where possible.

2.2.2. Production of exopolymer

Isolates were screened for their ability to produce exopolymer (exopolysaccharide) by plating onto NA/2 + 1% w/v additional carbon source. Eight different carbon sources; glucose, sucrose, maltose, mannose, glycerol, acetate, succinate, and citrate were examined. Growth and mucosity were scored qualitatively 0–3 on appearance for each parameter.

Selected strains were subsequently grown in continuous culture in a nitrogen- or sulphate-limited mineral salts medium [10] or nitrogen-limited (N-lim) synthetic pitting water medium supplemented with carbon source at 5–50 mM and trace elements where necessary.

The dilution rate (D) was maintained at 0.03 h^{-1}, pH at 7.0 ± 0.1, stirring speed at 250 rpm, and the temperature at 30°C. Aeration was by sparging with air at 6 vol. h^{-1} or by passive head-space gas exchange, depending on the culture.

Spent culture was collected in 20 L volumes and bacterial cells removed using a Pellicon tangential flow filtration system (Millipore) with a 0.22 μm membrane. The cells were discarded and the exopolymeric materials (EPS) concentrated from the cell-free filtrate using a second Pellicon system with a 10 k molecular weight cut-off membrane. The resulting concentrate was re-sterilised by filtration (Nalgene 0.22 μm membrane filter) and the polysaccharide fraction precipitated with 4 volumes of cold propan-2-ol with stirring. Precipitated EPS was collected by centrifugation at $5010 \times g$ for 30 min (MSE Coolspin) and either retained as a crude gel preparation for coating electrodes (see Wagner *et al.*, this volume p. 91), or dialysed against RO water and lyophilised to a white solid.

3. Results and Discussion

Direct plating of standing water from copper pipe systems gave a population of 1.385×10^6 mL^{-1}. From this population six strains, designated SW1–6, were distinguishable on colony morphology. The colony morphologies and the percentage populations are given in Table 1.

Table 1. Direct plating of standing water from copper pipe from a hospital in SW England

Isolate	Colony Morphology	% population
SW1	round, shiny, yellow, 1–2 mm	6%
SW2	round, shiny, white, 1 mm	6%
SW3	v. small, round, brown, clear, < 1 mm	20%
SW4	round, pink, shiny, 1 mm	2%
SW5	v. small, white, round, < 1 mm	28%
SW6	'pinprick' colonies	38%
Total population = 1.385×10^3 mL^{-1}		

3.1. Chemostat Enrichment

During the first experimental regime (STTW + glucose 1.1 mM + passive head space gas exchange) one organism (brown pigmented and subsequently identified as *Flectobacillus major*) predominated in the culture at 99% of the counted population. Other organisms were detected but at a very low proportion of the culture population. Changing the experimental regime by sparging the culture with air, but in all other respects maintaining the cultural conditions as before, had a dramatic effect. A four or five membered community became established in which the previously predominant organism was reduced to < 1% of the population or could not be detected (Table 2). The likely explanation for this is that the initially low dissolved oxygen tension (DOT) enabled the brown organism to become dominant. It is likely that the organisms predominating under aerated conditions had become established within the culture vessel — by colonising the internal surfaces — and thus escaping the selective pressure acting on the planktonic population. Once conditions became favourable these organisms were able to out-compete the brown strain which declined in numbers. Reverting to passive head-space gas exchange did not lead to re-establishment of the original predominating brown strain (Table 2) but rather to a subtle shift in the relative numbers of the mixed community, suggesting that factors other than DOT may be involved.

Changing the cultural conditions again, allowing the culture to become anaerobic by sparging with O_2-free nitrogen, produced an interesting effect. An *aerobic* population, or at least facultatively anaerobic, of the order of 10^6 mL^{-1} was maintained, typified by pin-prick colonies on aerobically incubated medium. However no growth was detected on similar medium incubated anaerobically.

Finally, re-introducing aerobic conditions and, at the same time, stressing the culture by the addition of 100 ppm Cu^{2+} ions in the medium (all other parameters maintained as before) enabled a three membered community to become established (Table 2).

In all the experiments described above, the plate counts and populations isolated on TSA4 and NA4 were treated separately.

Essentially similar results were obtained for each medium. However, it was notable that occasionally the populations were slightly different. Individual strains arose which did not appear on the alternative medium and, for example, during the last phase of the experiment isolates T14 and T15 were similar to each other but apparently different from isolates N11 and N12 (see characterisation results in Tables 3 and 4, pp.55–59).

In total, 29 strains of bacteria were isolated from the chemostat enrichment together with 6 from the direct plating and subjected to further testing to characterise them and to assign provisional identities (Tables 3 and 4). For the majority of the isolates it proved difficult to assign anything other than a tentative identity since many were unreactive in the characterisation tests. Where possible names were given but these should be regarded as provisional (Table 4 (iv)). A number of strains were sent to the National Collection of Industrial and Marine Bacteria, Aberdeen, for further testing; however, similar difficulties arose due to the unreactive nature of the strains.

3.2. Screening for Production of Exopolysaccharide

The isolates obtained by direct plating and from the chemostat enrichment were subjected to screening for the ability to produce polysaccharide as described earlier.

Table 2. Chemostat enrichment: steady state populations

Regime	Age (h)	Total pop^n (cfu mL^{-1})	Isolate	% pop^n	Medium
1. Passive head-space aeration	215	7.93 × E7	T1 (brown) T2 (yellow) others to 100%	99% 0.1%	TSA/4
		2.4 × E7	N1 (brown) others to 100%	>99%	NA/4
2. Sparged aeration	629	2.547 × E7	T4 (yellow) T5 (grey) T6 (white) T7 (grey) T8 (white)	65% 5% 8% 22% <1%	TSA/4
		1.161 × E7	N2 (grey) N3 (white) N4 (grey) N5 (white) others to 100%	12% 31% 56% <1%	NA/4
3. Passive head-space aeration	980	5.195 × E7	T9 (grey) T10 (white) T11 (white) T12 (grey) T13 (white)	7% 37% 39% 17% <1%	TSA/4
		5.945 × E7	N6 (grey) N7 (white) N8 (white) N9 (grey) N10 (white)	7% 29% 41% 22% <1%	NA/4
4. Anaerobic	1438	9.166 × E5	'pinprick' colonies aerobic plates only — no growth on anaerobic plates.		TSA/4
		3.308 × E6			NA/4
5. Aerobic + 100 ppm Cu^{2+}	2411	4.2 × E7	T14 (white) T15 (white) T16 (yellow)	52% 42% 6%	TSA/4
		4.9 × E7	N11 (white) N12 (white) N13 (yellow)	50% 34% 16%	NA/4

Table 3. Chemostat enrichment isolates — colony morphology

Isolate	Colony morphology
T1	small, round, brown, shiny, 1.5 mm
T2	round, raised, shiny, yellow
T3	pin-prick, white
T4	very small, round, yellow
T5	round/irregular, grey, matt
T6	very small, round, white, shiny
T7	pin-prick, grey
T8	round, white, shiny, 1–2 mm
T9	round/irregular, grey
T10	very small, white, round
T11	pin-prick, white
T12	pin-prick, grey
T13	round, white, shiny, 2 mm
T14	small, white, shiny
T15	very small/pin-prick, white
T16	round, shiny, yellow
N1	round, brown, shiny, convex, mucoid, 2–3 mm
N2	round/irregular, grey, matt
N3	pin-prick, white
N4	pin-prick, grey
N5	round, white, shiny, 1–2 mm
N6	round/irregular, grey
N7	very small, white, round
N8	pin-prick, white
N9	pin-prick, grey
N10	round, white, shiny
N11	small, white, shiny
N12	very small, white, shiny
N13	round, white, shiny

In addition, a further three strains, designated PAW, PAY and PAB, previously isolated from corroding copper pipe in a hospital in Germany [7], were also included in the screening. Growth and mucosity were scored on a subjective basis on appearance after 4 days and 7 days incubation at 30°C. Data for day 7 are given in Table 5.

No clear pattern emerged but strains T4, N1, SW2, SW3, PAW, and PAB, appeared to grow well and produce mucoid colonies on a number of carbon substrates. Strain PAB was particularly noteworthy in this respect.

Table 4. (i), (ii) *Characterisation tests on enrichment isolates*

CHARACTERISATION TESTS ON ENRICHMENT ISOLATES															ASSIMILATION										
(na = not available)																									
IS	G	S	CAT	OX	OF	AG	NO3	TRP	GLU	ADH	URE	ESC	GEL	PNPG	GL	AR	ME	MN	NG	ML	GT	CP	AD	MT	CT	PC	OX
T1	na																										
T2	-	F/R	+	+	O	vg	-	-	-	-	-	-	-	-	+	+	-	-	-	+	-	-	+	-	-	-	+
T3	-	R	±	+	O	vg	+	-	-	-	-	-	-	-	-	-	-	-	-	-	-	-	-	-	-	-	+
T4	-	R	-	-	F	yy	+	-	-	-	-	-	-	-	+	+	+	+	+	+	-	-	+	+	+	-	-
T5	±	R	-	-	F	gy	-	-	-	-	-	-	+	-	-	-	-	-	-	-	-	-	-	-	-	-	-
T6	-	R	+	+	O	vg	+	-	-	-	+	-	-	-	+	+	+	+	+	+	+	-	+	+	-	-	+
T7	-	R	+	+	O	bg	-	-	-	-	+	-	-	-	+	+	+	+	+	-	+	-	+	+	+	-	-
T8	+	R	+	-	O	vb	-	-	-	-	+	+	-	-	+	+	+	+	+	-	+	-	+	+	-	-	-
T9	-	R	+	-	OF	gyg	-	-	-	-	-	-	-	-	+	+	-	+	-	-	-	-	-	-	-	-	-
T10	-	R	+	+	O	vg	+	-	-	-	+	-	-	-	+	+	+	+	-	+	+	-	+	+	+	-	+
T11	-	R	+	+	O	vg	+	-	-	-	-	-	-	-	+	+	-	-	-	+	+	-	+	+	+	-	+
T12	-	R	+	+	O	vg	+	-	-	-	+	-	-	-	+	+	-	-	-	-	+	-	-	+	-	-	+
T13	+	R	+	-	O	vb	-	-	-	-	+	-	+	-	+	-	+	-	+	-	+	-	-	+	+	-	+
T14	-	R	+	+	O	vg	+	-	-	-	+	-	-	-	+	+	-	-	-	-	+	-	+	+	-	-	+
T15	-	R	+	+	O	vg	+	-	-	-	-	-	-	-	+	-	-	-	-	-	+	-	-	-	-	-	+
T16	na																										

CHARACTERISATION TESTS ON ENRICHMENT ISOLATES

(na = not available)

IS	G	S	CAT	OX	OF	AG	NO3	TRP	GLU	ADH	URE	ESC	GEL	PNPG	GL	AR	ME	MN	NG	ML	GT	CP	AD	MT	CT	PC	OX
																		ASSIMILATION									
N1	-	R	+	+	O	vv	-	-	-	-	-	+	-	+	-	-	-	-	-	-	-	-	-	-	-	-	+
N2	-	R	-	-	O/F	gyg																					
N3	-	R	+	+	O	vg	-	-	-	-	+	-	-	+	-	-	-	-	-	+	-	+	-	-	-	+
N4	+	R	+	-	O/F	gyg	-	-	-	-	-	-	-	-	-	-	-	-	-	-	-	-	-	-	-	-
N5	+	R	+	-	O	vb	-	-	-	-	+	+	-	+	+	-	+	-	+	-	-	-	-	+	-	-
N6	-	R	±	-	F	gy																					
N7	-	R	+	+	O	vg	+	-	-	-	+	-	-	+	+	-	-	-	-	+	-	-	-	-	-	+
N8	-	R	+	+	O	vg	+	-	-	-	+	-	-	-	+	-	-	+	-	-	-	-	-	-	-	+
N9	-	R	-	+	O	vg	+	-	-	-	+	-	-	+	+	-	-	-	-	+	-	-	-	-	-	+
N10	+	R	+	-	O	vb	-	-	-	-	+	-	-	+	+	-	+	-	+	+	-	+	-	+	-	-
N11	±	R	+	+	O	vg	+	-	-	-	+	-	-	-	-	-	-	-	-	-	-	-	-	-	-	+
N12	-	R	+	+	O	vg	+	-	-	-	+	-	-	-	-	-	-	-	-	-	-	-	-	-	-	+
N13	na																										

Table 4. *(iii) Characterisation tests on enrichment isolates (continued)*

CHARACTERISATION TESTS ON DIRECT PLATE ISOLATES																												
(na = not available)															ASSIMILATION													
Is	G	S	CAT	OX	OF	AG	NO3	TRP	GLU	ADH	URE	ESC	GEL	PNPG	GL	AR	ME	MN	NG	ML	GT	CP	AD	MT	CT	PC	OX
SW1	-	R	+	-	0	nr	-	-	-	-	-	+	-	+	+	+	+	-	-	+	+	-	+	+	+	-	-
SW2	-	R	+	-	O	vb	+	-	-	-	+	-	-	-	+	-	-	-	-	+	+	-	-	-	-	-	-
SW3	-	R	+	-	F	yy	-	-	-	-	-	+	-	+	+	-	+	-	+	+	+	-	+	-	-	-	-
SW4	-	R	+	-	O	vv	-	-	-	-	+	-	-	-	+	+	+	+	+	+	+	-	+	+	+	+	-
SW5	-	R	+	+	0	bg	+	-	-	-	+	-	-	-	-	-	-	-	-	-	-	-	-	-	-	-	+
SW6	-	R	+	+			+	-	-	-	+	-	-	-	-	-	-	-	-	-	-	-	-	-	-	-	+

Legend: IS = Isolate, G = Gram-reaction, S = cellshape (R = rod, C = coccus, F = filament), CAT = catalase, OX = oxidase, OF = oxidative/fermentative, AG = arginine-glucose medium (arginine dihydrolase), NO_3 = nitrate, TRP = tryptophan (indole), GLU = glucose, ADH = arginine dihydrolase, URE = urea, ESC = esulin, GEL = gelatin, PNPG = B-galactosidase, GL = glucose, AR = arabinose, ME = mannose, MN = mannitol, NG = N-acetyl-glucosamine, ML = maltose, GT = gluconate, CP = caprate, AD = adipate, MT = malate, CT = citrate, PC = phenyl-acetate, + = positve reaction, - = negetive, nr = no reaction, NA = not available, b = blue, g = green, y = yellow, v = violet (for slope/butt of tube).

Table 4. (iv) Chemostat enrichment isolates — preliminary identities

Isolate	Preliminary identity
T1	NA
T2	*Pseudomonas* sp.
T3	*Moraxella / Pasteurella*
T4	*Agrobacterium* sp.
T5	*Ochrobactrum anthropi*
T7	*Pseudomonas / Sphingomonas*
T8	unknown
T9	unknown
T10	*Ochrobactrum anthropi*
T11	*Pseudomonas* sp.
T12	*Pseudomonas* sp.
T13	unknown
T14	*Pseudomonas* sp.
T15	*Pseudomonas* sp.
T16	NA
N1	*Flectobacillus major*
N2	NA
N3	*Pseudomonas*--like organism
N4	*Flavobacterium*
N5	unknown
N6	unknown
N7	*Pseudomonas/Achromobacter*
N8	*Pseudomonas* sp.
N9	*Ochrobactrum anthropi*
N10	unknown
N11	*Moraxella* sp.
N12	*Moraxella* sp.
N13	NA
SW1	unknown
SW2	*Pasteurella* sp.
SW3	unknown
SW4	*Pseudomonas*-like organism
SW5	*Moraxella* sp.
SW6	*Moraxella* sp.

(NA = not available)

Table 5. Growth/mucosity of strains on NA/2 + 1% w/v carbon source

Isolate	gluc	sucr	malt	mann	glyc	ace	succ	citr
T1	0/0	0/0	0/0	0/0	0/0	0/0	0/0	0/0
T2	0/0	0/0	0/0	0/0	0/0	0/0	0/0	0/0
T3	1/1	1/0	1/1	0/0	2/0	1/0	1/0	1/0
T4	3/2	3/1	3/2	3/1	3/2	2/1	2/1	0/0
T5	0/0	0/0	0/0	0/0	0/0	0/0	0/0	0/0
T6	1/0	1/1	1/0	1/0	2/1	0/0	1/0	0/0
T7	1/0	1/0	1/0	1/0	1/0	1/0	1/0	1/0
T8	3/0	1/0	1/1	3/0	2/0	3/0	1/0	1/0
T9	0/0	0/0	0/0	0/0	0/0	0/0	0/0	0/0
T10	1/0	1/0	1/1	1/0	2/1	0/0	0/0	1/0
T11	1/0	1/1	1/0	0/0	2/1	1/1	1/0	0/0
T12	1/0	1/0	1/0	1/0	2/1	1/0	1/0	1/0
T13	3/0	1/0	2/0	3/0	2/0	3/0	1/0	1/0
T14	1/0	1/0	1/0	1/0	2/1	0/0	1/0	1/0
T15	1/1	1/0	1/0	0/0	2/1	0/0	1/0	0/0
T16	0/0	0/0	0/0	0/0	0/0	0/0	0/0	0/0
N1	0/0	2/3	1/1	0/0	2/2	1/1	2/1	0/0
N2	0/0	0/0	0/0	0/0	0/0	0/0	0/0	0/0
N3	1/0	2/0	1/0	1/0	1/0	1/0	1/0	1/1
N4	1/0	1/0	0/0	1/0	1/0	1/0	1/0	1/0
N5	3/0	2/0	2/1	3/0	2/0	3/0	2/1	2/0
N6	1/0	1/0	1/0	1/0	1/0	0/0	1/0	0/0
N7	1/0	1/0	1/0	1/0	2/1	0/0	1/0	1/0
N8	1/1	1/0	1/1	1/0	2/1	0/0	1/0	1/0
N9	1/1	1/0	1/1	0/0	3/1	0/0	1/0	1/0
N10	3/0	2/0	2/0	3/0	2/0	3/0	1/0	2/0
N11	1/1	1/1	1/0	0/0	2/1	0/0	1/1	1/0
N12	1/1	1/0	1/0	1/0	2/1	0/0	1/0	1/0
N13	0/0	0/0	0/0	0/0	0/0	0/0	0/0	0/0
SW1	3/1	2/0	3/1	3/0	2/1	2/0	2/0	2/0
SW2	2/2	2/1	2/1	2/2	2/1	1/0	2/1	0/0
SW3	2/2	2/1	2/2	2/2	0/0	0/0	0/0	0/0
SW4	2/2	2/1	1/0	2/0	3/0	0/0	0/0	0/0
SW5	1/0	1/1	1/0	0/0	1/0	1/0	1/0	1/0
SW6	1/0	2/0	2/0	1/0	1/0	0/0	1/0	0/0
PAW	3/2	2/0	2/1	1/0	3/1	1/1	1/1	1/0
PAY	3/0	3/1	2/0	2/0	3/0	3/0	2/0	1/0
PAB	3/2	2/2	3/3	2/1	2/1	2/3	2/3	1/1

Table 6. Carbohydrate polymer production in continuous culture

Organism	Regime	C-source	Population cfu mL^{-1}	polymer* µg mL^{-1}
PAB	0.1 S-lim	succ 37 mM	2.78 × E6	3
	1.0 S-lim	succ 37 mM	2.17 × E7	11
	1.0 S-lim	succ 50 mM	8.33 × E6	3
	2.5 S-lim	succ 50 mM	4.33 × E7	8
	2.5 S-lim	succ 150 mM	2.56 × E6	0
	1.0 N-lim	succ 100 mM	5.75 × E5	4
(high DOT)	N-lim STTW	malt 1 mM	5.0 × E6	592
(low DOT)	N-lim STTW	malt 1 mM	4.7 × E6	226
PAW	N-lim STTW	gluc 1.1 mM	5.75 × E7	3
	N-lim STTW	gluc 1.1 mM	3.58 × E7	102
	N-lim STTW	gluc 2.5 mM	2.5 × E7	992
	N-lim STTW	sucr 1.0 mM	3.83 × E6	424
N1	N-lim/D.02	gluc 1.1 mM	1.4 × E7	162
	N-lim/D.01	gluc 1.1 mM	8.8 × E6	200
	N-lim/D0.2	glyc 2.0 mM	8.5 × E6	0

Legend:

PAB, PAW, N1	=	isolates
S-lim	=	sulphate limited medium
N-lim	=	nitrogen limited medium
D	=	dilution rate (h^{-1})
STTW	=	synthetic pitting water medium
gluc	=	glucose
malt	=	maltose
sucr	=	sucrose
glyc	=	glycerol
succ	=	succinate
DOT	=	dissolved oxygen tension.

*polymer – measured as total carbohydrate present in culture supernatant.

3.3. Production of Exopolymer

On the basis of the above results, strains N1, PAW, and PAB have been examined further for their ability to produce exopolysaccharide. Chemostat cultures were established using a mineral salts medium either sulphur or nitrogen-limited [10]; or the synthetic pitting water medium (STTW), nitrogen-limited; with glucose, maltose or succinate as the sole carbon source.

To date the most productive regime has been nitrogen-limited STTW with glucose as the carbon source, nitrogen and glucose being at 1 and 50 mM respectively; growth conditions otherwise being as described earlier. Strain PAB appeared to grow better under conditions of reduced DOT (passive head-space gas exchange). The culture looked more dense and certainly the culture was markedly pigmented compared to growth under aerated conditions. However, the viable counts showed little difference (Table 6, preceding page). Strain PAW appeared to grow more densely under aerated conditions (6 vol.h^{-1}). These variations will be the subject of further investigation.

Spent medium from the cultures was collected to 20 L and then processed to recover exopolysaccharide. Initially low (0.3 mg L^{-1}) yields were improved by supplementing the original STTW formulation with trace elements when yields of the order of 1–2 mg L^{-1} were obtained.

Polymer produced in this way and retained as a gel prior to dialysis and lyophilisation, or, rehydrated was used within sintered glass discs (frits) and to coat electropolished copper electrodes to investigate its potentiating effect on the pitting corrosion of copper tube. These experiments will be discussed in other papers (Sequeira *et al.*, Chapter 4 and Wagner *et al.*, Chapter 5, this volume).

4. Conclusion

These experiments have demonstrated that a variety of bacteria are present within the standing water and biofilm of copper pipes exhibiting Type $1^1/_2$ pitting.
Which organism(s) will predominate at any given time will depend on the prevailing environmental parameters. It is, however, clear that many such organisms are able to withstand marked changes in environmental conditions, re-emerging when conditions become favourable once again.

The production of exopolymer as exopolysaccharide is a feature of many of these organisms but depends both on the availability of a suitable carbon-source and on the nature of the nutrient stress imposed by the water supply.

5. Acknowledgements

This work was funded through the EC BRITE/EURAM programme: Project No. BE-4088.

References

1. H. S. Campbell, *J. Appl. Chem.*, 1954, **4**, 633–637.
2. H. S. Campbell, Corrosion and dissolution of metals by water, in *Water Treatment and Examination*, (Ed. W.S. Holden), 1970, Churchill, London.
3. O. von Franque, D. Gerth, and B. Winkler,*Werkstoffe und Korros.*, 1975, **26**, 4.
4. C. W. Keevil, J. T. Walker, J. McEvoy and J. S. Colbourne, 'Detection of biofilms associated pitting corrosion of copper pipework in Scottish hospitals,' in *Biocorrosion* (Eds L. C. Gaylarde and L. H. E. Morton), 1989, The Biodeterioration Society, London.
5. W. Fischer, I. Hanbel and H. H. Paradies, 'First results of microbial induced corrosion on copper pipes', in *Microbial Corrosion — 1*, (Eds C. A. C. Sequeira and A. K. Tiller), 1988, Elsevier Applied Science, London and New York.
6. A. H. L. Chamberlain and P. Angell, 'Influences of microorganisms on pitting copper tube,' in *Microbially Influenced Corrosion and Biodeterioration*, (Eds N. E. Dowling, M. W. Mittelman and J. C. Danko), 1991, MIC Consortium, Knoxville, TN.
7. P. J. Angell, Microbial involvement in Type $1^1/_2$ pitting of copper. Ph.D. Thesis, University of Surrey, Surrey, UK, 1992.
8. A. A. Miles and S. S. Misra, *J. Hygiene*, 1938, **38**, 732.
9. D. J. Stewart, *J. Appl. Bacteriol.*, 1971, **34**, 779–786.
10. C. G. T. Evans, D. Herbert and D. W. Tempest, 'The continuous culture of microorganisms: 2. Construction of a chemostat,' in *Methods in Microbiology Vol. 2.*, (Eds J. R. Norris and D. W. Ribbons), 1970, Academic Press, London.

4

Membrane Properties of Biopolymeric Substances

C. A. C. SEQUEIRA*, A. C. P. R. P. CARRASCO*, D. WAGNER, M. TIETZ and W. R. FISCHER

Markische Fachhochschule, Laboratory of Corrosion Protection, Postbox 20 61, 58590 Iserlohn, Germany
*Instituto Superior Tecnico, Technical University of Lisbon, Av. Rovisco Pais 1096, Lisboa Codex, Portugal

ABSTRACT

Microbiologically influenced corrosion (MIC) of copper is accompanied by a 'biofilm' that influences transport processes on the copper surfaces. The membrane properties of this biofilm are relevant to the corrosion effects on copper tubes. These membrane properties have been established from measurements either with an 'artificial biofilm' composed of exopolymeric substances or with a 'biofilm' produced by microorganisms isolated from perforated copper tubes of an affected county hospital in Germany. The selectiveness of the ionic transport properties of these materials could be established.

1. Introduction

Microbiologically influenced corrosion (MIC) has been found to cause problems in drinking water installations made of copper tubes in main buildings [1–6]. The characteristic feature of this type of corrosion is the presence of a 'biofilm' of a primarily polysaccharide nature [4]. The biofilm matrix carries negative charges in a frequently patchy distribution. The chemical composition of the exopolymeric material has been investigated and elucidated in one case in detail [4].

A physico–chemical explanation of the manifestations of corrosion observed in the attacked pipes has been put forward taking into account membrane properties of this 'biofilm' and the varying distribution of the exopolymeric material on the surface of the copper pipes [7, 8].

One important goal to validate this physico–chemical explanation is to demonstrate experimentally the membrane properties and the ionic transport properties of these exopolymeric substances. A 'biofilm' produced by microorganisms isolated from the perforated copper pipes in an affected public building in Germany [9], and an artificial 'biofilm' composed of exopolymeric substances of similar chemical composition to those of the biofilm isolated from perforated pipes in this affected building, have been used for the validation [10]. Membrane potential measurements and cation transference number experiments have been chosen as powerful techniques to validate the proposed model.

2. Theoretical Considerations
2.1. Membrane Potential Measurements

Because the 'biofilm' matrix material carries negative charges the materials under consideration are expected to show a cation selective behaviour. Therefore, for a cation selective membrane, the potential difference can be measured for a potassium chloride electrolyte with a tenfold concentration difference taking into consideration a membrane contribution and a concentration contribution, by using an Ag/AgCl/Cl⁻ reference electrode:

$$\Delta U = \Delta U_{Ag/AgCl/Cl^-} + \Delta U_{membrane} = 118 \, mV$$

with

$$\Delta U_{Ag/AgCl/Cl^-} = (RT/zF)\log[a_1(Cl^-)/a_2(Cl^-)] = 59 \, mV$$

and

$$\Delta U_{membrane} = (RT/zF)\log[a_2(K^+)/a_1(K^+)] = 59 \, mV$$

where ΔU is the potential difference, R the gas constant, T temperature, z the ionic charge, F the Faraday constant, and a_1, a_2 are the ionic activities in the two compartments. Using saturated calomel reference electrodes ($Hg/Hg_2Cl_2/KCl_{sat.}$) the membrane potential difference only will be measured.

Table 1 shows examples of calculated potential differences for cation selective membranes using different reference electrodes and different concentrations for potassium chloride electrolytes.

Table 1. Calculated potential differences for a cation selective membrane in a potassium chloride electrolyte

Difference $\log a_1/a_2$	Potential Difference $Hg/Hg_2Cl_2/KCl_{sat.}$ (SCE) $\Delta U/mV$	Potential Difference $Ag/AgCl/Cl^-$ $\Delta U/mV$
1	59	118
2	118	236
3	177	354

2.2. Transference Number Measurements

The analytical method was devised by Hittorf in 1857, and is still widely employed. Consider the cell shown diagrammatically in Fig. 1, in which silver nitrate solution is electrolysed between a silver cathode and a silver anode. If a quantity, xF (F is the Faraday Constant), of electricity is passed through the cell, x g equiv. of silver will dissolve from the anode and pass into the solution as silver nitrate; simultaneously

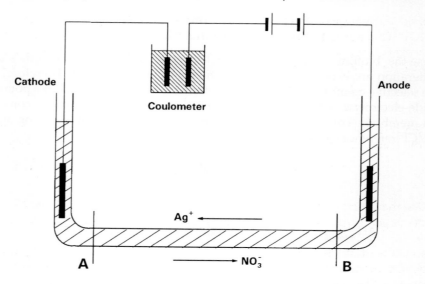

Fig. 1 Diagrammatic transport number apparatus for the performance of transference number measurements.

$x\,g$ equiv. of silver ions will be discharged and deposited at the cathode; and $96490\,x$ C of negative charge will move through the external circuit from anode to cathode, and neutralise the silver ions deposited there. The circuit is completed by the passage of $x\,F$ of charge through the silver nitrate solution by a process of electrolytic conduction; silver ions carry their positive charges towards the left hand electrode, and nitrate ions migrate to the right, and the sum of these complementary effects is the passage of $x\,F$ of negative charge through the cell from left to right (or of positive charge from right to left).

After the current has been passed for a moderate time it is found that the solution around the anode has gained in silver nitrate content, whereas the solution around the cathode has suffered a loss of silver nitrate. The greater bulk of intermediate solution is unchanged in concentration, as silver or nitrate ions migrating from a given volume are replaced by an equal number migrating into it.

Let A represent any cross-section through the solution such that the concentration at A is unchanged at the end of the experiment. That is, the decrease in silver nitrate round the cathode is all included in the cathode portion to the left of A (actual cells are designed so that this portion can be withdrawn for analysis after the experiment). This decrease in silver nitrate is fixed by (i) the cathode reaction, which is the loss of x equivalents of silver through deposition on the cathode, and, (ii) the movements, in or out of the cathode portion, of the silver and nitrate ions. These ions are present in equal numbers and carry equivalent charges, so the fractions of the current carried by each would be equal if they moved with equal velocities. They do not, however, as they differ in size, and in the resistance they meet in pushing through the solvent. If their speeds are written u for the cation and v for the anion, the fractions of the current carried by each across surface A are $u/(u + v)$ by Ag^+ and $v/(u + v)$ by NO_3^-. That is, $ux/(u + v)g$ equiv. of silver ion pass into the cathode portion, and $vx/(u + v)g$ equiv. of nitrate ion pass out. This loss of nitrate ions is accompanied by an equivalent net loss of silver ion (loss of x, deposited, minus gain of $ux/(u + v)$), so:

$$\frac{Loss\ of\ g\ equiv.\ of\ AgNO_3\ at\ cathode}{Quantity\ of\ electricity\ passed\ in\ Faradays} = \frac{v}{u+v} = t_a$$

By determining the two quantities on the left-hand side we can obtain the ratio t_a, called the anion transference number. The *cation transference number* then follows:

$$t_c = \frac{u}{u+v} = 1 - \frac{v}{u+v} = 1 - t_a$$

Notice that t_c cannot be called *the transference number of the silver ion*; transport numbers depend equally on v and u and are properties of the electrolyte solution as a whole.

The results of an electrolysis will not always be the deposition of cation at the cathode and the formation of cation at the anode, and the calculation of transport numbers must be based on a knowledge of what occurs at the electrodes.

It is also important to notice that in the argument given it is assumed that the water does not move, and the calculation is based on the *quantity* of electrolyte that a given quantity of solvent contains before and after passing current. The calculation cannot be based on *concentration* changes, because the cathode or anode portion analysed at the end of the experiment is not at an uniform concentration.

3. Experimental
3.1. Synthesis of Biopolymeric Materials

Synthetic biopolymeric materials consisting of single substances, binary mixtures and ternary mixtures of xanthan, agarose and alginate were purified and prepared as described by Wagner et al. [11]. Based on the analysis of the structure of a biofilm in an affected building, xanthan is used as a model substance, agarose as a matrix substance and alginate as a stabiliser substance. The measurement of membrane potentials and transference numbers requires a material of a stable gel-like constitution, and therefore considerable attention was paid to the swelling behaviour of each material.

The synthetic exopolymeric substances or their binary or ternary mixtures were dissolved in water to obtain concentrations of 3% xanthan, 1% agarose and 0.5% (w/w) alginate. To obtain the ternary mixture, (i) half of the final volume of distilled water was added to agarose in one beaker, and, (ii) half of the final volume of distilled water was added to a mixture of alginate and xanthan in a second beaker. Both solutions were heated in a water bath close to the boiling point with stirring. The agarose solution became transparent and liquid after 10 min, the xanthan/alginate solution became viscous after about 30 min. The obtained gelatinous biopolymer was covered and stored in a refrigerator for swelling until a stable constitution was obtained.

Biopolymeric materials produced by microorganisms were obtained from batch and continuous culture of a bacterium isolated from corroding copper pipe [9]. The organism was grown in batch and continuous-culture and the spent medium processed to recover extracellular material. Measurements were conducted with EPS # 930 406 B, EPS # 930 604 F1, EPS # 930 507 F1 and EPS # 930 820, the numbers indicating a laboratory code. Further information about these exopolymeric substances is given in reference [9].

This biopolymeric material was obtained either in a freeze-dried constitution or in a gelatinous water/2-propanol suspension. The different pretreatments of these materials are described in Section 4.2.

3.2. Membrane Potential Measurements

For membrane potential measurements a two-compartment cell as described in Fig. 2 was used. The compartments were filled with aqueous electrolytes prepared from one salt but with different concentrations. To keep the chemistry of the electrolyte as simple as possible, potassium chloride as a 1:1 aqueous electrolyte was chosen for these experiments since monovalent ions cause higher potential differences than bivalent ions.

The outer compartment was a beaker containing the electrolyte of the lower concentration. The inner compartment consisted of a flexible PVC tube with an inner diameter of 7 mm containing the electrolyte with the higher concentration. The ratio of the volumes was *ca.* 50 : 1 (lower to higher concentration) in the two compartments. Care had to be taken that the electrolytes in the two compartments had the same height to avoid the development of hydrostatic pressure.

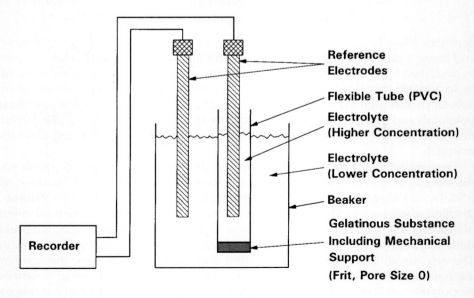

Fig. 2 Experimental set-up for membrane potential measurements.

The two compartments were electrically connected by the prepared gel-like exopolymeric material. Since the synthetic exopolymeric substances and the polymeric materials produced by microorganisms show a gel-like constitution they needed mechanical support when they were used as an ion-conducting connector in the two-compartment cell. Different materials were tested as a mechanical support for the biopolymers. The results of measurements carried out in concentration cells show that a sintered glass disc with pore size 0 is a suitable and cheap support. This support was filled with the gelatinous substance by applying a small vacuum.

Membrane potentials, established across the supported polymeric material, were measured as voltages between two electrodes dipped one into each compartment. As reference electrodes two $Ag/AgCl/Cl^-$ electrodes were used. This type of electrode is directly sensitive to the concentration of chloride within the compartment. The Ag/AgCl electrodes were prepared by polarising a silver wire in 1 M hydrochloric acid with a constant current density of 5 mA cm^{-2} for 30 min. Using two saturated calomel reference electrodes ($Hg/Hg_2Cl_2/KCl_{sat.}$) the membrane potential difference is measured only. Care has to be taken that the diffusion potential at the phase boundaries electrolyte (compartment)/electrolyte (reference electrode) is negligible.

3.3. Transference Number Measurements

Following the reasoning presented in Section 2.2, a cell has been devised for the transference number measurements as shown in Fig. 3. For most of the experiments, the cell was filled with 900 mL of diluted NaCl electrolyte solution, a current of 0.01–5 mA was passed for 1–5 h, and the porous frits (pore size 0) separating the central compartment from the electrode compartments were either (i) covered with a thin layer (< 2 mm) of an aqueous solution containing the diluted biopolymeric material, or (ii) covered with synthetic exopolymeric substances material (layer of *ca.* 2 mm thickness), which was sucked into the frits by pumping. Intermediate portions of the central solution, and electrode portions were withdrawn via taps located at the respective compartments. The several portions were weighed and analysed.

4. Results and Discussion
4.1. Membrane Potentials of Synthetic Exopolymeric Substances

On starting the experiments it was realised that parameters such as the pore size of the frit, the gel composition and location, and the electrolyte concentration, among others, needed to be optimised to enable proper membrane potential measurements to be made.

4.1.1. Frit selection
Frits of different pore sizes and a dialysis tube were tested to separate the two compartments containing 0.1 M and 10^{-3} M KCl. Due to the 100 fold concentration difference a potential difference of about 118 mV should have been observed at the start of the measurement. An experimental deviation of maximum 10 mV from this value

Fig. 3 Photo of the cell for transport number measurements.

was found as shown in Fig. 4. A strong decrease of this starting potential was observed in every case; the frit with the biggest pore size 0 showing the strongest decrease. This frit was chosen as a carrier for the exopolymeric substances under investigation.

4.1.2. Influence of gel location

These experiments were conducted using a ternary mixture consisting of 3% xanthan, 1% agarose, 0.5% (w/w) alginate and diluted electrolyte concentrations of potassium chloride with a 10-fold difference (10^{-4} M/10^{-5} M). When the gelatinous substance was placed on top of the frit in the silicon tube it took about seven hours to get the expected stable potential value of *ca.* 118 mV ± 10 mV for a time period of

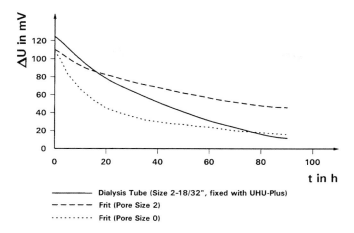

Fig. 4 Blind values of the potential difference obtained with different frits and a dialysis tube in 0.1 M/1 × 10⁻³ M potassium chloride electrolytes as a function of time.

maximum 100 h. A decrease of the potential values was then observed.

The time to reach a stable potential value of 115 mV ± 10 mV could be reduced to a few minutes by sucking the gel into the frit by applying a vacuum. This is shown in Fig. 5 for a 10^{-4} M /10^{-5} M potassium chloride system. Comparable potential values (110 mV ± 5 mV) have been obtained using sodium chloride electrolytes. Using the same concentration in both compartments leads to potential values of 0 mV ± 5 mV for the duration of the experiment.

It can be concluded that the stabilisation time of several hours is due to the location of the gel mixture on top of the frit and can be minimised to minutes by sucking the gel into the frit.

4.1.3. Influence of gel composition

In Section 4.1.2. the potential values obtained within a time scale of minutes were shown in Fig. 5. Results of long-term experiments using xanthan, agarose and alginate or their binary or ternary mixtures as separating membranes are given below.

The potential values obtained for the single components in 5×10^{-4} M/5×10^{-5} M potassium chloride are shown in Fig. 6. The expected cation selective behaviour was found for the model substance xanthan with values of 105 mV ± 7 mV. The matrix material agarose shows stable values of 85 mV. No stable values could be obtained for alginate, it showed the expected cation selective behaviour with a value of about 108 mV but, already, after a short time, a continuous decrease of the potential difference occurred. It could be seen during the measurements that this substance dissolved continuously into the solution, and this might be the cause for the observed decrease in the potential values.

Figure 7 shows the results obtained with the binary mixtures. The best results were obtained with the agarose/xanthan mixture, with constant values of *ca.* 120 mV for several days. The two mixtures containing alginate showed a decrease in the potential values immediately from the start (mixture with agarose), or after about

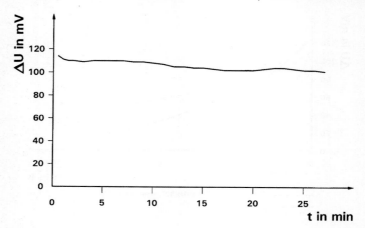

Fig. 5 *Potential difference obtained in 1×10^{-4} M/1×10^{-5} M potassium chloride electrolytes with the ternary mixture xanthan/agarose/alginate placed within the frit as an operating membrane as a function of time.*

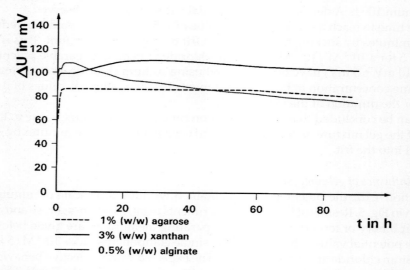

Fig. 6 *Potential difference values for the single synthetic exopolymeric substances in 5×10^{-4} M/ 5×10^{-5} M potassium chloride solutions as a function of time.*

30 h (mixture with xanthan), due to the continuous dissolution of alginate into the electrolyte.

The ternary mixture showed the expected stable values of 120 mV ± 5 mV at the beginning and a slight continuous decrease after about 20 h, due to the alginate, as reported above (Fig. 8). An artificial ageing of the gel mixture in an oven at 40°C for 24 h seems to have a small beneficial effect on the potential values, as shown in Fig. 8. Ageing for more than 24 h did not further affect the potential values.

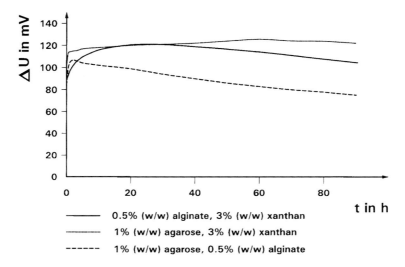

Fig. 7 Potential difference values for the binary mixtures of the synthetic exopolymeric substances in 5×10^{-4} M/5×10^{-5} M potassium chloride solutions at a function of time.

Summarising, the most stable potential values were achieved with the binary mixture xanthan/agarose, which were showing cation selective behaviour.

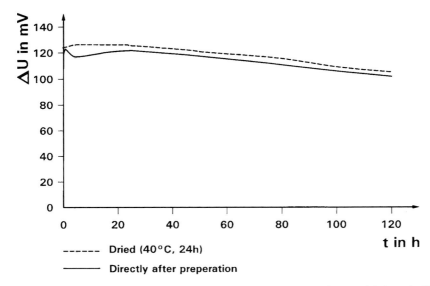

Fig. 8 Potential difference values for the ternary mixture xanthan/agarose/alginate in 5×10^{-4} M/5×10^{-5} M potassium chloride solutions as a function of time.

4.1.4. Separation of membrane and concentration contributions of the potential difference

The contribution of the membrane and of the concentration to the potential difference can be separated by the use of different reference electrodes as already described. This was investigated for the binary mixture agarose/xanthan using the potential values obtained after 40 h. The results are summarised in Table 2, for potential differences obtained as a function of different electrolyte concentrations of potassium chloride in the two compartments.

The contribution of the membrane decreases with increasing electrolyte concentration, while the contribution made by the concentration to the potential difference remains stable.

4.1.5. Effect of electrolyte concentration

Membrane potential measurements were made in electrolytes of various concentrations, with a 10-fold concentration difference and the binary mixture of xanthan/agarose. The results are shown in Fig. 9. The best values for the expected cation selective behaviour were obtained with 110 mV ± 5 mV for the 5×10^{-5} M/5×10^{-4} M potassium chloride system; 90 mV ± 5 mV were found in the 5×10^{-4} M/5×10^{-3} M system, and *ca.* 70 mV ± 5 mV for the system 5×10^{-3} M/5×10^{-2} M potassium chloride.

The separation of the membrane contribution from the concentration contribution by the use of the two different reference electrodes (as described in Section 4.1.4.) showed that with increasing electrolyte concentration, the membrane contribution to the measured potential difference decreases, due to the decrease of the space charge layer. The limiting values of 59 mV showing only the contribution of the concentration difference were obtained with the system 5×10^{-2} M/0.5 M potassium chloride.

A further decrease of the electrolyte concentration below 5×10^{-6} M potassium

Table 2. Potentials after 40 h obtained with 1% agarose/3% xanthan as a separating membrane

Concentration KCl	Potential difference Ag/AgCl/Cl$^-$ ΔU/mV	Potential difference Hg/Hg$_2$Cl$_2$/KCl$_{sat.}$ ΔU/mV	ΔU/mV
		Contribution of membrane	Contribution of concentration
5×10^{-2} M/ 5×10^{-3} M	59	1	58
5×10^{-3} M/ 5×10^{-4} M	76	21	55
5×10^{-4} M/ 5×10^{-5} M	98	44	54

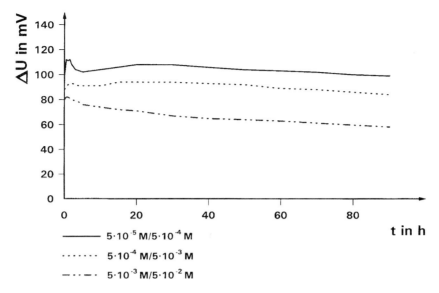

Fig. 9 Potential difference values for the binary mixture xanthan/agarose in potassium chloride electrolytes with different concentrations as a function of time.

chloride should not influence the expected potential difference of 118 mV with a 10-fold concentration difference. But, with these very dilute solutions again a decrease of the measured potential was observed. It is assumed that the high resistance within the system causes these problems.

A plot of the potential difference $\Delta U = \Delta U_{Concentration} - \Delta U_{Membrane}$ vs log c_1/c_2 with a concentration difference of 10-fold should show a slope of 118 mV/decade. But, due to the results obtained from the effect of the electrolyte concentration, this expected value could only be achieved with the system 5×10^{-5} M/5×10^{-4} M potassium chloride. All other systems with a 10-fold concentration difference showed smaller values.

To check the expected linearity of a ΔU vs log c_1/c_2 plot the concentration of 5×10^{-4} M potassium chloride was fixed while the concentration c_2 was varied to smaller values using the binary mixture xanthan/agarose. The linearity was verified for a concentration difference of one order of magnitude with the expected slope of 118 mV ± 5 mV/decade (Fig. 10). With higher concentration differences a strong deviation from the expected potential difference was noticed and was probably due to the high resistance of these diluted solutions.

To check the expected linearity over a reasonable concentration range, the higher concentration $c_1 = 0.005$ M potassium chloride was fixed while the concentration c_2 was varied to smaller values. It can be concluded from Fig. 9 for the system 5×10^{-3} M/5×10^{-4} M KCl that a slope of 90 ± 5 mV/decade can be expected in a ΔU vs log c_1/c_2 plot.

The results obtained for the binary mixture xanthan/agarose after different times are summarised in Fig. 11. For a concentration difference log $c_1/c_2 = 2.0$ a potential

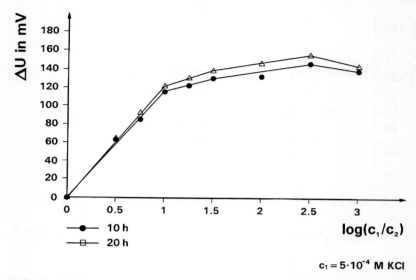

Fig. 10 Potential difference values as a function of the electrolyte concentration in different potassium chloride electrolytes for the binary mixture xanthan/agarose after different times.

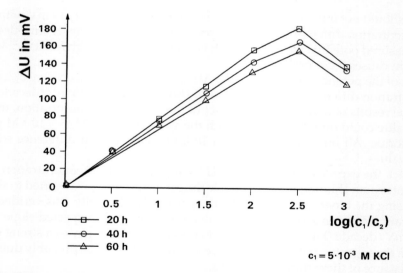

Fig. 11 Potential difference values as a function of the electrolyte concentration in different potassium chloride electrolytes for the binary mixture xanthan/agarose after different times.

value of 152 mV was observed after 20 h giving a slope of 76 mV/decade. This means that the expected linearity of this plot can be verified for a reasonable concentration range of two decades. A decrease of the potential values was observed at higher concentration differences. With increasing duration of the experiments, the maximum potential value decreased to 132 mV/2 decades after 60 h giving a slope of 66 mV/decade.

4.2. Membrane Potentials of the Biopolymers Produced by Microorganisms

Different pretreatments of the biopolymeric materials produced by microorganisms were tested before conducting the membrane potential measurements:

(i) The freeze-dried material was poured in a humid atmosphere for 30 h until a material of a stable gel-like constitution was obtained.

(ii) A sterilised gel-like water/2-propanol suspension of the material was used as delivered.

(iii) The material was dialysed for 2–4 days.

(iv) The effect of ageing of the dialysed material was investigated.

No significant differences in the measured potential differences were obtained with these various pretreatments within an experimental error of *ca.* 10 mV.

Figure 12 depicts this result with the example of the biopolymeric material # 930 406 B in the 5×10^{-5} M/5×10^{-4} M potassium chloride system. If the experiment is performed directly after dialysing the freeze-dried material, nearly stable potential values of 95 mV to 83 mV maximum could be observed. The difference of about 12

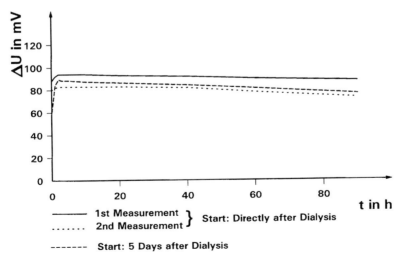

Fig. 12 Potential difference values for the dialysed biopolymeric material # 930 406 B in 5×10^{-5} M/5×10^{-4} M potassium chloride electrolytes as a function of time.

mV indicates the experimental error. Ageing of the biopolymeric material for 5 days had no significant influence (ΔU_{max} = 85 mV).

Two further important facts can be concluded from this figure:

(i) The biopolymeric substance produced by microorganisms showed the expected cation selectivity.

(ii) The obtained potential values remained nearly stable during the duration of the experiment.

Sample # 930 507 F1, delivered as a gelatinous colourless water/2-propanol suspension, yielded stable values of 80 mV and 82 mV (Fig. 13). The measurement was started 14 days after dialysis of 48 h. The reproducibility of two measurements is excellent. The effect of electrolyte concentration on the potential difference of this biopolymeric material was checked with the nondialysed suspension EPS # 930 507 F1 under the same experimental conditions as outlined in Fig. 11 to check the expected linearity in a reasonable potential range. The reference concentration c_1 was 5×10^{-3} M potassium chloride. In the system 5×10^{-3} M/5×10^{-4} M potassium chloride a stable potential value of 64 mV was measured (figure not shown) indicating the slope that has to be expected. Linearity of the plot was maintained up to two orders of magnitude of concentration difference with a slope of ca. 60 mV/decade taking the values after 20 h duration of the experiment (Fig. 14). The slope decreased to 51 mV/decade with increasing duration of the experiment (60 h). At higher con-

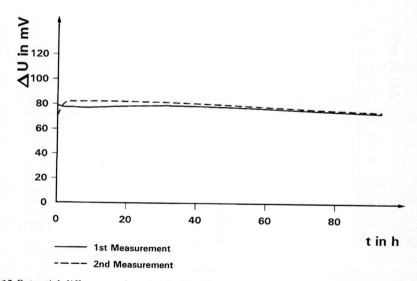

Fig. 13 *Potential difference values for the biopolymeric material # 930 507 F1 two weeks after dialysis of two days duration in 5×10^{-5} M/5×10^{-4} M potassium chloride electrolytes as a function of time.*

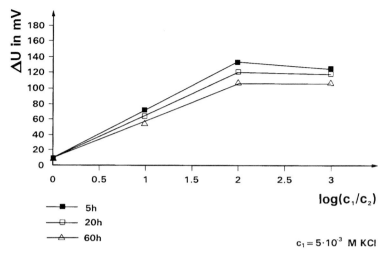

Fig. 14 Potential difference values as a function of the electrolyte concentration in different potassium chloride electrolytes for the biopolymeric material # 930 507 F1 as a gelatinous water/2-propanol suspension after different times.

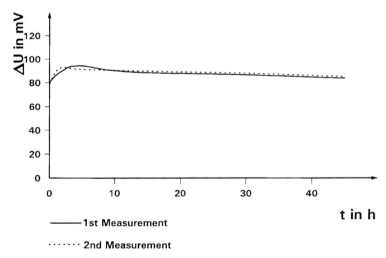

Fig. 15 Potential difference values for the biopolymeric material # 930 820 in 5×10^{-5} M/5×10^{-4} M potassium chloride electrolytes as a function of time.

centration differences a constant potential value is obtained. It may be assumed that the high resistance within the system caused these difficulties.

Figure 15 shows the membrane potential measurements as a function of time with the biopolymeric material # 930 820 delivered in a water/2-propanol suspension.

The obtained stable potential values of 91 and 92 mV show the expected cation selective behaviour and depict the excellent reproducibility.

It may be assumed that impurities consisting of proteins prevent the measurement of the expected value of 118 mV for a 10-fold concentration difference. Proteins in the neutral pH range are expected to show an anion selective membrane behaviour and therefore they may cause a certain compensation effect to the measured potential difference.

4.3. Transference Number Measurements

Using the Hittorf technique as described in the experimental section, the transference numbers of sodium chloride through the biopolymeric materials # 930 604 F1 (delivered in water/2-propanol suspension) and # 920 623 (delivered freeze-dried) were determined. Preliminary results at the anode compartment, for the # 930 604 F1 material, and $10^{-2}/10^{-1}$ M sodium chloride electrolytes, were as indicated in Table 3.

In the central compartment we did not detect concentration variations. The cation tranference number was obtained from the formula

$$\frac{Loss\ of\ NaCl\ at\ anode}{Faradays\ passed} = t_c$$

Our preliminary average cation tranference number was 0.4008, and therefore the average anion transport number was 0.5992. We have performed identical measurements with the substance EPS # 920 623, and in the absence of biopolymeric substances at the porous frits. We did not detect significant differences in the tranference number results obtained. Since no evidence for a cation-selective behaviour had been obtained, we decided to vary the experimental conditions, namely the concentration of the electrolyte, and the electrolysis parameters (current through the Hittorf cell, and duration of electrolysis).

The use of more dilute electrolytes had no significant effect on the measured cation tranference numbers (Table 3). For an electrolysis duration of 5 h, the increase of current also had no noticeable effect on the transference number (Table 3).

However, when we performed Hittorf measurements for a short duration, we were able to measure higher cation transference numbers. Moreover, and particularly for the less diluted solutions, it was possible to detect clearly the effect of the biopolymeric substance on the calculated transference number (Table 4). For this set of measurements, we used currents of the order of 0.01–0.2 mA for 10^{-4} M NaCl, 0.1–1.5 mA for 10^{-3} M NaCl, 2–10 mA for 10^{-2} M NaCl, and 3–15 mA for 10^{-1} M NaCl. However, for currents higher than 5 mA we noticed deterioration of the gelatinous substance within the frit at the cathode side, so most experiments were performed with currents < 5 mA. The noticeable effect of the time of electrolysis on the tranference number is probably due to diffusive mixing, so its minimisation requires electrolysis durations shorter than about 1 h.

The sodium chloride solutions in these experiments were analysed by precipitation titrations with silver nitrate, in which the chromate ion, was used as the indicator. During the quantitative analysis we found it increasingly difficult to obtain a reasonable good end point in the titration of Cl^- with Ag^+ when we moved to more dilute sodium chloride solutions. To overcome this difficulty complementary meth

Table 3. Cation transference numbers of sodium chloride through biopolymeric substance # 930 604 F1

	10^{-2} M NaCl		10^{-1} M NaCl	
Initial anode portion, g	301.6	300.5	300.3	301.1
Initial NaCl amount, g	0.1763	0.1756	1.755	1.760
Anode portion after electrolysis, g	29.77	30.13	28.96	30.26
NaCl amount after electrolysing	0.0088	0.0090	0.1563	0.1634
Current passed, mA	2	2	3	3
Time of electrolysis, h	5	5	5	5
Faradays passed, F	0.000373	0.000373	0.000560	0.000560
Cation transference number	0.3948	0.3964	0.3979	0.4142
	10^{-4} M NaCl		10^{-3} M NaCl	
Initial anode portion, g	299.6	304.8	301.1	297.1
Initial NaCl amount, g	0.0018	0.0018	0.0176	0.0174
Anode portion after electrolysis, g	29.54	30.25	30.95	31.10
NaCl amount after electrolysing	0.00012	0.000025	0.00145	0.0004
Current passed, mA	0.01	0.03	0.1	0.3
Time of electrolysis, h	5	5	5	5
Faradays passed, F	0.000002	0.000006	0.000019	0.000056
Cation transference number	0.4685	0.4355	0.4109	0.4311

Table 4. Cation transference numbers of sodium chloride as a function of time of electrolysis

Biopolymeric substance # 930 604 F1					
NaCl concentration	Time of electrolysis, h				
	5	4	3	2	1
10^{-4}	0.4685	–	0.4705		0.4719
10^{-3}	0.4109	0.4219	–	0.4341	0.4617
10^{-2}	0.3948	0.4891	0.4791	0.4815	0.4912
10^{-1}	0.3979	–	–	0.4796	0.4856
Blank values					
NaCl concentration	Time of electrolysis, h				
	5	4	3	2	1
10^{-4}	0.3812	–	0.4019		0.4017
10^{-3}	0.3925	0.4015	–	0.4091	–
10^{-2}	0.4005	–	–	0.4013	0.4156
10^{-1}	0.4011	0.4101	0.4008	–	–

Table 5. Cation transference numbers of sodium chloride through a synthetic biopolymeric substance of xanthan/agarose/alginate

	10^{-2} M NaCl		10^{-1} M NaCl	
Initial anode portion, g	300.6	299.1	298.9	302.3
Initial NaCl amount, g	0.1757	0.1748	1.7471	1.7669
Anode portion after electrolysis, g	29.12	31.363	31.330	29.718
NaCl amount after electrolysis, g	0.0151	0.01317	0.1801	0.1681
Current passed, mA	2	5	3	5
Time of electrolysis, h	1	1	1	1
Faradays passed, F	0.000075	0.000187	0.000112	0.000187
Cation transference number	0.4389	0.4735	0.4598	0.5115
	10^{-4} M NaCl		10^{-3} M NaCl	
Initial anode portion, g	300.1	299.1	305.2	302.3
Initial NaCl amount, g	0.0018	0.0018	0.0178	0.0177
Anode portion after electrolysis, g	31.21	35.25	28.26	31.65
NaCl amount after electrolysis, g	0.00012	0.000017	0.00108	0.00005
Current passed, mA	0.05	0.15	0.5	1.5
Time of electrolysis, h	1	1	1	1
Faradays passed, F	0.000002	0.000006	0.000019	0.000056
Cation transference number	0.5610	0.5913	0.5198	0.5499

ods of analysis were used, namely for the H^+ and Cl^- determination by means of pH and Cl^--ion selective electrodes.

The tranference number measurements shown in Table 5 clearly indicate a cation selective behaviour of the synthetic exopolymeric materials. However, the observed differences, with and without the gel at the frit, should be treated with some caution, giving that many variables are involved in the Hittorf cell.

5. Conclusions

The results obtained with the synthetic exopolymeric substances can be summarised as follows:

For all three single substances investigated (xanthan, agarose, alginate) a cation selective behaviour could be observed, xanthan giving the best values. Stable potentials could be achieved for xanthan and agarose for least 100 h. After this period of time the experiments were stopped. The potential values obtained with alginate decreased slowly, because this substance dissolved continuously into the solution. The xanthan/agarose binary mixture showed the expected and stable values. For the alginate/xanthan and alginate/agarose mixtures no stable potential values could be obtained due to the effect of the alginate. For the ternary mixture xanthan, agarose, alginate a slow decrease caused by alginate dissolving was detected. The experimental error during these experiments is in the range 5–10 mV. An accelerated ageing of the ternary mixture in an oven resulted in stabilisation of the gel-like material within the frit and did not cause a change of membrane properties. After this treatment, stable potential values could be obtained for about 250 h. This period of time is long enough to perform at least short term corrosion experiments under electrochemical control.

The contribution of the membrane effect and the concentration effect to the measured values could be clearly separated by varying electrolyte concentration and reference electrodes. The expected linearity of a $\Delta U/\log(c_1/c_2)$ plot could be verified over a concentration range of 2 orders of magnitude.

Identical results as described for the synthetic exopolymeric material have been obtained for the biopolymeric substances produced by microorganisms, but the measured potential differences were lower in the latter case. This may be caused by impurities consisting of proteins that provide a certain compensation effect to the potential difference due to their anion selectivity in the neutral pH-range.

For a mixture of synthetic exopolymeric material consisting of xanthan, agarose and alginate, the average cation transference number was 0.4123. This value is similar to transference number data collected in the absence of this substance at the frit. In the case of more dilute solutions it was possible to measure larger cation transference numbers. The results originating from the transference number measurements have to be treated as preliminary at this stage.

In diluted solutions it was possible to show an influence of the biopolymeric substance produced by microorganisms on the cation transference numbers by performing preliminary Hittorf experiments, but further study is necessary to clarify this transport phenomenon.

It can be stated that these model considerations hold true for both types of investigated exopolymeric materials. This means on one hand that the choice of the model substance xanthan is supported and on the other hand, it has been possible to produce exopolymeric substances from bacteria isolated from copper pipes affected by MIC, and importantly, both substances show the expected cation selective transport properties.

Finally, it can be concluded that the presence of microorganisms found in copper tube installations can lead to the formation of exopolymeric substances that have cation selective membrane properties. The electrolyte concentrations used in these membrane potential experiments are comparable to the concentration of these ions in potable water within pipework exhibiting corrosion. Therefore, cation selective behaviour of the exopolymeric substances occurring in this corrosion process is plausible.

6. Acknowledgement

The financial support of this work within the BRITE/EURAM Project No. 4088 ("New Types of Corrosion Impairing the Reliability of Copper in Potable Water Caused by Microorganisms") is gratefully acknowledged. We would like to thank A. H. L. Chamberlain, H. H. Paradies and N. Wardell for fruitful discussions and for the delivery of the biopolymeric substances.

References

1. G. G. Geesey, P. J. Bremer, W. R. Fischer, D. Wagner, C. W. Keevil, J. Walker, A. H. L. Chamberlain and P. Angell, in *Biofouling and Biocorrosion in Industrial Water Systems* (Eds G. G. Geesey, J. Lewandowski and H. C. Flemming), 1994, C. R. C. Press, Boca Ratom, USA, Ch. 16, pp. 243–263.
2. P. Angell, H. S. Campbell and A. H. L. Chamberlain, International Copper Association (ICA) Project No. 405, Interim Report, August 1990.
3. H. Campbell and A. H. L. Chamberlain, in *Corrosion and Related Aspects of Materials for Potable Water Supplies*, (Eds P. McIntyre and A. D. Mercer), 1993, The Institute of Materials, London (UK), p. 222.
4. W. Fischer, I. Hänßel and H. H. Paradies, in *Microbial Corrosion 1*, (Eds C. A. C. Sequeira and A. K. Tiller), 1988, Elsevier Applied Science, pp. 300–327.
5. W. Fischer, H. H. Paradies, D. Wagner and I. Hänßel, *Werkstoffe und Korros.*, 1992, **43**, 496.
6. W. R. Fischer, D. Wagner and H. H. Paradies, *Microbiologically Influenced Corrosion (MIC) Testing*, ASTM STP 1232, (Eds J. R. Kearns and B. Little), 1994, American Society for Testing and Materials, Philadelphia, pp. 275–282.
7. W. R. Fischer, D. Wagner and H. H. Paradies, in European Federation of Corrosion, Publication No. 8, *Microbial Corrosion* (Eds C. A. C. Sequeira and A. K. Tiller), 1992, The Institute of Materials, pp. 168–187.
8. A. H. L. Chamberlain, W. R. Fischer and H. H. Paradies, *3rd EFC Workshop on Microbial Corrosion*, Estoril, Portugal, March 1994, this volume, Chapter 1.
9. J. N. Wardell and A. H. L. Chamberlain, *3rd EFC Workshop on Microbial Corrosion*, Estoril, Portugal, March 1994, this volume, Chapter 3.
10. M. Thies, U. Hinze and H. H. Paradies, *3rd EFC Workshop on Microbial Corrosion*, Estoril, Portugal, March 1994, this volume, Chapter 2.
11. D. Wagner, H. Siedlarek, W. R. Fischer, J. N. Wardell and A. H. L. Chamberlain, *3rd EFC Workshop on Microbial Corrosion*, Estoril, Portugal, March 1994, this volume, Chapter 5.

5

Corrosion Behaviour of Biopolymer Modified Copper Electrodes

D. WAGNER, H. SIEDLAREK, W. R. FISCHER, J. N. WARDELL*
and A. H. L. CHAMBERLAIN*

Märkische Fachhochschule, Laboratory of Corrosion Protection, Postbox 20 61, 58590 Iserlohn, Germany
*School of Biological Sciences, Guildford, Surrey GU2 5XH, UK

ABSTRACT

Microbiologically influenced corrosion (MIC) causes problems in copper potable water installations in main buildings. The damage of copper tubes is accompanied by a 'biofilm' influencing transport processes on the copper surfaces. A physico–chemical explanation has been offered for the impairment observed on the basis of electrochemical experiments on copper electrodes in the presence of either an 'artificial biofilm' composed of exopolymeric substances or a 'biofilm' produced by microorganisms isolated from perforated copper tubes of an affected county hospital in Germany. Ionic properties of 'artificial biofilms' and 'biofilms', produced by microorganisms have been investigated using cyclic voltammetry.

Possible corrosion reactions of copper covered with exopolymeric substances in the presence of chloride and sulphate ions have been discussed.

1. Introduction

Microbiologically influenced corrosion (MIC) has been found to cause problems in drinking water installations made of copper tubes in main buildings [1–4]. The deterioration of metal surfaces in the presence of microorganisms is usually accompanied by a 'biofilm'. The particular 'biofilm' consists of different organic materials, e.g. polysaccharides, glycolipids, oligopeptides, and is believed to be involved in many corrosion processes [5, 6]. These 'biofilms' also include consortia of bacteria and fungi, and some algae, in addition to the anionic exopolymeric material to provide attachment and structural integrity [7, 8].

Many investigations dealing with MIC in potable water installations report on the influence of living microorganisms leading to the formation of a 'biofilm' which was discussed for the first time in 1948 [9] and more recently in 1990 [10]. This paper discusses a different approach where physicochemical properties of 'biofilms' play a decisive role in the MIC process only. These physical properties depend entirely on its chemical composition and are regarded as time-invariant during our electrochemical experiments.

The nature and chemical composition of the 'biofilm' detected on the inner surface of the copper tubes were investigated and elucidated for an affected county hospital in Germany [11, 12]. It was found that the biofilm plays a key role in this copper corrosion process [11]. The biofilm has been found on the bare copper surface of perforated copper pipes. All reaction layers consist of visible solid copper corrosion products which are located on top of this biofilm or mixed up with this biofilm [11, 12]. Parts of the chemical composition of this biofilm have been investigated [11]. Its structure is currently described as linear and/or crosslinked acidic or partly non-ionic polysaccharides. The primary composition consists of similar structures which have been found for xanthan, e.g. pyruvate residues, highly cross-linked and of high molecular weight. It also contains alginate like structures, a polysaccharide consisting of mannurate and galuronate residues arranged in a non-regular clockwise pattern along a linear chain. This type of biofilm is expected to be cation selective and permeable for water and oxygen, respectively [13], because of the accumulation of fixed negative charges associated with its major components — xanthan and alginate. The experimental validation of this model based on ionic transport properties by biopolymers and the explanation of relevant corrosion reactions are the main goal of an interdisciplinary approach to this problem [14]. These physical properties have been investigated with independent experimental techniques. The expected ionic transport properties were checked using transference number measurements and cyclic-voltammetry. The results of the transference number measurements are reported separately [15].

The results of the described biofilm analysis enable the role of the biopolymeric material in the corrosion process to be identified by using cyclic voltammetry. The experiments have been performed with two types of biopolymeric material:

(i) Xanthan has been chosen as a model substance, alginate as a stabiliser, and agarose has been used as a matrix to embed model substance and stabiliser. Because this polymeric material does not contain any microorganisms, the term 'biofilm' for this substance is not completely justified. However, this material used is very similar to the known exopolymeric substances (EPS). The terms 'artificial biofilm' or, alternatively, 'synthetic exopolymeric substances' have been chosen to specify this polymeric material.

(ii) The microorganisms used for the production of the biopolymeric material were isolated and cultured from perforated copper pipes taken from the affected county hospital in Germany [2]. The obtained 'biofilm', i.e. extracellular polymers, should be of the same comparable composition as the 'biofilm' detected in the perforated copper pipes [16].

This paper addresses the following questions:

(i) How do anions, and pH influence the corrosion reactions of a copper surface when coated with a 'biofilm' compared to a bare copper surface?

(ii) Does the existence of a 'biofilm' influence the precipitation of solid corrosion products on the metal surface when living microorganisms are absent?

(iii) Do the results reveal any differences when the copper electrode is coated either with an 'artificial biofilm' or with a biopolymeric material produced by microorganisms?

2. Theoretical Considerations

The cation selectivity and permeability of the exopolymeric substance (EPS) give rise to the following arguments:

(i) For a copper surface in contact with a dilute electrolyte, only cations and neutral components, e.g. water and oxygen, but not anions, can pass through the coating when this consists of exopolymeric substances with a high water content. This allows the formation of copper oxides or copper hydroxides underneath the coating, but not the deposition of salt layers, e.g. copper(I) chloride.

(ii) The electrochemical force for the transport of copper ions through the biopolymeric coating is caused by the gradient of the electrochemical potential of ions produced through the anodic partial reaction of the corrosion reaction, ie. copper oxidation.

(iii) A very thin layer consisting of copper(I) oxide is formed [17–20], when a bare copper surface comes into contact with an aqueous phase. The dissociated protons from the hydroxyls of the carboxo groups of the anionic polysaccharides are oriented towards the copper(I) oxide layer. Therefore, a low pH can be assumed at the phase boundary material of metal/EPS because of the values of the dissociation constants of the carboxylic acid groups. Typical pK_a values of interest here are in the range 4.2–5.6 according to titration experiments and conductivity measurements [21].

Furthermore, the electrolyte at the copper surface/EPS phase boundary becomes necessarily acidic, if an adhering deposit of copper(I) oxides or copper(I) hydroxides is formed through the electrochemical reactions (1 & 2):

$$2Cu + H_2O \rightarrow Cu_2O + 2H^+ + 2e^-$$

$$Cu + H_2O \rightarrow CuOH + H^+ + e^-.$$

Higher pH values in the range between 6.5 and 9.5 are established at the EPS/electrolyte phase boundary corresponding to the pH range of potable water. The pH range of 6.5 to 9.5 is required for potable water according to Federal German Potable Water Regulations [22]. These considerations infer the existence of a pH gradient across the EPS coating.

Cyclic-voltammetry is a powerful experimental technique:

(i) to prove whether the described physical properties are valid for a coating consisting of exopolymeric substances, and,

(ii) to investigate the influence of these properties on the precipitation of solid corrosion products.

3. Evaluation of the Influence of Ionic Transport Properties by Cyclic Voltammetry

The effect of the exopolymeric substance adhering to the surface of the copper can be determined, when the mechanism of the formation of reaction layers of solid corrosion products is well understood. Anions like chloride and sulphate, and protons are the main parameters influencing the deposition of reaction layers of solid corrosion products of copper when in contact with aerated cold water having the quality of potable water. In particular the influence of chloride and sulphate ions has been extensively investigated [23–27].

The present knowledge of the influence of the anions, ie. chloride and sulphate, can be summarised as follows:

In sulphate-containing electrolytes voluminous reaction layers of crystalline copper(I) oxide are formed at potentials higher than the threshold potential. The deposited layers do not significantly inhibit the anodic partial reaction. The manifestation of corrosion is a general attack. In electrolytes containing chloride two reaction layers are formed, namely copper(I) chloride underneath, and an amorphous film of copper(I) oxide on top of the copper(I) chloride. This amorphous copper(I) oxide is formed via hydrolysis of copper(I) chloride, and inhibits the anodic metal dissolution. The observed manifestation of corrosion is therefore repassivating pitting [23, 24].

Characteristic peaks occurring in cyclic voltammograms can be attributed to the formation and reduction of reaction layers in electrolytes containing chloride and sulphate ions [27, 28]. In an electrolyte containing chloride ions one oxidation peak and one reduction peak occur that can be attributed to the influence of these chloride ions within the corrosion process. In electrolytes containing sulphate no peaks are found that can be attributed to the influence of the sulphate ions.

The anions, e.g. chloride and sulphate, should be excluded from the electrolyte available at the phase boundary when the copper electrode is coated with an exopolymeric substance because of the cation selectivity of the latter. This implies that the two peaks caused by the influence of chloride ions should not be observed in a cyclo-voltammetric experiment with a coated copper electrode. In contrast, for the electrolyte containing sulphate ions the influence of the sulphate cannot be evaluated.

However, the sulphate ions as electrolyte are suitable to investigate the influence of pH. Since the pH at the copper/EPS phase boundary of the coated electrode is necessarily acidic, the cyclic voltammogram of the bare copper electrode when in contact with a neutral, sulphate ion-containing electrolyte reveals a characteristic peak due to the reduction of the reaction layer consisting of copper(I) oxide. A considerably smaller amount of copper(I) oxide should be formed underneath the coating of the EPS on the copper surface because the formation of copper(I) oxide is pH dependent. This implies that the integral charge of the reduction peak in a cyclic voltammogram with a coated electrode in an electrolyte containing sulphate should be considerably smaller than the correspondent integral charge obtained using a bare copper electrode.

4. Experimental
4.1. Electrochemical Set-up

Potentiodynamic measurements have been performed with a potentiostat (EG & G, Princeton Applied Research, Model 273 A) with automatic IR compensation. Experimental data have been collected and processed with the software package SOFTCORR II (EG & G, Princeton Applied Research).

Experiments were conducted in a Faraday cage in the absence of light. Measuring electrodes were prepared from hard copper tubes (German Standard: SF-Cu, F37; ISO-Standard: Cu-DHP; DHP: De-oxidised High Phosphorus) with a diameter of 22 mm according to DIN 1786 [29]. Rings with an outer surface area of 8 cm^2 were abraded with emery paper, polished with diamond paste and electropolished in orthophosphoric acid (70% w/w) for 30 s with an anodic current density of $i = 0.2$ A cm^{-2}. The rings were positioned in an electrode holder as shown in Fig. 1. Only the outer surfaces of these rings were polarized using a common three electrode arrangement (Fig. 2). A ring of platinum wire was positioned concentrically around the copper and was used as a counter electrode. A mercury sulphate electrode (Hg/Hg$_2$SO$_4$/K$_2$SO$_4$ sat.) or a calomel electrode (Hg/Hg$_2$Cl$_2$/KCl sat.) was used as a reference electrode. Electrode potentials were recalculated versus the standard hydrogen electrode (SHE).

A Haber–Luggin (HL) capillary was used to diminish the ohmic voltage drop. The solution was stirred using a magnetic stirrer, and temperature of the electrolyte was kept constant at 20°C. The solutions were prepared using distilled water and chemicals of pa. grade, i.e. sodium sulphate and sodium chloride electrolytes. The electrolyte within the cell (pH 6–6.5) was aerated with a mixture of 20 vol. % oxygen and 80 vol. % nitrogen.

4.2. Preparation of the Synthetic Biopolymeric Substances

Agarose was prepared by swelling in water in the presence of 5 mM EDTA at 50°C for 20 h. The material was washed with deionized water until a constant pH of 6.2 was obtained and EDTA was removed [21].

Partly deprotonated *alginic acid* was prepared from crude alginate (free acid) which was received as a cream-coloured powder from Hoechst (Frankfurt/M.); 10 g of alginate was dissolved in 0.1 M K$_2$HPO$_4$, pH 8 in the presence of 1 mM KOH, treated for 4 h at 70°C under continuous stirring. Normally, alginate in water forms a vis-

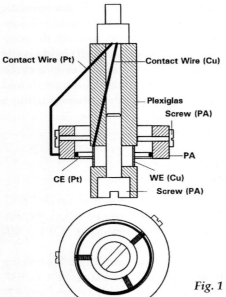

Fig. 1 Schematic diagram of an electrode holder.

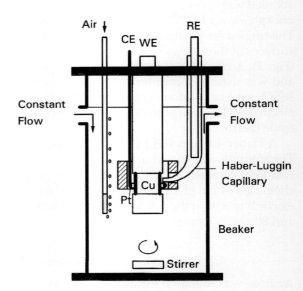

Fig. 2 Schematic diagram of an electrochemical cell.

cous colloidal solution which is not transparent. Upon heating to 70°C, and in the presence of HPO_4^{2-} and KOH, the kinematic viscosity increases reaching a maximum value of 1000 s^{-1}, and a transparent solution at alkaline pH is obtained. The K$^+$ (Na$^+$) salt of alginic acid was purified by precipitation with cold EtOH/water (55 w/45 w) at 4 °C, centrifuged at 10 000 xg and redissolved at 10°C in the presence of

1–5 mM K(Na)OH until a colourless, clear solution of alginic acid was obtained. Molecular weight measurements by static light scattering and turbidity measurements yielded values between 250 000–300 000, respectively [21].

'Native' xanthan (xanthan 1) which specifically means the pyruvated and N-acetylated form of the polysaccharide B-1459 produced by *Xanthomonas compestris*, was obtained from Sigma (St. Louis, MO, USA) as a cream-coloured powder; 100 mg of xanthan was dissolved in 1.5 L water, converted to the Na(K)-form by adding Na(K)HCO$_3$ (0.5 M) at 20°C under continuous stirring (800 rpm), keeping the pH $\cong$ 8.0. This solution (1.5 L solvent, 100 mg biopolymer) was heated to 40–50°C, and EDTA of 2.5 mM (2.0–3.0) was added, (pH 8.0,) and the solution was made up to 0.1 M NaCl by adding the appropriate amount of NaCl. After 4 h (4–6 h) this solution was cooled, dialysed against several charges of deionised water (20°C) and lyophilised. The colourless powder of xanthan is hygroscopic, the potassium salt more so than the sodium salt [21]. Stock solutions were prepared by dissolving alginic acid, (K(Na) potassium salt) in 0.1 M K(Na)Cl, pH 7.0–8.0, or pH 6.0–7.0 in 0.01 M acetate buffer, until a clear transparent solution was obtained (20°C). Continuous stirring is necessary when adding agarose for about 1 h so that the solution does not increase in viscosity. Under continuous stirring (800 rpm), and in the presence of 0.1 M–0.5 M NaCl xanthan (Na or K salt) was added. The addition of xanthan to the solution of alginic acid (sodium salt) and agarose has to be made over a period of 30 min [21]. The obtained solution was heated to 100°C in a steam bath under pressure (~ 2 atm) for 10 min.

Using the procedure described above, the following solutions were prepared to coat the copper electrodes:

(i) 0.3 % (w/w) xanthan/0.1% (w/w) agarose/0.05% (w/w) K-alginate, and

(ii) 0.3% (w/w) xanthan/0.1% (w/w) agarose.

4.3. Preparation of Biopolymers Produced by Microorganisms

Biopolymeric material was obtained from batch and continuous culture of a bacterium isolated from corroding copper pipe [16]. The organism was grown in batch and continuous-culture and the spent medium processed to recover extracellular polymer.

Measurements have been performed with the obtained EPS # 930 630 BF1 and EPS # 930 820. The numbers indicate a laboratory code. Further information about these exopolymeric substances is given in Ref. [16].

4.4. Techniques to Produce Coated Electrodes

Different methods have been employed to coat the copper surfaces with the biopolymeric material.

The electrode was covered with the synthetic biopolymeric material either by dipping the electrode in the prepared solution as described in Section 4.2., or by homogeneous distribution of the exopolymeric substance on the copper surface with a pipette. The latter technique was also used for the biopolymeric material produced by microorganisms.

Using another technique the exopolymeric substances were deposited directly on an electropolished copper surface using a rotating apparatus especially developed for this purpose. Electropolished copper electrodes (as described above) were mounted vertically on a silicon rubber disc attached to the drive shaft of a small d.c. electric motor running at 1–36 rpm. Approximately $^1/_3$ of the electrode circumference dipped into a water/propan-2-ol suspension of the microbial polymer. As the electrode rotated, a thin film of polymer was deposited on the surface of the electrode. Test electrodes were coated in this way for 18 h to allow a reasonable film thickness to develop.

5. Results
5.1. Bare Copper Electrodes

Current density potential curves with bare copper electrodes have been measured to explain the cyclo-voltammetric results obtained with electrodes coated with EPS. Figure 3 shows the voltammogram obtained in an aerated sodium chloride electrolyte with a very slow scan rate (0.01 mV s^{-1}). The measurement started at –950 mV$_H$.

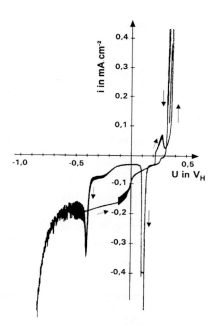

Fig. 3 *Potentiodynamic current density potential curve of copper in sodium chloride; $c_{(Cl-)}$ = 200 mg L^{-1}; pH 6–6.5; aeration; scan rate: 0.01 mV s^{-1}; start potential: –950 mV$_H$; reverse potential: +300 mV$_H$.*

The curve starts in the range of hydrogen evolution, then a first plateau can be seen. At more positive potentials the cathodic current shows a further decrease and a second plateau occurs in the range of 40–220 mV$_H$, followed by a small oxidation peak with the maximum current density at 270 mV$_H$. The threshold potential is found at 310 mV$_H$ and corresponds to the threshold potential obtained from potentiostatic measurements [23]. In the reverse scan two reduction peaks can be detected, one peak at +100 mV$_H$ and the second one at -450 mV$_H$, the latter showing a small shoulder at – 200–350 mV$_H$. The total cathodic current densities are considerably lower in the reverse scan in the potential range between these two reduction peaks.

Figure 4 shows the voltammogram obtained in the same electrolyte under the same experimental conditions with a faster scan rate of 1 mV s^{-1}. With this scan rate, the shoulder that can be seen in the reverse scan at −120 to −250 mV$_H$ is much more distinct with its maximum current density of −0.31 ± 0.01 mA cm^{-2}.

The first plateau described in Fig. 3, scanning from negative to positive potential values, can be attributed to the diffusion controlled current of the oxygen reduction. Experiments on platinum disk electrodes show that this current corresponds to the reaction [30–32]:

$$O_2 + 2H_2O + 4e^- \rightarrow 4OH^-$$

Within the range of the second plateau with small total cathodic current densities the formation of a reaction layer is thermodynamically possible at potentials far below the activation potential [33]. Corrosion products consisting of thin tarnish layers can be detected on copper surfaces after performing potentiostatic experiments in this potential range, where the copper electrode shows a passive behaviour [34]. The shoulder between −120 and −350 mV$_H$ in the reverse scan, dependent on the scan rate, can be attributed to the reduction of this tarnish layer [16]. In an oxygen saturated electrolyte this anodic partial reaction is superimposed by an oxygen reduction reaction consuming protons [30]:

$$O_2 + 4H^+ + 4e^- \rightarrow 2H_2O$$

The formation of the thin reaction layer itself causes an acidification of the electrode surface at neutral pH, because chloride containing solutions do not have any buffer capacity, and makes the occurrence of oxygen reduction in this potential range possible.

The oxidation peak at 270 mV$_H$ and its corresponding reduction peak at +100 mV$_H$ (see Fig. 3) can be attributed to the formation and reduction of copper(I) chloride. On top of this copper(I) chloride copper(I) oxide is formed via hydrolysis and is reduced at –450 mV$_H$ [27, 28].

The scan was reversed at a potential of 360 mV$_H$ to avoid the formation of copper(II) corrosion products that would be obtained at values more positive than the reverse potential [17, 18].

The results described in Figs 3 and 4 for chloride ion-containing electrolytes also hold true for sodium sulphate-containing electrolytes, with one exception (figures not shown). The two peaks corresponding to the oxidation and reduction of copper(I)

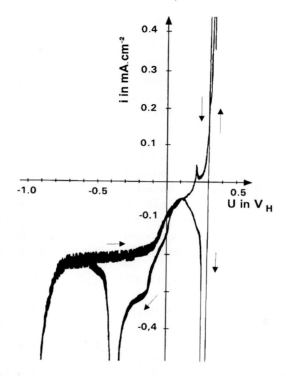

Fig. 4 *Potentiodynamic current density potential curve of copper in sodium chloride; $c_{(Cl^-)}$= 200 mg L^{-1}; pH 6–6.5 aeration, scan rate: 1 mV s^{-1}; start potential: –950 mV$_H$; reverse potential: +360 mV$_H$.*

chloride are missing, copper(I) oxide is formed directly on the copper surface leading to the differences in the physical and chemical properties of this copper(I) oxide described in Section 3 [23, 24].

In summary, copper has a passivity range of 40 to 300 mV$_H$. Within this range, tarnish layers, presumably anhydrous or hydrated modifications of copper(I) oxide, are formed. These tarnish layers inhibit the anodic and cathodic partial reaction of the corrosion process. The steep threshold in chloride- and sulphate-containing electrolytes is a corrosion process in the transpassive range with the oxidation of copper and the formation of reaction layers precipitated by the reaction of these copper ions with water.

5.2. Electrodes Covered with Biopolymeric Films

Cyclovoltammetric results obtained with copper electrodes covered with a synthetic biopolymeric film, consisting either of (i) 0.3% xanthan/0.1% agarose/0.05% (w/w) K-alginate, or (ii) 0.3% xanthan/0.1% (w/w) agarose do not show any differences. Therefore, these two films will not be differentiated in the following discussion.

Figure 5 shows the current density potential curve obtained in a sodium chloride

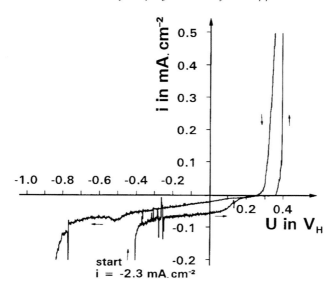

Fig. 5 Potentiodynamic current density potential curve of copper covered with 0.3% xanthan/ 0.1% agarose/0.05% (w/w) K-alginate in 1×10^{-3} mol L^{-1} sodium chloride; aeration; scan rate: 0.01 mV s^{-1}; start potential: -400 mV_H; reverse potential: $+400$ mV_H.

solution for an electrode coated with a synthetic biopolymer with a slow scan rate of 0.01 mV s^{-1}. The following differences compared to the results obtained with a bare copper electrode (see Fig. 3) can be seen:

(i) The two peaks described in Fig. 3 do not occur. So chloride ions cannot pass this coating and cannot form copper(I) chloride underneath the coating within the time scale of the experiment.

(ii) The total current densities corresponding to the oxygen reduction reaction in the first scan are considerably smaller on the coated electrode than on the bare copper electrode indicating that the coating is acting as a diffusion barrier.

(iii) No remarkable inhibition of the anodic partial reaction can be observed.

(iv) The integral charge of the peak corresponding to the reduction of copper(I) oxide at -520 mV is much smaller than that formed through the reaction with water under proton production at positive potentials. The amount of this copper(I) oxide underneath the coating remains nearly constant if the scan is stopped for 21 h at a reverse potential of 290 mV_H at positive total current densities. The reverse scan of this measurement shows a similar shape as in Fig. 5. This confirms the assumption that the pH underneath the coating is lower that in the bulk electrolyte.

Figure 6 shows the current density potential curve under the same experimental conditions as outlined for Fig. 5, but with a faster scan rate of 0.1 mV s^{-1}. Due to this faster scan rate the reduction peak at –520 mV$_H$ is more distinct, but with a comparable integral charge to that reported in Fig. 5. This is in confirmation of the results described above (ii–iv).

As outlined in Section 3, a voltammogram measured in sodium sulphate electrolytes under the same experimental conditions as described for Fig. 6, should yield the same shape of curve. This was verified as shown in Fig. 7. Only the peak corresponding to the reduction of the precipitated copper(I) oxide was obtained.

Figure 8 shows the current density potential curve obtained in sodium chloride electrolytes with an electrode covered with synthetic biopolymeric material under modified experimental conditions. The scan rate was increased to 1 mV s^{-1}, and the scan held at the reverse potential of 150 mV$_H$ for 5 min. Under these conditions it is possible to determine whether the tarnish layer described in Fig. 4 is also formed underneath the biopolymeric film. Figure 8 depicts the positive result. Only one reduction peak at –250 mV$_H$ corresponding to the reduction of the hydrated or unhydrated oxide can be seen. The size of the reaction layer is considerably lower than that obtained on a bare copper electrode (see Fig. 4).

Figure 9 shows a copper surface covered with a synthetic biopolymeric film after drying the electrode in a desiccator over silica gel following the experiment. It can be seen that the film has peeled off at random in some places, to reveal a bare copper surface showing pitting attack. The whole remaining film on the surface shows a positive stain with Gentian Violet as described in Ref. [35].

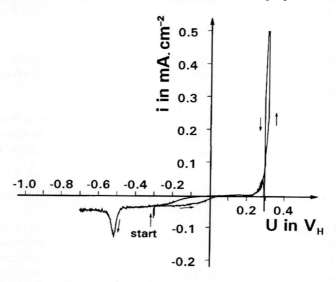

Fig. 6 *Potentiodynamic current density potential curve of copper covered with 0.3% xanthan/ 0.1% agarose/0.05% (w/w) K-alginate in 1×10^{-3} mol L^{-1} sodium chloride; scan rate: 0.1 mV s^{-1}; aeration, start potential: –300 mV$_H$; reverse current potential density: 0.1 mA cm^{-2}.*

The stain is positive especially at the transition between the film edge and the bare copper surface. When the electrode was wet, red copper(I) oxide could be seen on top of the biopolymeric substance. This layer peeled off after drying. This is the

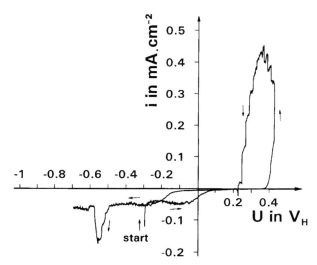

Fig. 7 Potentiodynamic current density potential curve of copper covered with 0.3% xanthan/ 0.1% agarose/0.05% (w/w) K-alginate in 1×10^{-3} mol L^{-1} sodium sulphate; scan rate: 0.1 mV s^{-1}; aeration; start potential: -300 mV_H; reverse current density: 0.1 mA cm^{-2}.

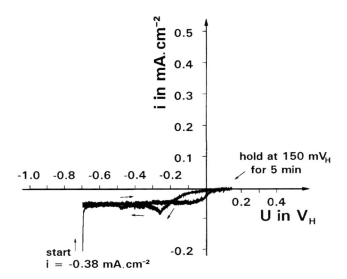

Fig. 8 Potentiodynamic current density potential curve of copper covered with 0.3% xanthan/ 0.1% (w/w) agarose in 1×10^{-3} mol L^{-1} sodium chloride; aeration; scan rate: 1 mV s^{-1}; start potential: -700 mV_H; stopped for 5 min at the reverse potential of 150 mV_H.

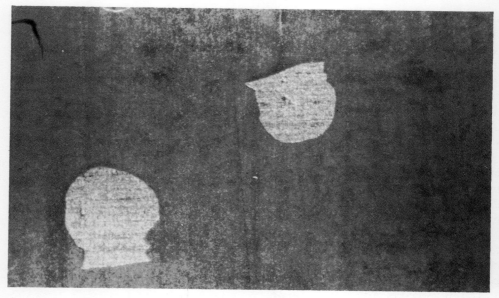

Fig. 9 Copper surface covered with 0.3% (w/w) xanthan/0.1% (w/w) agarose in 40× magnification after performing a potentiodynamic current density potential curve in 1×10^{-3} mol L^{-1} sodium chloride; aeration; scan rate: 10 mV s^{-1}; start potential: -700 mV_H; stopped for 21 h at the reverse potential of $+290$ mV_H.

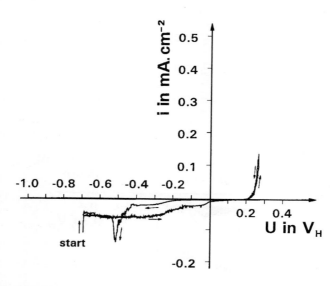

Fig. 10 Potentiodynamic current density potential curve of copper covered with EPS # 930 630 BF1 in 1×10^{-3} mol L^{-1} sodium chloride; aeration; scan rate: 0.01 mV s^{-1}; start potential: -700 mV_H; reverse current density: 0.1 mA cm^{-2}.

same amorphous shiny copper(I) oxide layer that is formed via hydrolysis on a bare copper electrode. Underneath this copper(I) oxide a slightly yellow layer can be detected that might consist of copper(I) chloride. These two reaction layers on top of the biopolymeric film cannot be reduced in the reverse scan of a cyclic voltammogram because this film is acting as an insulating barrier. The current density transient obtained when stopping the scan for 21 h at the reverse potential of 290 mV$_H$ to produce this electrode shows a strong decrease asymptotically towards zero. Transients obtained on bare copper electrodes in chloride containing solutions show a similar shape [23].

The electrochemical behaviour of a biopolymeric film produced by microorganisms on top of a copper surface was examined in sodium chloride electrolytes employing the experimental conditions outlined for Fig. 5, with a very slow scan rate of 0.01 mV s^{-1} (Fig. 10). All the results described earlier for the synthetic biopolymeric material (see Fig. 5) hold true for this type of exopolymeric substance as well. However, the reduction peak occurs at –520 mV$_H$ indicating the reduction of the precipitated copper(I) oxide, and the peaks for the formation and reduction of copper(I) chloride are missing compared to the scan of a bare copper electrode. The cathodic partial reaction was inhibited, but not the anodic partial reaction.

The influence of different pretreatments with exopolymeric substances on the coated electrodes, before performing the cyclic voltammetric experiment, have also been investigated. The following pretreatments were performed for both synthetic biopolymeric substances and biopolymers produced by microorganisms.

(i) Ageing of the exopolymeric substances for 164 h at 20°C, air contact, coating of electrode with aged material;

(ii) Exposure of the coated electrode at the free corrosion potential for 100 h, and

(iii) Cathodic protection of the coated electrode at 0 mV$_H$ for 24 h.

The results after performing pretreatment (i) are shown in Figs. 11 and 12 for both the synthetic (Fig. 11) and the microbial polymeric material (Fig. 12) using identical experimental conditions as described earlier (Figs. 5 and 10). No significant differences were obtained between these two materials. Only the reduction peak at –520 mV$_H$ can be seen in the reverse scan. Any effect of the pretreatment is not obvious (see Fig. 5, Fig. 10). After performing the two other pretreatments (ii) and (iii), only the reverse scan of the cyclic voltammogram was measured, starting at the free corrosion potential or at 0 mV$_H$, respectively. From the results obtained no influence of the pretreatment could be shown for both the synthetic and the microbial material.

5.3. Measurements at Platinum Electrodes

Voltammograms performed with coated copper electrodes in sodium chloride and sodium sulphate electrolytes showed at their start potential high cathodic current densities that decrease to very low values within a few minutes (see Figs. 5–12).

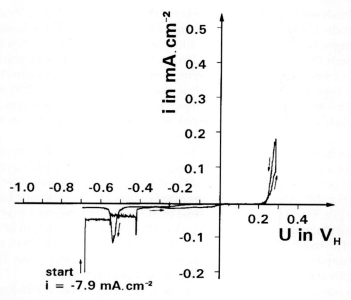

Fig. 11 Potentiodynamic current density potential curve of copper covered with 0.3% xanthan/ 0.1% (w/w) agarose in 1×10^{-3} mol L^{-1} sodium chloride; pH 6–6.5; aeration; scan rate: 0.01 mV s^{-1}; start potential: –700 mV$_H$; reverse current density: 0.1 mA cm^{-2}; pretreatment of biopolymeric material: ageing for 164 h at 20°C; air contact.

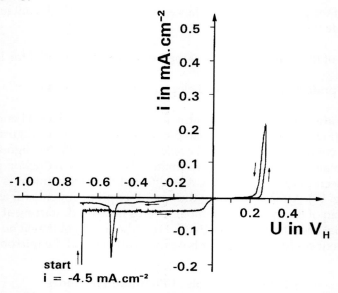

Fig. 12 Potentiodynamic current density potential curve of copper covered with EPS # 930 820 in 1×10^{-3} mol L^{-1} sodium chloride; pH 6–6.5; aeration; scan rate: 0.01 mV s^{-1}; start potential: –700 mV$_H$; reverse current density: 0.1 mA cm^{-2}; pretreatment of biopolymeric material: ageing for 164 h at 20°C; air contact.

To determine if these effects are to be attributed to processes involving the copper or processes within the biopolymeric material itself, measurements were made with a platinum electrode covered with the exopolymeric substance in a 1×10^{-3} mol L^{-1} sodium chloride solution. Figure 13 shows the obtained results. The high current at the start potential can be observed and no reduction peak occurs. This shows that these high cathodic currents do not affect the corrosion reactions involving copper species. It can be inferred that these currents are due to reduction processes within the biopolymeric coating itself. Consequently, they do not seem to play a relevant role in the corrosion mechanism.

6. Conclusions

Figure 14 summarises schematically the possible corrosion reactions of copper covered with biopolymeric material in electrolytes containing chloride ions.

The biopolymeric substance is cation selective and permeable for copper ions and hydrated protons. Chloride ions cannot diffuse through the coating. This discloses that at the Cu_2O/EPS phase boundary only corrosion reactions involving water can occur. Due to the high water content of the coating, sufficient water is available at the Cu_2O/EPS phase boundary. This leads to the formation of hydrated or unhydrated cuprous oxide, or cuprous hydroxide. But the acidic nature of the electrolyte at the Cu_2O/EPS phase boundary considerably limits the amount of oxide formed under the coating.

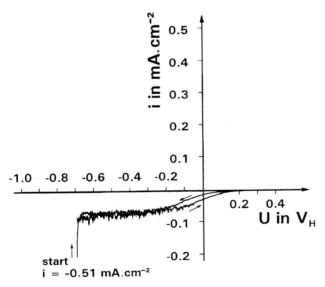

Fig. 13 Potentiodynamic current density potential curve of platinum covered with 0.3% xanthan/ 0.1% (w/w) agarose in 1×10^{-3} mol L^{-1} sodium chloride; aeration; scan rate: 1 mV s^{-1}; start potential –700 mV_H; reverse potential: +300 mV_H.

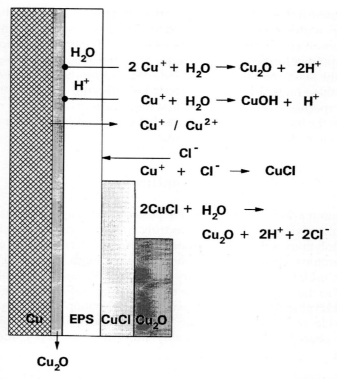

Fig. 14 *Possible corrosion reactions on copper covered with biopolymeric material in chloride ion containing electrolytes.*

Since the coating does not show conductive properties, only the reaction layers located directly on the copper surface, i.e. corrosion products formed via reaction with water, can be reduced. This barrier also inhibits the oxygen reduction reaction but has no significant influence on the anodic partial reaction.

In chloride-containing electrolytes the same reaction layers are obtained for a coated electrode as for a bare copper electrode. However, the precipitated copper(I) oxide formed via hydrolysis of the underlaying copper(I) chloride is now situated on top of the biopolymeric coating. Therefore, a reduction of these layers in the reverse scan of a voltammogram is not possible. Moreover, a comparable pitting attack as observed for bare copper electrodes can be observed at copper electrodes coated with EPS.

Finally, it has to be kept in mind that these model considerations are true for both the synthetic and the microbial polymeric substance on a copper surface. This means (i) that the choice of the model substance xanthan is justified; (ii) it has been possible to produce exopolymeric substances from bacteria isolated from copper pipes affected by MIC, and most importantly, (iii) both substances show the expected cation selective transport properties and have the same influence on the corrosion reactions of a copper surface in contact with an aqueous phase.

The application of cyclic voltammetry, in summary, is a powerful technique: (i) to prove the cation selectivity of the exopolymeric coating, and (ii) to show the influence of pH, as demonstrated in the experiments reported. A biopolymeric coating on a copper surface shows the expected cation selective properties and has a considerable influence on the corrosion reactions of a copper surface in contact with an aqueous phase, especially at the place where the precipitation of solid corrosion products will occur.

7. Acknowledgements

The financial support of this work within the BRITE-EURAM Project No. 4088 ('New Types of Corrosion Impairing the Reliability of Copper in Potable Water Caused by Microorganisms') is gratefully acknowledged.

References

1. G. G. Geesey, P. J. Bremer, W. R. Fischer, D. Wagner, C. W. Keevil, J. Walker, A. H. L. Chamberlain and P. Angell, in *Biofouling and Biocorrosion in Industrial Water Systems* (Eds G. G. Geesey, Z. Lewandowski and H.-C. Flemming), C. R. C. Press, Boca Raton, USA, 1994, Ch. 16, pp. 234–263.
2. P. Angell, H. S. Campbell and A. H. L. Chamberlain, International Copper Association (ICA) Project No. 405, Interim Report, August 1990.
3. H. Campbell, A. H. L. Chamberlain and P. Angell, in *Corrosion and Related Aspects of Materials for Potable Water supplies*, (Eds P. McIntyre and A. D. Mercer), 1993, The Institute of Materials, London, UK, 222–231.
4. W. R. Fischer, D. Wagner and H. H. Paradies, *Microbiologically Influenced Corrosion (MIC) Testing*, ASTM STP 1232, (Eds J. R. Kearns and B. Little), 1994, American Society for Testing and Materials, Philadelphia, PA.
5. B. Little, P. Wagner, S. M. Gerchakov, M. Walch and R. Mitchell, *Corrosion*, 1986, **42**, 533.
6. T. E. Ford, J. S. Moki and R. Mitchell, Corrosion '77, Paper 380, 9–13 March, 1977, San Francisco, CA, USA.
7. J. S. Nickels, R. J. Bobbie, D. F. Lott, R. F. Martz, P. H. Benson and D. C. White, *Appl. Environm. Microbiol.*, 1981, **41**, 1442.
8. W. G. Characklis and K. C. Marshall, *Biofilms*, Wiley Series in Ecological and Applied Microbiology, (Ed. R. Mitchell), J. Wiley & Sons Inc., New York.
9. R. F. Hadley, (Ed.), in *The Corrosion Handbook*, 1948, The Electrochemical Society Inc., New York N. Y., pp. 466–481, J. Wiley & Sons, Inc. N.Y.
10. J. T. Walker and C. W. Keevil, International Copper Association (ICA), Project No. 407, Final Report, 1990.
11. W. Fischer, I. Hänßel and H. H. Paradies, in *Microbial Corrosion 1* (Eds. C. A. C. Sequeira and A. K. Tiller), 1988, Elsevier Applied Science, pp. 300–327.
12. W. Fischer, H. H. Paradies, D. Wagner and I. Hänßel, *Werkstoffe und Korros.*, 1992, **43**, 496.
13. A. H. L. Chamberlain, W. R. Fischer and H. H. Paradies, 3rd European Federation of Corrosion Workshop on Microbial Corrosion, Estoril, Portugal, March 1994, this volume, pp. 1–16.

14. W. R. Fischer, D. Wagner, H. H. Paradies, M. Thies, U. Hinze, A. H. L. Chamberlain, J. N. Wardell and C. A. C. Sequeira, New types of corrosion impairing the reliability of copper in potable water caused by microorganisms, Brite-Euram Project No. BE-4088, Contract No. BREU-CT 91-0452, 1991–1994.

15. C. A. C. Sequeira, A. C. P. R. P. Carrasco, M. Tietz, D. Wagner and W. R. Fischer, *3rd European Federation of Corrosion Workshop on Microbial Corrosion*, Estoril, Portugal, March 1994, this volume, pp. 64–84.

16. J. N. Wardell and A. H. L. Chamberlain, *3rd European Federation of Corrosion Workshop on Microbial Corrosion*, Estoril, Portugal, March 1994, this volume, pp. 49–63.

17. C. D. Burke and T. G. Ryan, *J. Electrochem. Soc.*, 1990, **137**, 1358.

18. H. H. Strehblow and H. D. Speckmann, *Werkstoffe und Korros.* 1984, **35**, 512.

19. H. D. Speckmann, S. Haupt and H. H. Streblow, *Surface and Interface Analysis*, 1988, **11**, 148.

20. H. D. Speckmann, M. M. Lohrengel, J. W. Schultze and H. H. Strehblow, *Ber. Bunsenges. Phys. Chem.*, 1985, **89**, 392.

21. H. H. Paradies, personal communication.

22. Bundesgesetzblatt, Teil 1, No. 66, Bekanntmachung der Neufassung der Trinkwasserverordnung.

23. W. Fischer and B. Füßinger, *Proc. 12th Scand. Corros. Congr. & EuroCorr '92*, Dipoli (Finland), 1 (1992) 769.

24. H. Siedlarek, D. Wagner, M. Kropp, B. Füßinger, I. Hänßel and W. R. Fischer, in *Corrosion and Related Aspects of Materials for Potable Water Supplies* (Eds P. McIntyre and A. D. Mercer), 1993, The Institute of Materials, London, UK, p. 122.

25. M. Billiau and C. Drapier, *Materiaux et Technique*, 1976, **1**, 27.

26. M. Pourbaix, *J. Electrochem. Soc.*, 1976, **123**, 27.

27. F. M. Al-Kharafi and Y. A. El-Tantawy, *Corros. Sci.*, 1982, **22**, 1.

28. J. G. N. Thomas and A. K. Tiller, *Brit. Corros. J.*, 1972, **7**, 256.

29. DIN 1786, 'Copper tubes for plumbing, seamless drawn', Beuth Verlag Berlin, 05/1980.

30. W. Fischer and W. Siedlarek, *Werkstoffe und Korros.*, 1979, **30**, 695.

31. L. Muller and L. N. Nekrassov, *Electrochim. Acta*, 1964, **9**, 1015.

32. L. Muller and L. N. Nekrassov, *Electrochim. Acta*, 1965, **9**, 282.

33. M. Pourbaix, *Atlas of Electrochemical Equilibria in Aqueous Solutions*, 1966, Pergamon Press, New York.

34. H. Siedlarek, B. Füßinger, I. Hänßel and W. R. Fischer, *Werkstoffe und Korros.*, 1994, **45**, 654–662.

35. A. H. L. Chamberlain, P. Angell and H. S. Campbell, *Brit. Corros. J.*, 1988, **23**, 197.

Part 2

Microbial Corrosion:
Mechanisms
and General Studies

6

Contribution of Microbiological Phenomena in the Localised Corrosion of Stainless Steels

F. COLIN, M. J. JOURDAIN, G. D'AMBROSIO, A. POURBAIX*
and D. NOEL[†]

University of Nancy, IRH Génie de l'Environnement, 11 bis rue Gabriel Péri, 54500 Vandoevre, France
*CEBELCOR, Avenue Paul Heger, Grille 2B 1050 Bruxelles, Belgium
[†]EDF DER/EMA Les Renardières, Ecuelles 72250 Moret sur Loing, France

ABSTRACT

Pitting corrosion observed on 304L stainless steel pipes transporting water led us to suspect considerable induction or acceleration of corrosion processes due to the development of microorganisms. Discriminant experiments were carried out to quantitatively determine the extent of purely physico–chemical processes compared with purely biological processes. An experimental pilot loop, working with two identical and independent 30 L circuits, was built to monitor electrochemical measurements and to characterise the biofilm.

One circuit was fed with natural surface water (river water with high chloride content); the other with the same, but sterilised river water. The whole range of contact conditions encountered under real conditions (stagnation, flow and water renewal) was investigated, by means of the loop geometry (vertical, horizontal and oblique zones, enabling immersed areas, temporarily immersed areas, and dead zones to be simulated). The surface condition of the stainless steels covered a wide range of observed cases: soldered and unsoldered zones, as-received and passivated surfaces. The 300 day experiment was characterised by cycle simulations (6 in total), including flow, stagnation and renewal of circuit water (10% in each cycle). Besides electrochemical and biofilm characterisation measurements, the analytical monitoring of circuit water enabled variations due to bacterial growth or to corrosion to be shown. At the end of the experiment, a complete dismantling of both circuits enabled a visual comparison to be made of the surfaces affected by corrosion deposits and biofilm, in the non-sterile and sterile circuits. A similar comparison was made between the electrochemical and microbiological measurements in the sterile and non-sterile circuits. Corrosion was clearly more prevalent in the presence of biological activity.

1. Objectives

Pitting corrosion observed on 304L stainless steel pipes transporting river water suggested a large induction and/or acceleration of corrosion processes due to the developments of microorganisms.

Discriminating experiments were carried out to compare quantitatively purely physico–chemical processes with purely biological processes.

2. Experimental Procedures
2.1. The Experimental Set-up

An experimental pilot loop, working with two identical and independent 30 litre circuits, was constructed, in order to monitor electrochemical measurements, and to characterise the biofilm (Figs 1–3).

One circuit was fed with natural surface water (river water with a high chloride content 300–500 mgL^{-1}); the other with the same water, after sterilisation. The whole range of contact conditions encountered under real conditions (stagnation, flow and water renewal) was investigated by modifying the geometry of the loop (vertical, horizontal and oblique zones), enabling immersed areas, temporarily immersed areas and 'dead zones' to be established.

Moreover, the surface condition of stainless steel covered a wide range of observed cases: soldered and unsoldered zones as-received, and passivated surfaces.

The pipes were made of AISI 304L stainless steel of varying diameters: 20, 60 and 100 nm, with a thickness of 1.5–2 mm.

The stop valves were made of AISI 316L stainless steel.

2.2. Preparation of Materials

2.2.1. Cleaning and passivation

The part of the loop (Fig. 3) marked with the letter P was cleaned and passivated with a paste (AVESTA) containing 5% hydrofluoric acid (HF) and 22% nitric acid (HNO_3) The paste was applied for a period of one hour, and the loop was then rinsed with distilled water.

2.2.2. Sterilisation

The various parts of the loop were sterilised under different conditions, depending on their nature.

The stainless steel material was sterilised for 4 h at 180°C in dry heat, and for materials that would be damaged at high temperature, a tyndallisation process was used.

After assembling the various parts of the loop, it was disinfected using a solution of 10% formaldehyde over a period of 24 h. It was then rinsed three times with sterilised distilled water, the final water used being checked to ensure that it contained no bacteria or traces of formaldehyde.

2.2.3. Conditions of set-up

The experiment was carried out over a period of 300 days. It was characterised by cycle simulations; six in total, with flow, stagnation, and renewal of circuit water at a rate of 10 % for each cycle. After renewal, the water flowed for one hour. In the sterile loop, the procedure was the same, but with sterilised water.

After the sixth cycle, the two loops were maintained in conditions of stagnation for 4 months (110 days).

Fig. 1 General view of the experimental loop.

Fig. 2 View of the transluscent control during experiment.

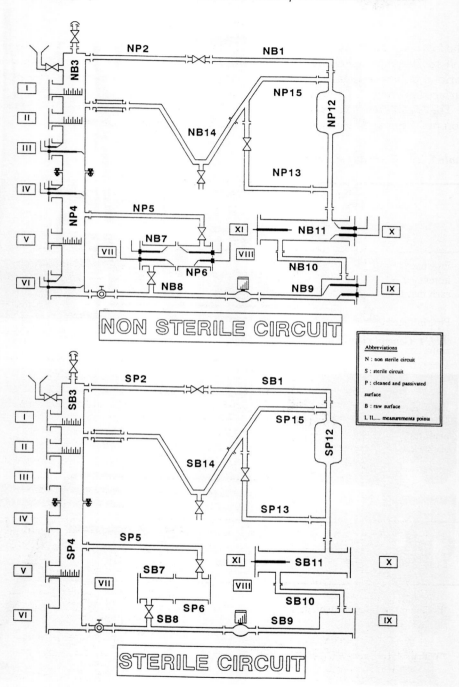

Fig. 3 Experimental set-up.

2.3. Instrumentation

The loop was instrumented with 16 measurements of corrosion potential (15 on the non-sterile loop; one on the sterile). Figure 3 and Table 1 show the establishment of various points of measurement according to the stagnation zone and the type of surface (as-received or cleaned and passivated surface).

The 'dead zone' was also instrumented in order to permanently measure corrosion potential and to measure corrosion galvanic current when required.

Table 1. Identification of measurement points

	Surface B : raw P : cleaned passivated		Number of measurement points	Measurements
Pipe temporarily immersed (1 h each month)	B	I	–	2 exposed steel surfaces for bacteriological analysis
		II	–	
Pipe immersed in stagnant water for 1 month		III	1–2	Electrochemical potential on and outside soldered zone
		IV	3–4	
	P	V	–	2 exposed steel surfaces
Water flows for 1 hour at beginning of each cycle		VI	5–6	Electrochemical potential on and outside soldered zone
		IX	15–16	
	B	X	11*–12*	potential under biofilm
		XI	13**–14**	potential corrosion and galvanic current
Dead zone	P	VII	7–8	Electrochemical potential on and outside soldered zone
		VIII	9–10	

*Even-numbered channels indicate measurements taken on soldered zone and odd-numbered channels outside from soldered zone.

**With a solid electrode Ag/AgCl (13 non-sterile loop) (14 in sterile loop).

2.4. Measurements

The analytical monitoring of the circuit enabled us to pinpoint variations due to bacterial growth, modifications of water characteristics and corrosion. These variations are shown in Table 2.

3. Results and Discussion
3.1. Bacteriological Analysis of the Water

For the non-sterile loop, the number of bacterial types was compared at the start of the experiment or on renewal of the circuit water (10% in each cycle) with the number after one month of stagnation. Table 3 shows, for example, the first and the fifth cycles.

No significant difference was observed for either aerobic or anaerobic bacteria.

Table 2. Measurements

	Water
At the beginning and at the end of each cycle	• *physico–chemical parameter* – pH value – conductivity – hardness – anions: HCO_3^-, Cl^-, SO_4^{2-} – cations – total organic carbon – dissolved oxygen – sulphide • *bacteriological parameters* – total bacteria – metabolic active bacteria – aerobic bacteria (colony-forming units in culture medium) – aerobic bacteria – anaerobic bacteria (SRB) – yeast and fungi – ferrobacteria
At the end of each cycle	**Exposed steel surface** – corrosion – quantitative bacteria in the biofilm
During experiment	– continuous measurement of electrochemical potential of stainless steel

Table 3. *Bacteriological analysis of the water (examples: 1st and 5th cycles)*

		At the beginning of the cycle		At the end of the cycle	
		1	5	4	5
No. per mL	Total bacteria	8.40 10^6	1.80 10^6	4.30 10^6	1.50 10^6
	Active bacteria	2.10 10^5	9.00 10^4	1.80 10^5	1.40 10^5
	Aerobic bacteria	5000	5100	2600	700
	Anaerobic bacteria	40 000	14 000	6000	4300
	Yeast and fungi	680	80	42	670
	Ferrobacteria	no	presence	no	presence
No. per L	SRB after 2 months of incubation at 30°C	36	750	0	2400

3.2. Physico–chemical Analysis

Table 4 shows measurements of pH value, dissolved oxygen and sulphide at the end of the experiment.

The pH value, and levels of dissolved oxygen and sulphide changed between the beginning and the end of the experiment. A decrease was noted in the pH value: from 7.7 to 6.8 in stagnant parts. There was no significant difference between the sterile and non-sterile circuits. A strong sulphurous odour was detected only in the non-sterile 'dead zone'. This was probably due to SRB activity.

3.3. Corrosion of Material

3.3.1. Electrochemical potential data
Table 5 shows the measurements of electrochemical potential.

Modifications in electrochemical corrosion were occasionally found just after water renewal. This was probably because the water was aerated and therefore had an

Table 4. *After experimentation, measurements of pH value, dissolved oxygen and sulphide*

Loop	Water	pH	Dissolved oxygen (gL^{-1})	HS$^-$ (gL^{-1})
Sterile	Stagnation	6.85	–	0
	Renewal	7.05	5	0
Non-sterile	Stagnation	6.70	–	3.2
	Renewal	7.45	1.9	0

Table 5. Electrochemical potential at the beginning and at the end of experimentation

	Surface B : as-received P : cleaned passivated	Number of measurement points		Potential value (Ag/AgCl mV)		
				Start	End of the 6th cycle	End of the experiment
Pipe temporarily immersed (1 h each month)		I	–	–	–	–
Pipe immersed in stagnating water for 1 month	B	II	–	–	–	–
		III	1 2	–100 –100	–70 –70	–100 –100
		IV	3 4	–100 –100	–70 –200	–100 –230
Water flows for 1 h at beginning of each cycle	P	V	–	– –	– –	– –
		VI	5 6	–100 –100	–70 –70	–220 –100
	B	IX	15 16	–60 –60	–250 –250	–250 –250
		X	11 12	–200 –200	–200 –200	–250 –250
		XI	13 14	–300 –150	–350 –120	–300 –120
Dead zone		VII	7 8	–20 –20	–400 –400	–350 –320
	P	VIII	9 10	–20 –20	–400 –400	–400 –400

increased oxidising power. At the end of the experiment, the higher potential was situated near the aerated area, the lower potential in the non-aerated area.

In the sterile circuit, the electrochemical potential remained stable during the experiment (channel 14 in measurement point XI). An electrochemical effect was found even when no bacteria were present. In the non-sterile circuit, no difference was recorded between soldered and not-soldered surfaces. The electrochemical potential detected in the aerated area was higher than the pitting potential. This did not, however, reach breakdown potential.

3.3.2. Visual examination

On dismantling the two circuits at the end of the experiment we carried out a visual examination. Corrosion deposits and biofilm were observed particularly on the surfaces of the non-sterile circuit.

3.3.3. The interior

- Sterile circuit (Figs 4, 5 and 8). No biofilm was observed (Fig. 4) but there were very small corrosion points at the soldered (Fig. 5) zones, and no corrosion in the dead zones (Fig. 8, p.118).

- Non-sterile circuit (Figs 4, 6 and 7). Extensive gelatinous rust-coloured biofilm was found in all zones (as-received or cleaned and passivated surfaces). The points of corrosion were larger and more numerous than in the sterile circuit.

- The dead zone had a sulphurous odour and was covered with a dark deposit with bright points (pitting).

Fig. 4 View of the transluscent control after experiment (sterile and non-sterile circuit).

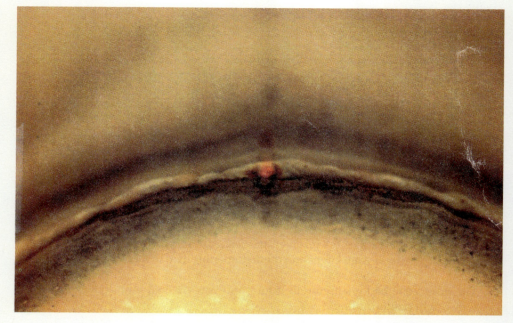

Fig. 5 *Appearance of corrosion of SP12 element of sterile circuit (S) after test.*

Fig. 6 *Appearance of NP12 element of non-sterile circuit (N) after test. (a): External lateral view.*

Microbiological Phenomena in the Localised Corrosion of Stainless Steels 117

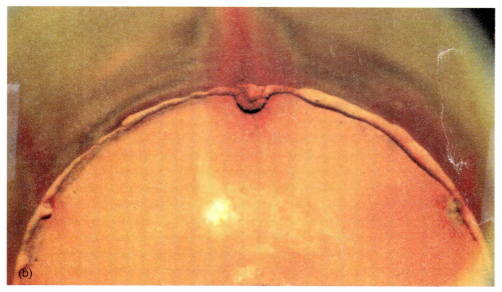

Fig. 6 Appearance of NP12 element of non-sterile circuit (N) after test. (b): Internal lateral view.

Fig. 7 Appearance of NB3 element after test.

Fig. 8 *Appearance of SB11 element after test.*

4. Conclusions

With this experimental set-up it was possible to compare the purely chemical processes with the purely biological processes. After 10 months, pitting corrosion was present in both circuits but was more significant in the non-sterile loop.

The observations and electrochemical measurements suggest that the more severe corrosion in the non-sterile circuit was due to the presence of larger amounts of polysaccharides, debris and biofilms in that circuit. Such deposits are assumed to induce local chemistries (higher chloride concentration, lower pH) that enhance local attack.

The role of the metallurgical structure and of the composition of the different phases of the weld on the initiation of local corrosion must still be investigated.

7
Attachment of *Desulfovibrio vulgaris* to Steels: Influence of Alloying Elements

D. FERON

Commissariat a l'Energie Atomique, CEREM-SCECF, BP No 6, 92265 Fontenay-aux-Roses Cedex, France

ABSTRACT

In the present paper, the influences of iron, chromium, molybdenum, nickel, vanadium and copper on the development of *Desulfovibrio vulgaris* are investigated and related to the attachment of these bacteria on seven different steels ranging from carbon steel to stainless steels. The results when related to dissolved metallic species suggest that iron (III) has a beneficial effect on the growth of *D. vulgaris* while molybdenum (VI), chromium (III), vanadium (V), and copper (II) limit the growth when the concentrations of these ions reach 0.1 mM. Experiments performed with metallic coupons show that the toxic effect of the alloying elements on bacterial growth is limited by the beneficial effect of dissolved iron. The attachment experiments were performed with one austenitic steel (18% Cr, 11% Ni, 2% Mo) and six ferritic steels containing 0–17% Cr, 0–2% Mo and up to 0.3% V.

On carbon steel, without any alloying element, the whole surface is colonised after 24 h. On steels containing chromium only, 13% Cr is needed to decrease the surface colonisation to about 50%, but with 17% Cr, only 10% of the surface is covered, mainly with polysaccharides, after 3 weeks. The toxic effect of molybdenum and vanadium is observed when these elements are together (from 1% Mo and 0.2% V). Similar observations are made on coupled coupons, with a small increase in the number of attached bacteria. The attachment of *D. vulgaris* seems to be mainly related to the steel composition and so to the physico–chemistry of the interface.

1. Introduction

Although several mechanisms have been proposed to elucidate the part played by Sulphate Reducing Bacteria (SRB) in the corrosion of steels, little evidence is available concerning the role of alloying elements in the prevention of Microbiologically Influenced Corrosion (MIC). The alloying elements which are added to metals to increase their properties and specially their corrosion resistance, may also alter or improve their susceptibility to MIC [1]. General considerations on the influence of alloying elements have been reviewed for stainless steels, copper, nickel, aluminium and titanium alloys [2]. The schematic diagram in Fig. 1 shows the main interactions taking place in the MIC processes. It outlines the role of metallic ions generated by the alloy degradation with respect to their influence on bacterial behaviour: metallic ions may influence bacterial growth, the attachment of bacterial cells to steel surface and so the build up of a biofilm.

As alloying elements (e.g. chromium, nickel, molybdenum, vanadium, copper ...)

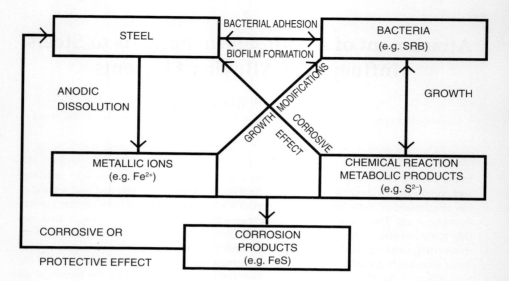

Fig. 1 Main interactions in the MIC of steels.

are added to steels for the improvement of their properties, these elements may also be dissolved together with iron during corrosion processes. The study of their influence on bacterial growth and adhesion is of importance in increasing the understanding of MIC mechanisms.

The sulphate reducing bacteria have long been known to be involved in corrosion under anaerobic environments. Previous studies on the initial settlement of SRB on steel surfaces revealed that different degrees of surface colonisation depend on the composition of the substratum while other authors reported that the corrosion behaviour depends on surface colonisation [3–8].

This paper summarises the results on bacterial attachment obtained during experiments that were undertaken to account for the role of alloying elements on the MIC of steels (from carbon steel to stainless steels and including low alloyed steels) exposed to *Desulfovibrio vulgaris*. The impact of metallic species (iron, chromium, nickel, molybdenum, vanadium, copper) on bacterial growth are reported as well as the bacterial adhesion on carbon steel, low alloyed and stainless steels.

2. Materials and Methods
2.1. Culture Conditions

The sulphate reducing bacterium *Desulfovibrio vulgaris*, strain Hildenborough (NCIMB 8303), was grown in a standard deaerated lactate–sulphate medium consisting of: NH_4Cl 1; $MgSO_4, 7H_2O$ 2; Na_2SO_4 4; K_2HPO_4 0.5; sodium lactate 2.7; yeast extract 1; ascorbic acid (when added) 0.1 g L^{-1} and mineral element solution 1 mg L^{-1}. The influence of dissolved metallic species was tested by adding soluble metallic species to the standard culture medium during its preparation. The culture medium was sterilised by autoclaving at 122°C for 20 min or by filtration at 0.22 µm. Deaeration was by boiling and by argon or nitrogen bubbling. Three to five days old cultures were used as a source of bacterial inoculum. After inoculation (10% v/v inoculum) the cultures were incubated at 32°C. Growth was followed using optical density measurements at 450 nm and chemical analyses of lactate and acetate (ion exclusion chromatography), sulphate (ion affinity chromatography) and sulphite (calorimetry method). Reported experiments were performed in 10 mL Hungate tubes, 70 mL or 150 mL flasks, or in 5 L fermenters.

2.2. Specimens

The tested steels, the chemical compositions of which are reported in Table 1, include:

- one carbon steel (mild steel), referred to as A42 in Table 1;
- one low alloy steel (3Cr1Mo) containing 2.25% of Cr and 1% of Mo;
- two low alloy steels with 9% Cr but containing 1% Mo (9Cr1Mo) or 2% Mo (9Cr2Mo);
- two ferritic stainless steels with 13% Cr (13Cr) or 17% Cr (17Cr);
- one austenitic stainless steel containing 17% Cr, 11% Ni and 2% Mo (316L).

Steel coupons (10 × 10 × 2 mm or 10 × 70 × 2 mm) were manufactured from plate or stub materials. During the manufacturing process, all coupons were given an identical, rough surface finish by polishing with a 'grade 80' paper (mean grain size: 200 µm). Coupled coupons were made of carbon steel (A42, anode) thermally welded with an alloy steel (cathode). All steel coupons were then washed with ethanol, rinsed with deionised water and dried before sterilisation and use.

Table 1. Chemical composition of tested steels (wt %, balance is Fe)

Steel designation	Cr	Mo	V	Ni	C
A42	–	–	–	–	0.2
3Cr1Mo	2.25	0.9/1.1	–	–	0.15
9Cr1Mo	9.5	1.05	0.25	0.2	0.12
9Cr2Mo	9.0	2.1	0.27	–	0.15
13Cr	13.0	–	–	0.15	0.14
17Cr	16.32	–	–	0.15	0.02
316L	17.09	2.03	–	11.23	0.016

2.3. Biofilm Observations

Biofilms on steel surfaces were observed under the scanning electron microscopy (SEM). The freshly withdrawn steel coupons, removed from the culture medium, were immersed in a 2.5% glutaraldehyde solution for 24 h at 4°C. The coupons were dehydrated by passing through a graded series of water–ethanol solutions (30–100% ethanol), then immersed in a 'PELDRI II' solution and then put in the primary vacuum chamber. Dried coupons were coated with gold (10 nm) and examined under SEM.

3. Influence of Metallic Ions

Three types of tests were performed in order to study the influence of metallic ions on the growth of *D. vulgaris*:

- the effect of dissolved metallic species was investigated with additions of dissolved metallic compounds at various concentrations in the culture medium;

- experiments were also run to observe whether metallic coupons themselves may limit or enhance bacterial growth;

- the influence of anodic dissolution products was studied by anodic polarisation of steel electrodes in the culture medium.

3.1. Dissolved Metallic Species

The test matrix is presented in Table 2: iron (II), molybdenum (VI), chromium (III), copper (II), nickel (II) and vanadium (V) were added separately to the standard lactate–sulphate medium at various concentrations up to 0.1 or 0.2 mM. Typical results of optical density measurements are presented in Fig. 2 where the points are mean values and the bars cover the extreme values of at least 3 experiments. The development observed during incubation of *D.vulgaris* in the standard medium is given as a reference. The same development, i.e. an increase of optical density from 0.15–1.4 within 30 h, was obtained with nickel (II) chloride concentrations up to 0.2 mM and with all the other metallic ions when their concentrations did not exceed 0.01 or 0.02 mM. With higher concentrations, the data plotted in Fig. 2 show that:

- with iron (II) citrate concentration of 0.2 mM, higher optical densities (up to 1.7) are reached within a shorter time (24 h). Complementary experiments performed with 0.2 mM of citric acid instead of iron citrate, prove that citrate alone has no significant influence on the bacterial growth.

- in the presence of 0.1 mM sodium molybdate or copper (II) chloride, or with 0.2 mM of sodium metavanadate, optical density hardly increases; this is in accordance with the lactate and sulphate concentrations which do not change;

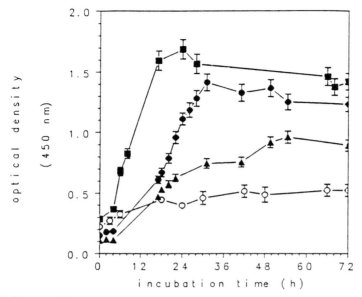

Fig. 2 Influence of the metallic ions on the growth of D. vulgaris. (■) 0.2 mM iron (II), (●) standard culture medium, (▲) 0.2 mM chromium (III), ( ) 0.1mM molybdenum (VI) or 0.2 mM vanadium (V).

Table 2. Test matrix of the influence of dissolved metallic species

Element	Compound	Concentrations (mol L^{-1}) in the medium		
		Standard	Tested	Inhibitory
Iron	Fe(III) citrate	10^{-5}	$10^{-6}, 2 \times 10^{-5}, 2 \times 10^{-4}$	–
Chromium	CrCl$_3$	–	$2 \times 10^5, 5 \times 10^5, 2 \times 10^{-4}$	2×10^{-4}
Copper	CuSO$_4$	10^{-7}	$10^{-7}, 10^{-6}, 10^{-5}, 10^{-4}$	2×10^{-4}
Molybdenum	Na$_2$MoO$_4$	10^{-7}	$10^{-7}, 10^{-6}, 10^{-5}, 10^{-4}$	–
Nickel	NiCl$_2$	10^{-8}	$2 \times 10^{-7}, 2 \times 10^{-6}, 2 \times 10^{-4}$	–
Vanadium	NaVO$_3$	–	$2 \times 10^{-5}, 2 \times 10^{-4}$	2×10^{-4}

- addition of 0.2 mM chromium (III) chloride slowly modifies the culture evolution which reaches its maximum (only 1.0 of optical density) after 50 h.

These results suggest that iron (II) has a beneficial effect on the growth of *D.vulgaris*, which is consistent with the high iron requirements for growth described in the literature [10]. They clearly show the inhibitory role of 0.1 mM copper (II) or molybdenum (VI) and of 0.2 mM vanadium (V), which is in accordance with previously published results for molybdenum [11]. The role played by chromium (III) is less obvious as its addition to the medium induced the formation of a precipitate: the growth limitation could be due either to a toxic effect of chromium or to the withdrawal of growth factors from the solution during the precipitation. It can be pointed out that the inhibitory effect of molybdenum and of chromium has been observed on a marine strain of SRB [11]. These experiments demonstrate that additions of metallic ions can markedly modify the growth of *D. vulgaris*.

3.2. Metallic Coupons

D. vulgaris were grown in standard culture medium with metallic coupons are added before the inoculum. The test matrix include all the alloys described in Table 1 and two sizes of coupons in order to vary the ratio between the steel surface and the culture volume: $2 \text{ cm}^2 \text{ mL}^{-1}$ with the small coupons ($10 \times 10 \times 2$ mm) and $20 \text{ cm}^2 \text{ mL}^{-1}$ with the large coupons ($70 \times 10 \times 2$ mm).

The bacterial growths observed with these different metallic coupons were the same as with the standard culture medium without any coupon, except in two cases:

- with large coupons of the unalloyed steel, A42, the bacterial growth was faster than in the standard conditions;

- 3Cr1Mo slowly modified the culture growth which reached its maximum after some 40 h instead of 30 h in the standard culture medium. This result has been reported previously and is based not only on optical density measurements, but also on chemical analyses of substrates and metabolisms [12].

These results are in accordance with the corrosion observations: except A42 and 3Cr1Mo, the other materials do not corrode in the previous conditions, and so do not influence greatly the bacterial growth in the bulk medium solutions [14]. With large coupons of 2Cr1Mo, the influence on the bacterial growth could be more surprising: at the beginning of the exposure, the corrosion of this material is not important and so few molybdenum or vanadium ions would be dissolved in the medium [14]. It could be a synergistic effect of these two compounds. Anyway, the bacterial growth was not inhibited but only slowly modified.

3.3. Dissolution Products

In order to force metallic compounds from steel to be dissolved in the culture medium, anodic polarisations of A42, 2Cr1Mo, 9Cr1Mo and 9Cr2Mo was conducted to produce more significant quantities of corrosion products. All the polarisations wee performed in 150 mL culture medium with an anodic current of 5 mA for 10 min. If only the following reaction of iron dissolution is considered

$$Fe \rightarrow Fe^{2+} + 2e^-$$

the imposed polarisation would correspond to a concentration of 10^{-4} mol L^{-1} of iron (II) in the culture medium.

The growth results obtained with the above four materials are exactly the same as in the standard culture medium without anodic dissolution products: there is no major effect of the anodic dissolution product on the growth of *D. vulgaris*. This is probably due to the fact that the toxic effects of alloying elements are compensated by the beneficial effect of dissolved iron.

4. Bacterial Adhesion
4.1. Experimental

Two sorts of specimens were used to study the bacterial adhesion on steels as shown in the test matrix of Table 3 where the crosses (×) indicate that the test was conducted:

- composition of standard coupons ranged from carbon steel to stainless steel as shown in Table 3; these experiments were performed to study the influence of alloying elements on bacterial adhesion after 24 and 48 h, two and three weeks of incubation with *D. vulgaris* in the standard culture medium;

Table 3. Test matrix of the bacterial adhesion on steels

Steel	Duration			
	24 h	48 h	2 weeks	3 weeks
Alone				
A42	×	×	×	×
3Cr1Mo	×	×	×	×
9Cr1Mo	×	–	×	×
9Cr2Mo	×	–	×	×
13Cr	×	–	×	–
17Cr	×	×	×	×
316L	–	×	–	×
Coupled with A42 (anode (A)/cathode (C) area ratio)				
A42	–	×	–	×
3Cr1Mo(1)	–	×	–	×
9Cr1Mo (1)	–	×	–	×
17Cr (1)	–	×	–	×
316L (1)	–	×	–	×
316L (5)	–	×	–	×
316L (0.2)	–	×	–	×

- coupled coupons were made of carbon steel thermally welded with an alloy steel; on these coupons, the anodic area (A 42) and the cathodic area (alloyed steel) were well defined and the ratio between these two areas was equal to 1; with 316L stainless steel as a cathode, ratios of 0.2 and 5 were also tested; if the bacterial adhesion is a function of the anodic or cathodic areas, it should be easily observed in these experiments.

All the tested coupons were examined under SEM after the preparation reported above (Section 2.3.). So only what is often called 'irreversible' adhesion was studied. Although the SEM observations give good qualitative results, quantitative information is more difficult to obtain. The attachment results are expressed in two ways:

- number of bacteria per surface unit (100 μm^2) which is quite easy to determine when few cells are visible on surfaces, but more difficult when there are heterogeneities, when biofilms are thicker and include high densities of cells;

- percentage of covered surface which includes not only bacteria cells but also polysaccharides, corrosion products and deposits; this percentage was initially determined 'manually' with graph paper, and is now calculated using image analysis, as shown below.

4.2. Attachment Results

Tables 4 and 5 show the results obtained on the standard coupons (uncoupled coupons). They are expressed respectively in terms of bacterial densities (number of cells per 100 μm^2) and of covered surface (percentage). The results for the coupled coupons are reported in Table 6.

On carbon steel (A 42), the whole surface was covered with bacterial cells after 24 h as shown in Fig. 3 and as reported in Tables 4 and 5: cells were numerous on the

Table 4. Number of bacteria cells per 100 μm^2 on uncoupled coupons

Steel	Test duration			
	24 h	48 h	2 weeks	3 weeks
A42	> 50	> 50	> 50	> 50
3Cr1Mo	> 50	> 50	> 50	> 50
9Cr1Mo	0–2	–	10–20	20–40
9Cr2Mo	0–2	–	3–20	20–30
13Cr	30–50	–	40– > 50	–
17Cr	0–5	0–5	10	0–5
316L	–	0–2	–	0–15

Table 5. Percentage of covered surface on uncoupled coupons

Steel	Test duration			
	24 h	48 h	2 weeks	3 weeks
A42	100%	100%	100%	100%
3Cr1Mo	40–60%	40–60%	60–100%	60–100%
9Cr1Mo	< 5%	–	10–30%	30–40%
9Cr2Mo	< 5%	–	20–30%	30–40%
13Cr	30–50%	–	40–60%	–
17Cr	< 5%	< 5%	10%	5–15%
316L	–	0–2	–	0–15

Table 6. Bacterial adhesion on coupled steels (number of bacterial cells per 100 µm²).(Anode: carbon steel.)

Cathode	Ratio A/C	Number of bacteria after		% of covered surface after	
		48 h	3 weeks	48 h	3 weeks
A42	1	> 50	> 50	100%	100%
3Cr1Mo	1	> 50	> 50	100%	100%
9Cr2Mo	1	10–30	20–40	10%	20/40%
17Cr	1	0–5	10–30	5/15%	20/50%
316L	1	0–5	5–15	5/15%	10/30%
316L	0.2	0–10	0–20	10%	20%
316L	5	0–5	0–15	5/15%	10/20%

steel surface and free extracellular polysaccharides (EPS) are visible. After two or three weeks, the biofilm is thick and desquamated in some areas as shown in Fig. 3. The same observations were made on standard coupons and on coupled coupons whatever the cathodic material (Table 6).

On 3Cr1Mo, many bacterial cells were present on the surface after 24 h (> 50 cells/100 µm²), but the whole surface was not covered as illustrated in Fig. 4. In that case, image analysis was conducted as also shown in Fig. 5 where the digitalised image is compared to the SEM image, and the percentage of covered surface calculated: 45% of the 3Cr2Mo surface was covered with cells after 24 h in that example. In fact, it varies between 40 and 60% depending on the examined surface area, as reported in Table 4. Similar results were obtained at 48 h, but on coupons coupled with carbon steel, the whole 3Cr2Mo surfaces were covered after 48 h (Table 6).

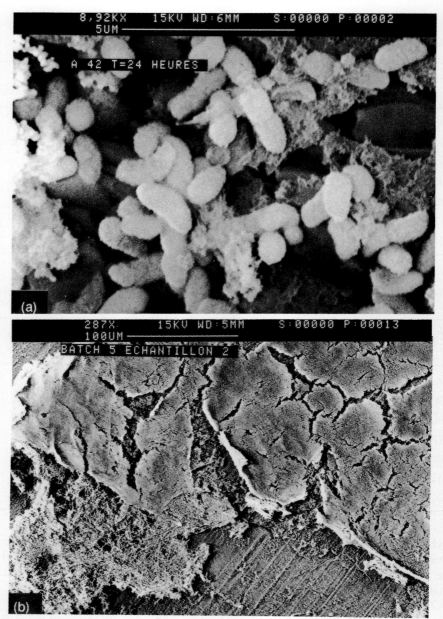

Fig. 3 D. vulgaris *adhesion on carbon steel. (a) After 24 h; (b) After 2 weeks.*

On 9Cr1Mo and 9Cr2Mo, very few bacteria cells were observed after 24 h as shown in Fig. 6. The percentage of covered surface was also very low: 1% in the case of Fig. 6, less than 5% reported in Table 5. Notice that, after 2 weeks at 32°C in the incubated culture medium, the number of bacteria is still low on the surfaces of these materials (Fig. 7): between 10 and 20 bacteria per µm², but EPS are visible and covered about

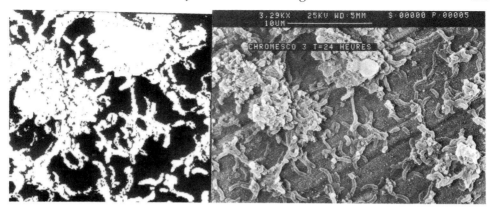

Fig. 4 D. vulgaris *adhesion on 3Cr1Mo (after 24 h, SEM and digitised images).*

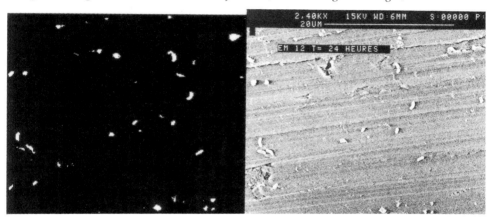

Fig. 5 D. vulgaris *adhesion on 9Cr2Mo (after 24 h, SEM and digitised images).*

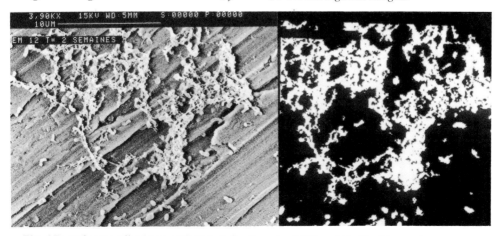

Fig. 6 D. vulgaris *adhesion on 9Cr2Mo (after 2 weeks, SEM and digitised images).*

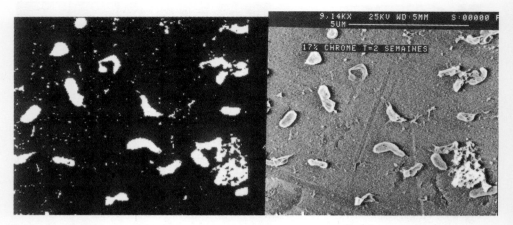

Fig. 7 D. Vulgaris *adhesion on 17Cr stainless steel (after 2 weeks, SEM and digitalised images).*

20–30% of the surface. On 9Cr2Mo coupled coupons, the number of cells and the percentage of covered surface was more significant at 48 h, but the same after 3 weeks as on uncoupled coupons.

On 13Cr stainless steel, a large number of cells were seen after 24 h, while on 17Cr stainless steel, very few cells were observed (Table 4). Even after 2 or 3 weeks, the number of cells attached on 17Cr steel was low (about 10 cells per 100 μm^2 (Fig. 7)) and few EPS are seen (the ratio of covered surface does not exceed 15%). On these two stainless steels (13Cr and 17Cr), the observations were similar to those with coupled coupons, the only difference being a little more bacteria on surfaces at the beginning of the exposure (Table 6).

On 316L stainless steel, the same observations as on 17Cr could be made: i.e. very few cells on the surfaces, even after 3 weeks (Tables 4 and 5). There was no difference between uncoupled coupons and 316L coupons coupled with carbon steel: the number of cells and the covered surface were always low and there was no connection with changes in the surface ratio between the anode and the cathode (Table 6).

As reported previously, a large amount of bacterial cells on the anode (carbon steel) of coupled coupons, is always seen and the cathode colonisation is a function of the alloyed steel. The fusion line which has the same roughness as the other parts of the coupons, is rather less colonised than the cathode, and so attachment of bacteria is less important at these parts.

4.3. Discussion

These results demonstrate unequivocally the influence of alloying elements on the amount of biofilm developed on metallic coupons in pure cultures of *D. vulgaris*.

On carbon steel, without alloying element, the whole surface is colonised by bacterial cells and free extracellular polysaccharides (EPS) even after only one day of the exposure. The ability of *D. vulgaris* to produce EPS has been demonstrated years ago [10]. It has also been observed that carbon steel can stimulate the excretion of polysaccharides from *Desulfovibrio desulfuricans* [13]. The whole colonisation of carbon steel surfaces has been reported also with *D. desulfuricans*, but with *Pseudomonas fluorescens*, the biofilm was reported to be less important [7]. The results of this study obtained with carbon steel are in accordance with the reported results in the literature. They confirm the rapid colonisation of carbon steel by *Desulfovibrio vulgaris* species but literature data remind us that the ability of SRB to produce biofilm is also a function of the bacteria species [3, 4, 7, 15].

On steels containing chromium only, 13% Cr is needed to decrease the surface colonisation to about 50% (same percentage of covered surface after one day and after two weeks), but on 17% Cr, only 10% of the surface is covered after two weeks. The difference in biofilm formation after two weeks in pure cultures of *D. vulgaris* is shown in Fig. 8. On both materials (13Cr and 17Cr), some EPS are seen but only after two or three weeks of exposure. The biofilm decrease as a function of the chromium content increase is also illustrated from a comparison of the results obtained on 3Cr1Mo and on 9Cr1Mo.

The comparison of bacterial adhesion on 13% chromium steel (13Cr) and 9% chromium steels (9Cr1Mo and 9Cr2Mo) shows that the biofilm is less important on materials containing only 9% chromium and also molybdenum (1 or 2%) and vanadium (0.2%) in the alloy steel. These observations demonstrate that the bacterial adhesion and development are more difficult when molybdenum and vanadium are used as alloying elements. It is in accordance with the inhibitory action of molybdenum and vanadium on bacterial proliferation as shown above (Section 3).

On stainless steels (17Cr and 316L), very few bacteria are seen after 24 h and the biofilm covers less than 15% of the surface of the coupons after 3 weeks. The fact that the amount of biofilm is less important on stainless steel than on carbon steel, seems to be a general finding with sulphate reducing bacteria: it has been reported for other SRB, *Desulfovibrio desulfuricans* and *Pseudomonas fluorescens* [5, 7]. It could be related to the toxic effect of alloying elements on bacteria growth. But on stainless steels, the metallic cation concentrations are probably low, even at the interface (no corrosion of the material, passive layer). The explanation could be related to the electrostatic bonding between hydroxyl groups on EPS and metal sites. It has been suggested that free EPS in the environment may saturate binding sites on solid substrate, leaving fewer available sites for bacterial adhesion [3, 9].

At the beginning of bacterial attachment, on alloyed steels with 9% chromium or more, few bacteria are present, but no EPS are seen. So the attachment of EPS occurs after initial attachment of *D. vulgaris* to steel surfaces. This is in accordance with an observation reported in the literature which indicated that the initial attachment of SRB was associated, not with EPS, but with lipopolysaccharides (LPS) on the outer cell membrane [3, 9]. This is consistent with the fact that the attachment of SRB cells is a function of the SRB species.

Similar observations are made on coupled and uncoupled coupons. There is only a small increase in the number of attached bacteria on coupons coupled with carbon steel, mainly at the beginning of the exposure. So it seems that the presence of bacterial cells on metallic coupons is not greatly influenced by anodic or cathodic areas. A recent work which reported MIC on carbon steel in seawater, noticed that there was

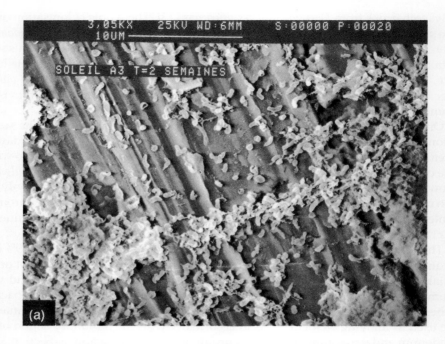

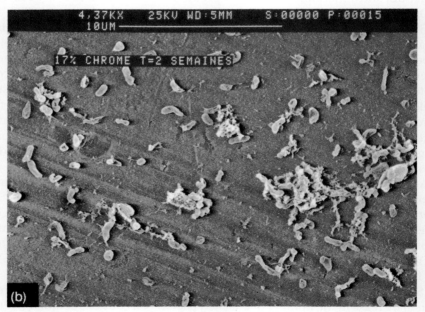

Fig. 8 D. vulgaris *adhesion on 13Cr and 17Cr stainless steels (after 2 weeks).*
(a) Stainless steel with 13% chromium (13Cr); (b) Stainless steel with 17% chromium (17Cr).

the same amount of SRB on corroded and non corroded areas [16]. That observation is consistent with the results presented in this paper.

5. Conclusion

The attachment of *D. vulgaris* to alloyed steels in pure culture medium is mainly related to the steel composition, in accordance with the toxic effect of alloying elements. It is not related to the anodic or cathodic areas, neither to a previous adhesion of EPS. This study emphasises the role played by the chemical composition of materials on the initial attachment of bacterial cells to metallic surfaces and so on the physico–chemistry of the interface. The interactions between metallic elements or ions and bacterial cells could play an important part in the process of cell adhesion to steels and need further investigation.

6. Acknowledgements

The author wishes to thank K. Bellamy, V. Ferrante, P. Fouchet, I. Guillerme and F. Pierot for their contributions to this work, and Y. Lefevre for the SEM observations and image analyses.

References

1. G. Beranger, J. Guezennec, G. Hernandez and C. Lemaitre, 'Prévention de la biocorrosion des aciers inoxydables par modification de la composition chimique', *Matériaux & Techniques*, Numero special Biocorrosion (ISSN 0032-6895), 1990, 25–33.
2. P. Wagner and B. Little, 'The impact of alloying on microbiologically influenced corrosion — a review', *Mat. Perform.*, 1993, **32**(9), 65–68.
3. A. H. L. Chamberlain, 'Biofilms and corrosion', in *Biofilms — Science and Technology* (Eds Melo *et al.*), Kluwer academic publishers, Netherlands, 1992, 207–217.
4. J. Guezennec, 'Le biofilm et ses conséquences sur le comportement des matériaux en milieu naturel', *Océanis*, **19**(3),1993, 1–6.
5. W. Lee and W. G. Characklis, 'Corrosion of mild steel under anaerobic biofilm', *Corrosion*, 1993, **49**(3), 186–199.
6. S. G. Berk, R. Mitchell, R. J. Bobbie, J. S. Nickels and D. C. White, 'Microfouling of metal surfaces exposed to seawater', *Int. Biodeterior. B.*, 1984, **17**, 29–35.
7. I. B. Beech and C. C. Gaylarde, 'Attachment of *Pseudomonas fluorescens* and *Desulfovibrio desulfuricans* to mild and stainless steel — First step in biofilm formation', in *Microbial Corrosion*, European Federation of Corrosion Publications No. 8 (Eds C. A. C. Sequeria and K. Tiller), The Institute of Materials, London, UK, 1992, 61–67.
8. A. Pederson, G. Hernandez-Duque, D. Thierry and M. Hermannsson, 'Effects of biofilms on metal corrosion', in *Microbial Corrosion* (Eds C. A. C. Sequeria and K. Tiller), European Federation of Corrosion Publications No. 8, The Institute of Materials, London, 1992, 165–168.
9. I. B. Beech and C. C. Gaylarde, 'Microbial polysaccharides and corrosion' *Int. Biodeterior. Bull.*, 1991, **27**, 95–107.

10. J. R. Postgate, *The Sulphate Reducing Bacteria*,1984, Cambridge Univ. Press, London, UK.
11. I. B. Beech, S. A. Campbell and F. C. Walsh, 'The role of surface chemistry in SRB influenced corrosion of steel' to be published.
12. D. Feron, V. Ferrante and S. Le Cavelier, 'Use of ion chromatography in microbiolocally influenced corrosion studies', in *Microbial Corrosion* (Eds C. A. C. Sequeria and K. Tiller), European Federation of Corrosion Publications No. 8, The Institute of Materials, London, UK, 1992,146–154.
13. I. B. Beech, C. C. Gaylarde, J. J. Smith and G. G. Geesey, 'Extracellular polysaccharides from *Pseudomonas fluorescens* and *Desulfovibrio desulfilricans* in the presence of mild and stainless steel', *Appl. Microbiol. Biotechnol.*, 1991, **35**, 65–71.
14. V. Ferrante and D. Feron, 'Microbially influenced corrosion of steels containing molybdenum and chromium: a biological and electrochemical study', in *Microbially Influenced Corrosion and Biodeterioration* (Eds Dowling, Mittleman and Danko), The University of Tennessee, Knoxville, TN, USA, 1991, occasional publication; ISBN: 0-9629856-0-0, 355-365.
15. D. White, R. Jack and N. Dowling, 'The microbiology of MIC', in *Microbially Influenced Corrosion and Biodeterioration* (Eds Dowling, Mittleman and Danko), The University of Tennessee, Knoxville, TN, USA, 1991, occasional publication; ISBN: 0-9629856-0-0.
16. N. Rollet-Benbouzit, 'Influence des bactéries sulfato-réductrices sur la corrosion d'aciers en milieu marin', Thesis, Université de Bretagne Occidentale, Brest (France), October 1993.

8
Electrochemical and Surface Analytical Evaluation of Marine Copper Corrosion

B. J. LITTLE, P. A. WAGNER, K. R. HART, R. I. RAY, D. M. LAVOIE,
W. E. O'GRADY* and P. P. TRZASKOMA*

Naval Research Laboratory, Stennis Space Center, MS 39529-5004, USA
*Naval Research Laboratory, Washington, DC 20375-5000, USA

ABSTRACT

X-ray absorption near edge structure (XANES) techniques can be used to differentiate Cu^{1+} and Cu^{2+} species within biofilms attached to surfaces. Copper ions could not be demonstrated with XANES within a marine biofilm of *Oceanospirillum* on a corroding copper surface. Furthermore, Cu^{2+} concentration cells do not appear to be a significant mechanism for microbiologically influenced corrosion in marine environments.

1. Introduction

An aerobic, gram-negative, marine bacterium, *Oceanospirillum*, was isolated from several copper-containing surfaces exposed in marine environments. When grown on copper, the organism produces copious amounts of extracellular polymer and accelerates corrosion of copper metal [1]. The organism with associated polymer has been shown to bind copper ions from solution. Geesey *et al.* [2] demonstrated that exopolymers produced by adherent bacterial cells promoted deterioration of copper. The authors developed a conceptual model for microbiologically influenced corrosion (MIC) in fresh water that required the formation of exopolymer-bound copper concentration cells. Our experiments were designed to determine whether or not the copper-binding properties of the exopolymer from a marine bacterium were important in the corrosion process. We attempted to detect the presence and valence state of copper ions in a marine biofilm and to relate the spatial distribution of bound copper species with localised corrosion.

2. Methods and Materials

Biofilms of *Oceanospirillum* were grown on 90:10 copper–nickel foils and on glass slides in batch and semi-batch cultures of nutrient-rich (AVS) [3] and nutrient-deficient (glutamate) seawater [4] media for six and ten weeks, respectively. Cultures maintained in batch cultures were not replenished with nutrients over time while medium in semibatch cultures was replaced biweekly. Glass slides colonised by *Oceanospirillum* were exposed to separate solutions containing Cu^{1+} and Cu^{2+}. Cu^{1+} in solution was maintained in an anaerobic condition to prevent oxidation to Cu^{2+}.

Corrosion rates were determined from polarisation curves using POLFIT software [5]. Copper–nickel foils were transitioned from culture medium through filtered seawater to distilled water and examined wet using environmental scanning electron microscopy (ESEM) to document the horizontal distribution of cells and localised corrosion [6]. Thin sections of epoxy-embedded foils were examined with transmission electron microscopy (TEM) and ESEM coupled with energy-dispersive X-ray spectroscopy (EDS) to resolve the relationship between bound metals and cells.

Biofilms were removed from copper substrata and bound copper concentrations determined using atomic absorption spectroscopy (AA) and X-ray photoelectron spectroscopy (XPS) [7]. X-ray absorption near edge structure (XANES) was used to determine the speciation of copper within biofilms on copper surfaces [8]. The electrochemical impact of copper concentration cells as defined by Geesey *et al.* [2] was evaluated using a dual-cell corrosion-measuring device [9] with galvanically coupled 99% copper electrodes in tap water and artificial seawater (3.5%) [10]. Identical electrodes were allowed to equilibrate for 16 h to stabilise the galvanic current. Cu^{2+} was added to one individual half-cell as cupric chloride (0.3 mM) and the resulting current measured. In an additional experiment, the dual cell was used to evaluate the electrochemical significance of a Cu^{1+}/Cu^{2+} concentration cell in artificial seawater. One half-cell was deaerated with bubbling nitrogen while the other half-cell was aerated and galvanic current was measured. Cu^{1+} (0.15 mM) was then added to the deaerated half-cell, Cu^{2+} (0.15 mM) to the aerated half-cell, and the resulting galvanic current measured.

3. Results

XANES spectra for Cu^{1+} and Cu^{2+} ions bound from solution within an *Oceanospirillum* biofilm grown on glass slides were unique (Fig. 1). Oxidation of Cu^{1+} bound within the biofilm was not observed during exposure to air. The corrosion rate of copper colonised by *Oceanospirillum* depended on the seawater medium and the rate at which nutrients were replenished. The highest corrosion current densities were measured in nutrient-deficient glutamate medium under semi-batch conditions (Fig. 2). Cells in association with copious amounts of polymer were distributed in patchy areas on all surfaces exposed to bacteria in both glutamate and AVS media (Fig. 3). Localised intergranular corrosion was documented on surfaces colonised in glutamate medium (Fig. 4). Attempts to demonstrate copper bound within *Oceanospirillum* biofilms grown on copper surfaces using TEM/EDS, XANES, XPS and ESEM/EDS were unsuccessful. Small amounts of copper (50 ppb) within biofilms from both media were determined with AA. The addition of 0.3 mM Cu^{2+} as cupric chloride to fresh water in one-half of the dual-cell corrosion measuring device resulted in a maximum galvanic current of 0.4 µA cm^{-2}. Under the same experimental conditions, the addition of 0.3 mM Cu^{2+} as cupric chloride to one half-cell of the dual-cell corrosion measuring device containing artificial seawater, no galvanic current could be measured. Results of galvanic current measurements with differential aeration produced a galvanic current of 3.1 µA cm^{-2}. Differential aeration coupled with copper speciation cells produced a maximum current of 9.5 µA cm^{-2}.

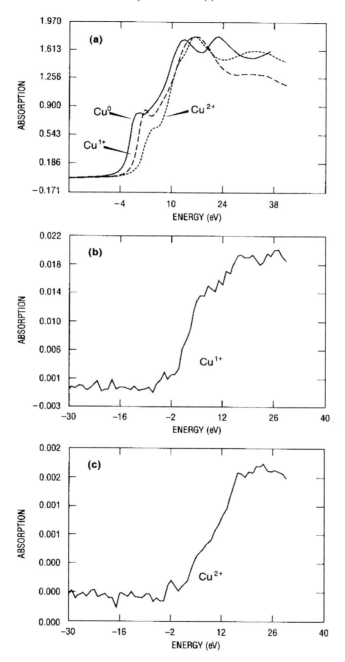

Fig. 1 XANES spectra (a) copper foil (Cu^0), cuprous oxide (Cu^{1+}), and cupric oxide (Cu^{2+}); (b) Cu^{1+} bound from solution within a biofilm of Oceanospirillum grown on a glass slide; and (c) Cu^{2+} bound from solution within a biofilm of Oceanospirillum grown on a glass slide.

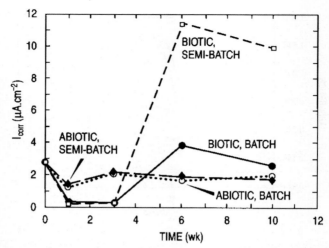

Fig. 2 I_{corr} *vs time for 90:10 copper–nickel in nutrient-deficient glutamate medium colonised with* Oceanospirillum *compared to abiotic controls for batch and semi-batch cultures.*

4. Discussion

Copper alloys are vulnerable to MIC in the form of pitting, crevice or underdeposit attack [11–13]. During seawater exposure, biofilms form on copper surfaces within hours [14]. Bacterial exopolymers are known to bind heavy metals from corroding metal substrata [15] and from solution [16]. Metallic ions associated with biofilms can be solubilised, incorporated into inorganic molecules or adsorbed onto internal or external portions of cells. Metal binding to cell envelopes of gram-negative bacteria [17], accumulation of copper within intracellular lysosomal structures [18] and immobilisation of copper ions by extracellular polymers [19] have been previously demonstrated. Valence states of metal ions associated with biofilms are largely unknown.

Surface analytical tools have been used to resolve questions related to bound metals within biofilms. For example, EDS analyses are excellent tools for demonstrating the presence of metal ions within biofilms but cannot be used to determine the speciation of metal ions. Several investigators are attempting to determine the speciation of bound metals within cultures grown in liquid media using XPS. However, XPS cannot be used to evaluate metals bound within biofilms attached to surfaces. High flux X-ray beams produced by synchrotron light sources are useful for probing local environments of metal atoms and can be used to investigate gases, liquids, solids, solutions, and gels. XANES* provides information on metal site symmetry, oxidation state, and the nature of the surroundings, and the absorption fine structure (EXAFS) provides details about the type, number, and distances of atoms in the vicinity of the absorber. Several studies have investigated Cu–N, Cu–O and Cu–S bonding in proteins [22, 23]. Similar bonding sites are likely to be found in marine biofilms.

---

*See Addendum on p. 142 for a more detailed description.

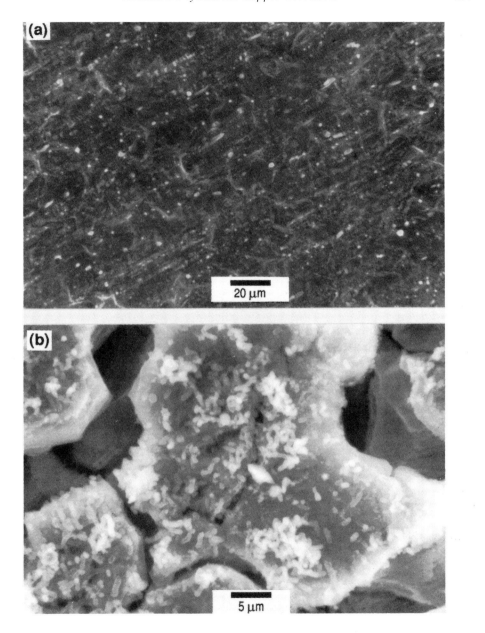

Fig. 3 ESEM micrographs of 90:10 copper–nickel surfaces after 10 weeks exposure to glutamate medium under semi-batch conditions (a) abiotic control and (b) Oceanospirillum culture.

The role of bound metals in accelerating MIC has not been clearly defined. Scotto et al. [24] attributed ennoblement of corrosion potential in natural seawater to accel-

Fig. 4 Intergranular corrosion under Oceanospirillum *biofilm shown in Fig. 3(b).*

eration of the oxygen reduction reaction by organometallic catalysts formed within biofilms. The presence of organometallic compounds formed between bacterial exopolymers and metals, either from a corroding metal surface or from an electrolyte, has never been demonstrated. One mechanism proposed for MIC of copper-containing alloys is related to the binding capacity of microbial exopolymers. The conceptual model for corrosion proposed by Geesey *et al.* [2] requires the formation of copper concentration cells in which Cu^{2+} generated from the corroding copper substratum is selectively bound within adjacent exopolymers having differential affinities for Cu^{2+}. In the model, the exopolymers are excreted from two different organisms.

The corrosion rate of copper colonised by *Oceanospirillum* for ten weeks depended on the seawater medium. The highest corrosion rates were measured in semi-batch cultures in nutrient-deficient glutamate. Cells in association with copious amounts of polymer were distributed in patchy areas on all surfaces exposed to *Oceanospirillum*. Attempts to demonstrate copper bound within biofilms grown on copper surfaces using XANES, XPS and ESEM/EDS were unsuccessful. Small amounts of copper (50 ppb) within biofilms from both media were determined with AA. Electrochemical data indicate that a galvanic current is generated in tap water by the formation of Cu^{2+} concentration cells. No current was generated in artificial seawater. In our investigations we were able to document Cu^{1+} and Cu^{2+} bound within biofilms grown on glass slides exposed to media containing the specific ions. Once Cu^{1+} was bound within the biofilm under anaerobic conditions, exposure to air did not result in further oxidation. Galvanic current measurements indicate that differential binding of Cu^{1+} and Cu^{2+} within adjacent aerobic and anaerobic regions within marine biofilms may be a significant mechanism for MIC.

5. Conclusions

XANES appears to be an excellent technique for detecting copper ions and their speciation *in situ* within biofilms. Based on surface analytical and electrochemical data, it is unlikely that the formation of Cu^{2+} concentration cells is a mechanism for MIC of copper alloys in marine environments. The electrochemical impact of Cu^{2+} concentration cells varies with the electrolyte and may be significant in fresh water systems and on surfaces that have adjacent aerobic and anaerobic areas within biofilms.

6. Acknowledgements

This work was supported by the Office of Naval Research, Dr. Michael Marron, Code 1141MB. NRL contribution number NRL/PP/7333—93-0026.

References

1. P. A. Wagner, B. J. Little and A. V. Stiffey, 'An Electrochemical Evaluation of Copper Colonized by a Copper-Tolerant Marine Bacterium,' *CORROSION '91,* 1991, Paper No. 109, NACE, Houston, TX, USA.
2. G. G. Geesey, M. W. Mittleman, T. Iwaoka and P. R. Griffiths, 'Role of Bacterial Exopolymers in the Deterioration of Metallic Copper Surfaces,' *Mat. Perform.* 1986, **23** (2), 37.
3. A. V. Stiffey and R. V. Lynch, 'Isolation and Characterization of a Bioluminescent Bacterium Obtained from a "Milky Sea" Area in the Arabian Sea,' *Int.Conf. on Marine Science of the Arabian Sea,* 1986, Paper No. 47, Karachi, Pakistan.
4. S. M. Gerchakov, B. Little and P. Wagner, 'The Role of Microorganisms in Electron Transport,' Argentina/USA Workshop on Biodeterioration, 1985, Sao Paulo, Brazil, Aquatec Quimica.
5. H. Shih and F. Mansfeld, *ASTM STP 1154, 174,* Philadelphia, PA, 1992.
6. B. Little, P. Wagner, R. Ray, R. Pope and R. Scheetz, 'Biofilms: An ESEM Evaluation of Artifacts Introduced During SEM Preparation,' *J. Indust.Microbiol.*,1991, **8**, 213–222.
7. C. P. Clayton, G. P. Halada, J. R. Kearns, J. B. Gillow and A. J. Francis, 'Spectroscopic Study of Sulfate-Reducing Bacteria–Metal Ion Interactions Related to Microbiologically Influenced Corrosion,' *ASTM STP 1232,* Philadelphia, PA, 1994, 141–152.
8. A. Bianconi and A. Marcelli, 'Synchrotron Radiation Research: Advances in Surface and Interface Science,' in *Techniques* (Ed. R. Z. Bachrach), Plenum Press, New York, 1992, p. 63.
9. B. Little, P. Wagner, S. M. Gerchakov, M. Walch and R. Mitchell, 'The Involvement of a Thermophillic Bacterium in Corrosion Processes,' *Corrosion,* 1986,**42** (9),533.
10. B. Little, P. Wagner and D. Duquette, 'Microbiologically Induced Cathodic Depolarization,' *CORROSION '87,* 1987, Paper No. 370, National Association of Corrosion Engineers, Houston, TX, USA.
11. D. H. Pope, A *Study of Microbiologically Influenced Corrosion in Nuclear Power Plants and a Practical Guide for Countermeasures,* 1986, Research Project 1166-6, Electric Power Research Institute, Palo Alto, CA.
12. I. Alanis, L. Berardo, N. Decristofaro, C. Moina and C. Valentini, 'A Case of Localized Corrosion in Underground Brass Pipes,' *Biologically Induced Corrosion* (Ed. S. C. Dexter), NACE, Houston, TX, 1986, 102.

13. B. Little, P. Wagner, R. Ray and M. McNeil, 'Microbiologically Influenced Corrosion in Copper and Nickel Seawater Piping Systems,' *Mat. Perform.*, 1990, **24** (3), 10.
14. B. Little, 'Factors Influencing the Adsorption of Dissolved Organic Materials from Natural Waters,' *J.Colloid Interface Sci.*, 1985, **108** (2), 331.
15. T. E. Ford, J. P. Black and R. Mitchell, 'Relationship Between Bacterial Exopolymers and Corroding Metal Surfaces,' *CORROSION '90*, 1990, Paper No. 101, NACE, Houston, TX, USA.
16. H. E. Jones, P. A. Trudinger, L. A. Chambers and N. A. Pyliotis, *Zeitschrift für Allg. Mikrobiologie*, 1976, **16** (6), 425.
17. T. J. Beveridge and S. F. Koval, 'Binding of Metals to Cell Envelopes of *Escherichia coli* K-12,' *Appl. Environm. Microbiol.*, 1981, **42** (2), 325.
18. A. V. S. de Reuck and M. P. Cameron (Eds), *Lysosomes*, Boston, MA, Little, Brown and Co., 1963.
19. M. W. Mittleman and G. G. Geesey, 'Copper-Binding Characteristics of Exopolymers from a Freshwater-Sediment Bacterium,' *Appl. Environm. Microbiol.*, 1985, **49** (4), 846.
20. J. R. Kearns, C. R. Clayton, G. P. Halada, J. B. Gillow and A. J. Francis, 'The Application of XPS to the Study of MIC,' *CORROSION '92*, 1992, Paper No. 178, NACE, Houston, TX, USA.
21. J. Lumsden, personal communication, 1992.
22. F. W. Lytle, 'Experimental X-Ray Absorption Spectroscopy,' *Appl. Synchrotron Radiation*, Beijing, China, 1988.
23. S. P. Cramer and K. O. Hodgson, *Prog. Inorg. Chem.*, 1979, **25** (1).
24. V. Scotto, R. Dicintio and G. Marcenaro, 'The Influence of Marine Aerobic Microbial Film on Stainless Steel Corrosion Behaviour,' *Corros. Sci.*, 1985, **25** (3), 185.

Addendum

X-ray absorption (XAS) is a technique for the investigation of electronic structure and local environment of specific atoms in liquids, solids, gases, solutions and gels. Although XAS has been used to study electronic states and local environments of atoms for over a half century, it has only become an important tool for structural investigations with the recent development of synchrotron radiation sources. An abrupt increase in absorption is observed when the X-ray energy is sufficient to liberate inner shell electrons. This is called the absorption edge and occurs at energies of several kiloelectronvolts for 1s electrons. The energy region surrounding the absorption edge, characterised by XANES, is less than the ionisation potential or threshold and contains information about charge density of the absorbing atom. Above the threshold, the absorption spectrum of an isolated atom gradually decreases monotonically as the X-ray photon energy increases. For atoms involved in chemical bonding, the absorption spectrum above the threshold is characterised by a structure called extended X-ray absorption fine structure (EXAFS). XAS depends on precise measurement of the absorption cross section in the neighbourhood of characteristic absorption edges. The effect is analogous to low energy electron diffraction except that the source of electrons is a specific type of atom within the sample. A unique feature of XAS is the element specificity which occurs because of the separation in energy of the absorption edges of different elements. In summary, XANES provides information on metal site symmetry, oxidation state and the nature of the surroundings, and the EXAFS provides details about the type, number, and distances of atoms in the vicinity of the absorber.

9

Microbial Degradation of Fibre Reinforced Polymer Composites

P. A. WAGNER, R. I. RAY, B. J. LITTLE and W. C. TUCKER*

Naval Research Laboratory, Stennis Space Center, MS 39529, USA
*Naval Undersea Warfare Center Division, Newport, RI 02841, USA

ABSTRACT

Two fibre reinforced polymer composites were examined for susceptibility to microbiologically influenced degradation. Composites, resins, and fibres were exposed to sulphur/iron-oxidising, calcareous-depositing, ammonium-producing, hydrogen-producing and sulphate-reducing bacteria (SRB) in batch culture. Surfaces were uniformly colonised by all physiological types of bacteria. Epoxy and vinyl ester neat resins, carbon fibre, and epoxy composites were not adversely affected by microbial species. SRB degraded the organic surfactant on glass fibre and preferentially colonised fibre-vinyl ester interfaces. Hydrogen-producing bacteria appeared to disrupt bonding between fibers and vinyl ester resin and to penetrate the resin at the interface.

1. Introduction

Fibreglass/polymer and carbon/polymer composite materials are used in many aquatic environments. With high strength to weight ratios and improved stiffness for high performance, these materials surpass conventional metals and alloys for many structural applications. Unfortunately, little attention has been paid to environmental degradation. It was long believed, for example, that fibreglass boat hulls would not suffer the corrosion, biofouling or deterioration found in conventional materials. However, it is now recognised that all engineering materials become colonised by microorganisms, including bacteria, within hours of exposure to natural waters [1]. Microorganisms grow and produce a viscoelastic layer or biofilm. The environment at the biofilm/material interface is radically different from the bulk medium in terms of pH, dissolved oxygen, and organic and inorganic species [2]. Furthermore, polymeric composites are subject to degradation from moisture intrusion and osmotic blistering [3]. Although the problems of moisture intrusion and blistering have been studied [4] and can be eliminated by proper manufacturing and maintenance procedures [5], repair costs and safety risks are high.

Polymeric composites are subject to many kinds of environmental degradation. Tucker and Brown [6] showed that carbon/polymer composites galvanically coupled to metals are degraded by cathodic reactions in seawater. Jones et al. [7] demonstrated that epoxy and nylon coatings on steel were breached by mixed cultures of marine bacteria. Pendrys [8] reported that P-55 graphite fibres were attacked by a mixed culture of *Pseudomonas aeruginosa* and *Acinetobacter calcoaceticus,* common soil isolates. Possible mechanisms for microbial degradation of polymeric composites include: direct attack of the resin by acids or enzymes, blistering due to gas evolution, enhanced cracking due to calcareous deposits and gas evolution, and polymer destabilisation by concentrated chlorides and sulphides.

2. Experimental Procedure
2.1. Identity and Maintenance of Bacterial Cultures

A sulphur/iron oxidising bacterium, *Thiobacillus ferroxidans,* Leathan strain, obtained from Dr. Norman Lazaroff, State University of New York, Binghamton, NY, was maintained in 9K medium [9] containing 3.0 g $(NH_4)2SO_4$, 0.1 g KCl; 0.5 g K_2HPO_4, 0.5 g $MgSO_4 \cdot 7H_2O$, 0.01 g $Ca(NO_3)_2$ dissolved in 700 mL distilled water previously acidified to pH 2.5 with H_2SO_4. The salt solution was sterilised at 250°C and 150 psi for 15 min. Three hundred mL of an iron solution containing 44 g $FeSO_4 \cdot 7H_2O$ in pH 2.5 H_2SO_4 were filter sterilised and added to the salt solution.

Pseudomonas fluorescens, a calcareous-depositing bacterium, was obtained from the American Type Culture Collection (ATCC #17571), Rockville, MD. The organism was originally isolated from polluted seawater. *Ps. fluorescens* was maintained in a medium containing 0.25 g calcium acetate, 0.4 g yeast extract, 1.0 g glucose, dissolved in 100 mL distilled water, and adjusted to pH 8.0 using NaOH [10].

Lactococcus lactis subsp. *lactis,* ATCC #19435, an ammonium-producing bacterium, was maintained in brain heart infusion medium [11]. *Clostridium acetobutylicum,* ATCC #824, a bacterium previously shown to produce copious amounts of hydrogen from fermentation of sugars, was maintained in a growth medium described by Ford *et al.* [12]. Sulphate-reducing bacteria (SRB), isolated as a mixed culture of facultative microorganisms from a corrosion failure of a carbon steel waster piece on a surface ship [13, 14] were maintained in Postgate B growth medium [15].

2.2. Exposure Conditions

Triplicate coupons (2.5 × 2.5 × 0.6 cm) of two fibre reinforced polymer composites — a carbon fibre (T-300) reinforced epoxy (NARMCO-5208/T-300, BASF)* and a glass (S-2) and carbon (T-300) reinforced vinyl ester (Derakane 411-45)† were exposed to microbiological cultures for 161 days. The epoxy was cured in a vacuum bag autoclaved at 121°C (250°F). Vinyl ester resins were post-cured at 100°C for 8 h. Additionally, carbon fibres, glass fibres, vinyl ester, and epoxy resins were individually exposed for 90 days to SRB and hydrogen-producing bacteria. Glass fibres had been treated with organofunctional Silane A-172**, a vinyl tris (2-methoxyethoxy) silane. All cultures were maintained at room temperature and were periodically refreshed with new media. Triplicate uninoculated controls were maintained under the same exposure conditions.

---

*Structural Materials, Anaheim, CA
†Dow Chemical, Midland, MI
**®Union Carbide, Danbury, CT

2.3. Moisture Uptake

Samples were weighed before and after exposure. Moisture uptake was calculated after the biofilm had been removed with a cotton swab containing acetone and the sample reweighed.

2.4. Surface Analysis

Samples were examined before and after exposure using an environmental scanning electron microscope (Electroscan Corporation, Wilmington, MA) coupled with an energy-dispersive X-ray analysis system (NORAN, Middleton, WI) (ESEM/EDS). The ESEM uses a secondary electron detector capable of forming high resolution images at pressures in the range of 0.1 to 20 torr. At these pressures, specimen charging is dissipated into the gaseous environment of the specimen chamber, enabling direct observation of uncoated, nonconductive specimens, including polymeric composites. If water vapour is used as the specimen environment, wet samples can be observed. Wet biofilms can be imaged directly without fixation, dehydration or metal coating, and EDS data can be collected at the same time sample morphology/topography is photographed [16, 17]. Exposed coupon samples were fixed in 2% glutaraldehyde, rinsed to distilled water, examined wet for evidence of degradation resulting from microbial activity, and compared to uninoculated controls.

3. Results and Discussion

In all cases, composite, neat resin and fibre surfaces were colonised by all microbial types. Neither the epoxy nor the vinyl ester composites were adversely affected by calcareous-depositing or ammonium-producing bacteria. There was no evidence of attack of resins and fibres remained embedded within both resins.

Composites exposed to sulphur/iron-oxidising bacteria (Fig. 1) were covered with crystalline deposits containing iron and sulphur in addition to microbial cells. All surfaces exposed to SRB were black due to the deposition of iron sulphides. No damage to the epoxy composite, epoxy neat resin, carbon fibres or vinyl ester neat resin could be attributed to the presence and activities of SRB and hydrogen-producing bacteria.

SRB grew preferentially at fibre/resin interfaces on the vinyl ester composite (Fig. 2). Figure 3 shows unexposed glass fibres (Fig. 3(a)), glass fibres exposed to uninoculated medium (Fig. 3(b)), and glass fibres exposed to SRB in culture medium (Fig. 3(c)). Glass fibres exposed to SRB lost all rigidity after the 90-day exposure so that the weave pattern was no longer evident. Control glass fibres remained rigid and maintained the original weave pattern. Glass fibres are routinely treated with an organic surfactant used to size the fibres and to facilitate handling. The silane surfactant promotes adhesion between the vinyl ester resin and glass fibres. Microbial degradation of the surfactant by SRB was further demonstrated with ESEM/EDS dot maps of silicon distribution (Fig. 4). Dot maps of control fibres exposed to uninoculated media showed concentrations of silicon within the core of each fibre

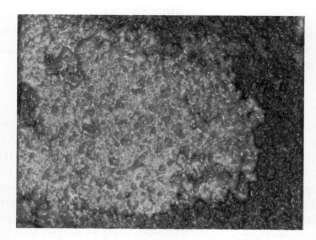

Fig. 1 Sulphur/iron-oxidising bacteria with crystalline deposits on surface of fibre reinforced polymeric composite (3 ×).

Fig. 2 SRB at fibre/resin interfaces of vinyl ester composite.

and small amounts of silicon along the length of each fibre. Similar maps for fibres exposed to SRB showed increased amounts of silicon along the length of the fibre. Many microorganisms are known to degrade organic polymers [18, 19]. The mixed anaerobic culture containing SRB used in this work has been shown previously to degrade marine caulks and traditional polymeric coatings [7, 13, 14].

Microbial Degradation of Fibre Reinforced Polymer Composites 147

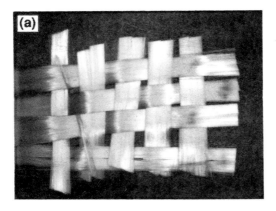

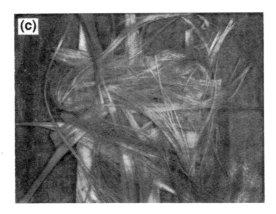

Fig. 3 Light microscope micrographs of glass fibres (2 ×) (a) unexposed, (b) exposed to culture medium, and (c) exposed to SRB in culture medium.

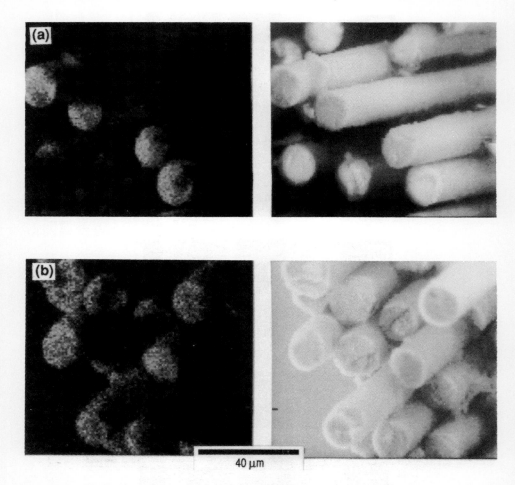

Fig. 4 EDS dot map of silicon for glass fibres (a) exposed to culture medium and (b) exposed to SRB in culture medium.

Hydrogen-producing bacteria appeared to disrupt bonding between fibres and vinyl ester resin (Fig. 5). The organisms penetrated the resin and disruption of fibres and resin may be due to gas formation within the composite.

Previous work published for moisture uptake for vinyl ester neat resin and the carbon vinyl ester composite indicate that the materials should be saturated after 90 days [20]. The neat resin and the carbon vinyl ester composite are typically saturated at 0.78 and 2.25% weight gain, respectively. In the presence of biofilms, moisture uptake was typically 0.1 and 0.9% for the neat resin and composite, respectively. It appears that biofilms may act as a diffusion barrier for water, retarding moisture uptake.

Fig. 5 Hydrogen-producing bacteria at disrupted interfaces between fibres and vinyl ester resin.

4. Conclusions

Epoxy resin and carbon fibres, either individually or in composite, were not degraded by sulphur/iron-oxidising, hydrogen-producing, calcareous-depositing, or SRB. Bacteria did colonise resins, fibres and composites, but did not cause damage. SRB preferentially colonised vinyl ester composites at the fibre–resin interfaces. SRB did not degrade neat vinyl ester resin. SRB degradation of the organic surfactant on glass fibres was demonstrated with ESEM/EDS. Hydrogen-producing bacteria appear to have disrupted the fibre–vinyl ester resin bonding with penetration of the resin.

5. Acknowledgements

This work was supported by the University of Rhode Island Seagrant Foundation, Kingston, RI 02881 and by the Offshore Technology Research Center, Texas A&M University, College Station, TX 77845. Naval Research Laboratory Contribution Number NRL/PP/7333—93-0025.

References

1. B. J. Little and P. Wagner, 'Factors Influencing the Adhesion of Microorganisms to Surfaces,' *J. Adhesion*, 1986, **20** (3), 187–210.
2. B. Little, R. Ray, P. Wagner, Z. Lewandowski, W. C. Lee, W. G. Characklis and F. Mansfeld, 'Impact of Biofouling on the Electrochemical Behaviour of 304 Stainless Steel in Natural Seawater,' *Biofouling*, 1991, **3**, 45–59.
3. R. Davis, J. S. Ghota, T. R. Malhi, and G. Pritchard, 'Blister Formation in RP: The Origin of the Osmotic Process,' 38th *Annual Conference, SPI Reinforced Plastics/Composites Institute*, 1983, Paper No. 17-B, New York, NY, Society of Plastics Industry.
4. 'Guidance Notes for the Manufacture of Glass Fiber Reinforced Polyester Laminates to be Used in Marine Environments,' *British Plastics Federation Report*, 1978.
5. 'Repairs to Blisters in Glass Fiber Hulls,' *British Plastics Federation*, Report No. 224/1.
6. W. C. Tucker and R. Brown, 'Blister Formation on Graphite/Polymer Composites,' *J. Composite Materials*, 1989, **23** (4), 389–395.
7. J. M. Jones, M. Walsh and F. G. Mansfeld, 'Microbial and Electrochemical Studies of Coated Steel Exposed to Mixed Microbial Communities,' CORROSION '91, 1991, NACE, Houston, TX, USA,Paper No. 108.
8. J. P. Pendrys, 'Microbiologically Induced Degradation of P-55 Graphite Fibers,' *J. Electrochem.Soc.*, 1989, **136**, 113C.
9. M. P. Silverman and D. G. Lundgren, 'Studies on the Chemoautotrophic Iron Bacterium *Ferrobacillus Ferrooxidans*. I. An Improved Medium and a Harvesting Procedure for Securing High Cell Yields,' *J. Bacteriol.*, 1959, **77**, 642–647.
10. E. Boquet, A. Boronat and A. Ramos-Cormenzana, 'Production of Calcite (Calcium Carbonate) Crystals by Soil Bacteria is a General Phenomena,' *Nature*, 1973, **246**, 527–528.
11. Difco Manual, Difco Laboratories, Detroit, MI.

12. T. E. Ford, P. C. Searson, T. Harris and R. Mitchell, 'Investigation of Microbiologically Produced Hydrogen Permeation Through Palladium,' *J. Electrochem.Soc.*, 1990, **137**, 1175–1179.
13. J. Jones-Meehan, K. L. Vasanth, R. K. Conrad, B. Little and R. Ray, 'Corrosion Resistance of Several Conductive Caulks and Sealants from Marine Field Tests and Laboratory Studies with Marine, Mixed Communities Containing Sulfate Reducing Bacteria,' *Int. Symp.on Microbiologically Influenced Corrosion (MIC) Testing*, 1992, Miami, FL, American Society for Testing and Materials, Philadelphia, PA, USA.
14. J. Jones-Meehan, M. Walch, B. Little, R. Ray and F. Mansfeld, 'ESEM/EDS Studies of Coated 4140 Steel Exposed to Marine, Mixed Microbial Communities Including SRB,' *8th Int.Congr. on Marine Corrosion and Fouling*, 1992, Taranto, Italy. Oebalia, 1993, **XIX**, Suppl.: 267–275.
15. J. R. Postgate, *The Sulphate Reducing Bacteria*, Cambridge University Press, London, UK, 1979, p. 26.
16. B. Little, P. Wagner, R. Ray, R. Pope and R. Scheetz, 'Biofilms: An ESEM Evaluation of Artifacts Introduced During SEM Preparation,' *J. Ind. Microbiol.*,1991, **8**, 213–222.
17. P. Wagner, B. Little, R. Ray and J. Jones-Meehan, 'Investigations of Microbiologically Influenced Corrosion Using Environmental Scanning Electron Microscopy,' *CORROSION '92*, 1992, NACE, Houston, TX, Paper No. 185.
18. B. F. Sagar, 'The Mechanism and Prevention of Microbial Attack on Polyurethane Coatings,' *Shirley Institute Conference*, 1981, Manchester, UK, Shirley Institute Publication S41.
19. F. J. Upsher, 'Microbial Attack on Materials,' *The Royal Australian Chemical Institute*, 1976, **43-44** (6), 173-176.
20. W. C. Tucker and R. J. Brown, 'Moisture Absorption of Graphite/Polymer Composites Under 2000 Feet of Seawater,' *Composite Materials*, 1989, **23**, 787–797.

10

The Modelling of Microbial Soil Corrosion on Iron Oxides and Hydroxides

T. S. GENDLER*, A. A. NOVAKOVA and L. E. ILYINA

Moscow M.V. Lomonosov State University, 117234 Moscow, Russia
*Institute of Physics of the Earth, Russian academy of Sciences

ABSTRACT

This work is devoted to studying by means of Mossbauer spectroscopy the effect of microflora activity on the transformation of amorphous iron hydroxide (AIH) and fine dispersed amorphous iron oxide (AIO) on the surface of soil clay mineral kaolinite. The incubation of AIO + kaolinite in sucrose solution resulted in partial dissolution of particles 4–5 nm in diameter and in reduction of Fe(III) to Fe(II). In the case of AIH + kaolinite incubation only part of the Fe(III) ions passed into the incubation solution and reduced to Fe(II), the major part of the Fe(III) remained on the surface participating in gradual formation of an organo–iron chelate (evidently Fe(III)-citrate).

1. Introduction

It is widely known that ^{57}Fe Mossbauer spectroscopy is especially useful in determining the nature of iron compounds as well as the chemical or electronic state of iron involved in complex biogeochemical processes.

One such phenomena is the transformation and migration of iron oxides and hydroxides during the soil corrosion process. Fe(III) is the most abundant and potential electron acceptor in most soils [1]. Only sulphate provides a greater potential for oxidation of organic matter than Fe(III)[2].

Reduction of Fe(III) to ferrous iron Fe(II) can be simply expressed as:

$$Fe^{3+} + e^- \rightarrow Fe^{2+}$$

But the actual reaction in soils is more complex. It has a biological basis and is directly related to the metabolism of microorganisms. Numerous organisms that reduce Fe(III) have been reported in an excellent review of D. Lovley [3]. As is clear from that work there is strong evidence for microbial catalysis of Fe(III) reduction.

2. Experimental

The purpose of the present work is the modelling of one of the possible stages of such reduction, i.e. the influence of nonspecific microflora vital activity on the transformation of amorphous ferric hydroxide (AIH) precipitated on a clay mineral surface. Finely dispersed kaolinite — $Al_4(Si_4O_{10})(OH)_8$ — was chosen as a support for AIH because this mineral occurs widely in soils and also has a very extensive surface area and strong absorption capacity.

The scheme of the experiments was as follows:

(i) Kaolinite was incubated under a solution of $^{57}FeCl_3$ (pH 5) for 1 month so that it was deposited on the surface of the finely dispersed AIH to give AIH + K. At pH1 no Fe^{3+} sorption on the kaolinite surface was observed.

(ii) Incubation of IAH + K under 1% sucrose solution at room temperature from 1 to 3 months, then drying the precipitate and collecting the solution.

(iii) Annealing part of the AIH + K particles at 573K for forming fine dispersed amorphous iron oxide (AIO) + K.

(iv) Incubation of AID + K under 1% sucrose solution in the same way as AIH + K from 1 to 3 months.

As is shown in our previous work [4] AIH synthesised from $^{57}FeCl_3$ solution and precipitated on the surface of finely dispersed kaolinite formed fine layers with thickness $d < 5.0$ nm. It must be mentioned that these data were obtained by means of electron microscopy and Mossbauer spectroscopy because the films were X-ray amorphous.

These layers possessed positive charge and were attracted by strong coulombic forces to the negatively charged Si–O basal planes of the kaolinite tetrahedrons. Thus the amorphous ferric hydroxide coated on clay has a more exposed surface area for microbial reduction than that of free AIH particles.

In recent work we analysed the transformations of these very thin layers under microflora vital activity during the incubation in 1% sucrose solution. At each stage of our research the samples dried after incubation and frozen incubation solutions were analysed by means of Mossbauer spectroscopy. The spectra are displayed in Fig. 1.

The AIH + K spectrum (Fig. 1(a)) was fitted in the model of two doublets with parameters: $\delta_1 = 0.46$ mm s^{-1}; $\Delta_1 = 0.63$ mm s^{-1} and $\delta_2 = 0.42$ mm s^{-1}; $\Delta_2 = 0.90$ mm s^{-1}. This denotes the existence of two different Fe(III) states in the structure of fine AIH films on the kaolinite surface (with lower and greater degrees of cubic symmetry distortion respectively).

The spectrum of this sample after incubation under sucrose solution for a period of 1 month (Fig. 1(b)) has a very similar form, but is less resolved and may be fitted in the same model only with addition of a third doublet. Its parameters were $\delta_3 = 0.49$ mm s^{-1} and $\Delta_3 = 0.25$ mm s^{-1}. This fact indicates the formation of a new Fe(III)-compound on the kaolinite surface and according to the parameters it may be identifies as iron chelate, i.e.-Fe(III)-citrate.

After 3 months of incubation drastic changes in the spectrum were observed (Fig.

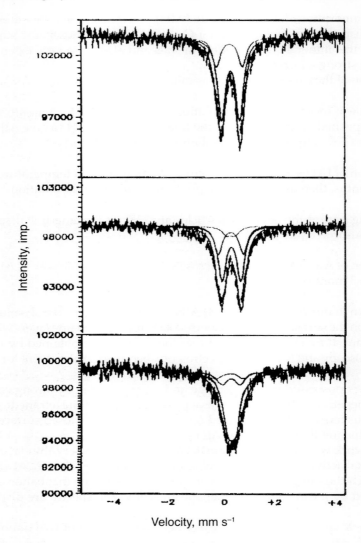

Fig. 1 Mossbauer spectra of AIH on the kaolinite surface after incubation in sucrose solution: (a) initial spectrum (before incubation), (b) 1 month of incubation, (c) 3 months of incubation. T = 300 K.

1(c)). The main part of it consisted of the Fe(III)-citrate doublet and the total spectrum area decreased. This denotes both changes in the atomic bonds inside a new compound and the dissolving of part of the Fe^{3+} ions in the incubation solution. The spectrum of this frozen solution is shown in Fig. 2(a). It has the form of a doublet with parameters $\delta_3 = 1.49$ mm s^{-1} and $\Delta_3 = 3.22$ mm s^{-1} characteristic for Fe(II) compounds.

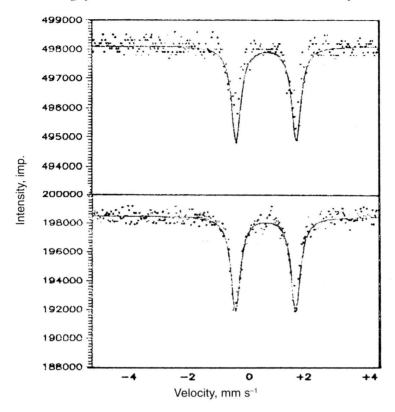

Fig. 2 *Mossbauer spectra of frozen (T = 80 K) sucrose solution after 3 months of incubation: (a) AIH + K, (b) AIO + K, resulted in appearance of the same Fe(II)-compound, but its quantity in the case of AIH is much smaller.*

After annealing AIH + K particles at 573K they transform into films of superparamagnetic haematite (α-Fe_2O_3) with the range of particle sizes 3–15 nm on the surface of kaolinite (AIO + K). The central doublet part of its Mossbauer spectrum (Fig. 3(a)) is due to particles of 3–9 nm, and the six peaks part is due to particles 9 nm.

The incubation of AIO + K in sucrose solution resulted in partial dissolution of the particles 3–5 nm in diameter (we can see the gradual disappearance of the central part in Figs 3(b) and 3(c)) and the transition of Fe^{3+} into Fe^{2+} in the incubation solution (Fig.2(b)). No evident effect of microorganisms on the haematite particles of a size more than 9 nm is observed; the spectra having hyperfine magnetic splitting with the same parameters, but there is a little change in the ratio of intensities between the components. These small changes in the spectra of 'insoluble' haematite particles are currently under research.

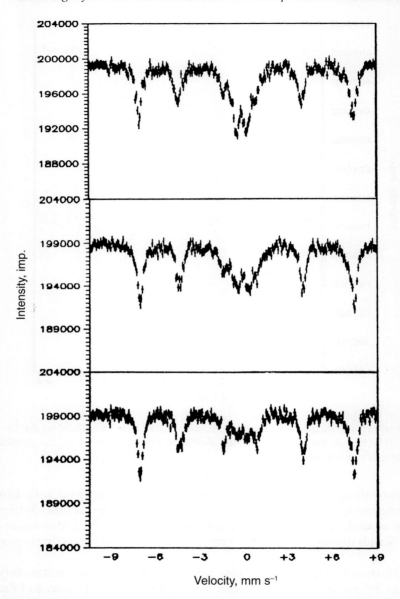

Fig. 3 *Mossbauer spectra of AIO + K after incubation in sucrose solution: (a) initial spectrum, (b) 1 month of incubation, (c) 3 months of incubation. T = 300 K.*

3. Discussion and Conclusions

As was shown above, in both cases (AIH + K and AIO + K) the fraction of Fe(III) dissolved in the incubation sucrose solution in the form of Fe(II) increased with increasing incubation time. The possible reaction [5] in the solution — according to the Mossbauer spectra parameters (Fig. 2) — is:

$$Fe^{2+} + 2\ org^- \rightarrow Fe(org)_2\ salt.$$

In the case of AIH + K incubations the changes of relative contribution of AIH, Fe(III)-citrate and Fe(org)$_2$ salt as a function of incubation time are shown in Fig. 4. It is quite evident from this picture that the Fe(III)-citrate formation process is activated. Our experiments have proved that the microbial reduction of amorphous ferric hydroxide requires an interim period, i.e. for the formation of more soluble Fe(III)-compounds (in our case Fe(III)-citrate) which is chelated by the products of microflora vital activity.

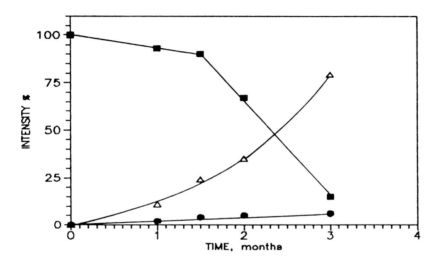

Fig. 4 Relative contributions of AIH (■) and Fe(III)-citrate (Δ) on the kaolinite surface and Fe(II)-compound (•) in the sucrose solution in dependence on incubation time.

References

1. F. N. Ponnamperuma, 'The Chemistry of Submerged Soils,' *Adv. Agron.*, 1972, **24**, 29–96.
2. W. S. Reebergh, 'Rates of Biogeochemical Processed in Anoxic Sediments,' *Ann. Rev. Earth. Planet Sci.*, 1983, **11**, 374–382.
3. D. R. Lovley, 'Organic Matter Mineralization with the Reduction of Ferric Iron,' *Geomicrobiol. J.*, 1987, **5**, (3/4), 375–399.
4. T. S. Gendler, L. S. Ilyina, L. O. Korpachevsky and A. A. Novakova, 'Formation of Hematite on the Kaolinite Surface,' *Dokl. Akad. Nauk USSR*, 1981, **259**, 199–204 (in Russian).
5. A. K. Tiller, 'Aspects of Microbial Corrosion,' in *Corrosion Processes* (Ed. R. N. Parkins), Applied Science Publishers, London and New York, 1982.

11

The Influence of Corrosion Experiments on Microorganisms and Biofilms in the Sessile Phase

D. WAGNER, J. T. WALKER*, W. R. FISCHER and C.W. KEEVIL*

Märkische Fachhochschule, Laboratory of Corrosion Protection, Frauenstuhlweg 31, 58644 Iserlohn, Germany
*Centre for Applied Microbiology and Research, Research Division, Porton Down, Salisbury, Wilts SP4 0JG, UK

ABSTRACT

A microbiologically influenced corrosion process occurring in a potable water installation of a German county hospital was accelerated drastically under well defined experimental conditions with special regard to microbiology and corrosion using the polarisation potential as an accelerating parameter. Episcopic differential interference contrast (DIC) with UV fluorescence microscopy was chosen to prove the establishment of microorganisms and biofilms on the electrode surface after negative or positive polarisation of the electrode. These results correlate to results obtained (i) from samples of a laboratory loop after one year exposure, and (ii) from samples of the installation of the affected county hospital.

1. Introduction

The deterioration of metal surfaces in the presence of microorganisms is normally accompanied by a 'biofilm'. The particular 'biofilm' is comprised of different organic materials, e.g. polysaccharides, glycolipids, oligopeptides and is believed to be involved in many corrosion processes [1, 2]. For example, this kind of microbiologically influenced corrosion (MIC) has been found to cause problems in potable water installations in main buildings [3–6].

The 'biofilm' of corroded copper pipes taken from a potable water installation of a county hospital in Germany affected by MIC was investigated in detail using episcopic differential interference contrast (DIC) with UV fluorescence microscopy [7]. Microorganisms such as diatoms and bacteria were rapidly and reliably detected on and in pipe surfaces. The observed biofilm possessed a structure that was neither homogeneous nor confluent over the surface and which resembled a heterogeneous mosaic of microcolonies.

It was possible to simulate the MIC process occurring in an affected potable water installation in test rigs installed in the affected county hospital and in a laboratory loop [8, 9]. The slow corrosion process was accelerated drastically using the polarisation potential as an accelerating parameter under well defined experimental conditions with special regard to microbiology and corrosion [10]. The manifestations of corrosion including the formation of corrosion products that were obtained in a period of two years in test rigs and laboratory loops [9] could be simulated in six days in potentiostatic tests.

This paper addresses the following questions:

(i) How does the polarisation potential influence the establishment of microorganisms and biofilms on the electrode surface, i.e. the sessile phase?

(ii) Do the obtained results correlate with the results obtained from a laboratory loop after one year exposure?

The method of choice to investigate the sessile phases was episcopic differential interference contrast (DIC) with UV fluorescence microscopy. The suitability of this method was shown in detail in Ref. [7].

2. Experimental
2.1. Electrolyte

Potable water taken from the installation of the affected county hospital in Germany was used to operate the laboratory loop and to perform the electrochemical experiments. The chemical composition of this potable water is described in Ref. [11].

2.2. Laboratory Loop

Figure 1 shows a schematic diagram of a laboratory loop consisting of copper (German Standard: SF-Cu, F37; ISO-Standard: Cu-DHP). The fittings in the loop were protected by sealings made of propylene to avoid the dissolution of zinc by dezincification into the electrolyte. Only the stop cock, flow control and valves consisted of copper/zinc alloy because no alternative materials were available. Connection between the copper pipes was by soft soldering when fittings could not be used [12].

The water was circulated within the loop using a 30 L storage tank and a pump operated by a time control. Intermittent operating conditions were applied as shown in Fig. 2. Analytical measurements within the loop were performed after the last flow period of the water at 9 o'clock. The flow rate of the water was regulated < 1.2 L min^{-1} to maintain a laminar flow according to the German Standard DIN 1988 [13]. The flow rate was checked by a flowmeter. The air was removed from the system by a ventilating valve; 20 L of the electrolyte were exchanged every 15–20 days to avoid an increase of the concentration of corrosion products and a decrease of nutrient concentration. The electrolyte was aerated continuously in the storage tank. Carbon dioxide was removed from the air by passing through 2M NaOH to maintain the buffer capacity of the electrolyte.

Analytical measurements were performed after the last flowing period of the water to guarantee a quantitative mixture of the water in the copper pipes and the water in the storage tank.

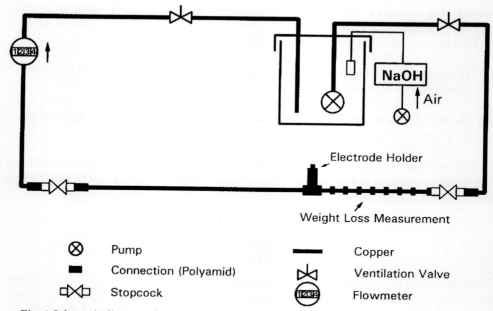

Fig. 1 Schematic diagram of a laboratory loop.

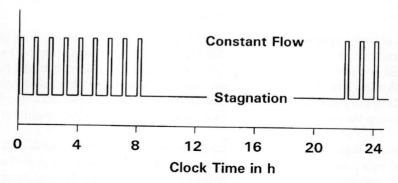

Fig. 2 Operating conditions in the laboratory loop (schematically).

Oxygen concentration, pH, conductivity, buffer capacity as well as copper and zinc activity were determined according to methods that have been published [14]. Tube samples were removed at regular time intervals. Once removed, a rubber silicon bung was placed in one end and the pipe filled with the electrolyte before another bung was placed in the outer side to ensure that the pipe surface remained wet. Samples were then stored at 5 °C until analysis. After removing the silicone bung from one end of the pipe two incisions 0.5 cm apart were cut horizontally along the tube for 1 cm. Another incision was then made 1 cm from the end of the pipe to release a section of tube with a limited curvature.

2.3. Potentiostatic Experiments

Potentiostatic experiments were performed in a Faraday cage in the absence of light. Measuring electrodes were prepared from hard copper tubes (SF-Cu, F37) with a diameter of 22 mm. Rings with an outer surface area of 8 cm² were pretreated with emery paper, polished with diamond paste and electropolished in ortho-phosphoric acid (70 % w/w) with an anodic current density of $i = 0.2$ A cm^{-2}. The rings were positioned in an electrode holder as shown in Fig. 3. Only the outer surfaces of these rings were polarised using a common three electrode arrangement (Fig. 4). A ring of a platinum wire was positioned concentrically around the copper and used as a counter electrode. A mercury sulphate electrode (Hg/Hg$_2$SO$_4$/K$_2$SO$_4$ sat.) was used as a reference electrode. All potentials were referred to the standard hydrogen electrode. A Haber–Luggin capillary was used to diminish the ohmic voltage drop. The electrolyte was stirred and the temperature was kept constant at 20 °C.

The electrolyte within the cell was exchanged at a rate of 1.5 L d^{-1} while performing the potentiostatic tests. The electrolyte was aerated as described in Section 2.2. Since very dilut solutions were used and ohmic voltage drops could not be avoided completely, cut-off experiments were performed at the end of immersion.

The electrodes were stored in the electrolyte at 5 °C until analysis. A part of the ring of *ca.* 0.5 cm² was cut and was used for the microscopical investigations.

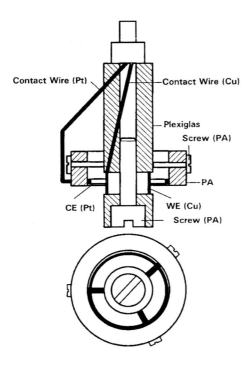

Fig. 3 *Schematic diagram of an electrode holder.*

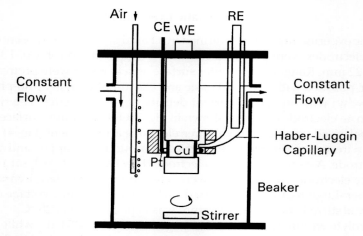

Fig. 4 Schematic diagram of an electrochemical cell.

2.4. Microscopy

Microscopical examination of the pipe sections was carried out using a Nikon Labophot-2 which combines both epifluorescence and episcopic differential interference contrast as shown in Fig. 5 [15]. A B-2A filter block which has a 510 nm dichroic mirror, an excitation filter of 450–490 nm and a barrier filter of 520 nm were used for the epifluorescence. An immuno gold staining (IGS) block was used for DIC and the objective lenses were non-contact M Plan Apo 40, 100 and 150 times with 0.5, 0.8 and 0.95 numerical apertures, respectively. Light source was a 100 W halogen lamp. Neutral density filters were used where required to suppress high background fluorescence. Photography was carried out using a Nikon F-801 35 mm camera and 160 ASA tungsten Kodak transparency film.

2.5. Staining

Acridine orange (BDH, Eastleigh, UK) was dissolved (0.2 % w/w) in sterile distilled water and maintained at 5 °C as a stock solution. The tiles removed from the copper samples were bathed for 2 min in 0.02 % (w/w) filtered (0.2 µm polycarbonate filter) acridine orange (AO) stock solution and then gently rinsed by immersion in 10 mL sterile distilled water. Copper surfaces were also stained for the presence of polysaccharide using periodic acid-Schiffs reagent (PAS)[16].

3. Results

Figures 6 and 7 show the same area of a copper surface removed from the laboratory loop after 345 days operating time using the episcopic DIC mode (Fig. 6) and the UV epi-fluorescence mode (Fig. 7). Figure 6 shows a copper surface with a layer of cor-

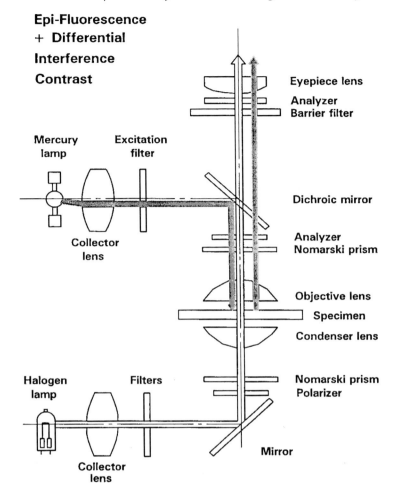

Fig. 5 Schematic diagram of the episcopic differential contrast (DIC) with UV fluorescence microscopy.

rosion products with some debris apparent. The 'biofilm' is visualised as a heterogeneous mosaic structure on the copper surface. This material showed a positive PAS stain reaction of other portions of the sample indicating that the 'biofilm' contains polysaccharide. On viewing with UV-Fluorescence (Fig. 7) small groups of bacterial microcolonies were observed on top of the copper corrosion products and in their polymer matrix. Individual diatoms with no obvious extracellular polymer were also present on other parts of this sample (Figure not shown).

Figure 7 may give some idea of the viability of the cells being visualised, as it has been suggested that slow growing cells have a tightly coiled DNA and a low RNA content and so fluoresce in the red yellow part of the spectrum [17].

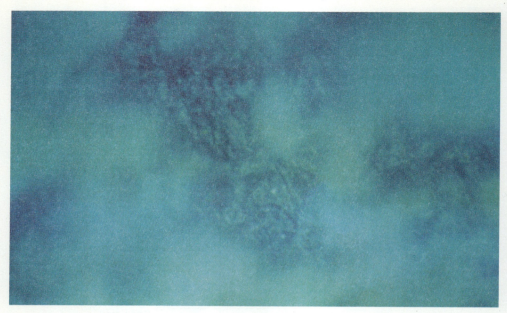

Fig. 6 DIC mode of a copper sample extracted from the laboratory loop after 345 days operating time at 1500 × magnification.

Fig. 7 UV fluorescence mode of a copper sample extracted from the laboratory loop after 345 days operating time at 1500 × magnification, same surface area as Fig. 6.

Figure 8 shows the UV-fluorescence mode of a copper surface polarised for 48 h at -300 mV$_H$, i.e. in the region of O$_2$ limiting diffusion currents, in the potable water taken from the affected county hospital. A high number of diatoms and bacteria were detectable including some organic debris giving a positive PAS stain. Figure 9 depicts a copper surface using the DIC mode polarised at the same potential for 140 h. This surface also shows the voluminous structure of the heterogeneous 'biofilm'. Anodic polarisation at 650 mV$_H$ (to promote dissolution of copper) for 48 h could not prevent the bacteria settling on the corroded copper surface as depicted by Fig. 10 which shows a high number of slow growing bacteria. The biofilm obtained on this copper electrode showed a positive PAS-stain.

4. Discussion

The attachment of bacteria to form a biofilm has been extensively reviewed due to the cost implications in biodeterioration of the substrata [18] and also in terms of health of the general public [19]. The visualisation of bacteria on surfaces has been well studied using light microscopy methods where the substrata transmit light [20].

However, as metallic materials do not transmit light, DIC was used in the episcopic mode to determine the presence or absence of bacteria. Tube sections themselves present other problems when one considers visual examination of the surface. The samples were cut from a tube or a ring and so the circumference of the section re-

Fig. 8 *UV fluorescence mode of a copper surface polarised for 48 h at -300 mV$_H$ (400 × magnification).*

Fig. 9 *DIC mode of a copper surface polarised for 140 h at −300 mV$_H$ (1000 × magnification).*

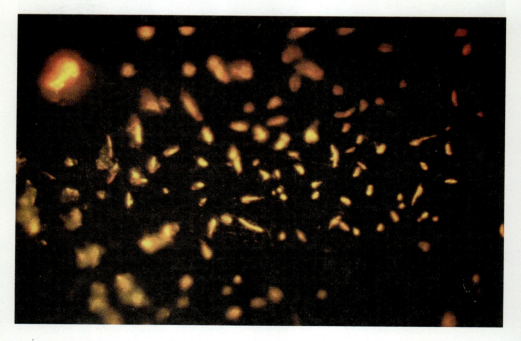

Fig. 10 *UV fluorescence mode of a copper surface polarised for 48 h at 650 mV$_H$ (1000 × magnification).*

sulted in a curved sample and because of this non-contact long working distance objectives were used. Such objectives negated the use of coverslips, which may have otherwise compressed the 'biofilm' resulting in artefacts. The techniques used enabled the height of the biofilm to be assessed as approximately 500 µm at its maximum point. Importantly, the biofilm has been demonstrated to be composed of a series of microcolonies and polysaccharides which extend into the water phase while creating accelerated zones of corrosion at the substrata. Physically the microcolonies would form differential aeration cells while the metabolic activity of the cells would create acidified zones on the metallic substrate resulting in increased corrosion due to the presence of the biofilm [21].

It was shown that the electrochemical potential is a suitable parameter to accelerate this MIC process in the laboratory with respect to the detection of microorganisms and biofilm in the sessile phase. The establishment of microorganims and biofilm is not affected by negative or positive polarisation of the material. The results correlate to results obtained from a laboratory loop after one year exposure. They are also in agreement with the results obtained with samples taken from the installation of the affected county hospital [7].

5. Acknowledgement

Parts of this work have been funded by the International Copper Association, which is gratefully acknowledged.

References

1. B. Little, P. Wagner, S. M. Gershakov, M. Walch and R. Mitchell, *Corrosion*, 1986, **42**, 533.
2. T. E. Ford, J. S. Moki and R. Mitchell, *Corrosion '77*, Paper No. 380, 9–13 March, Moscone Center, San Francisco, CA, USA.
3. G. G. Geesey, T. Iwaoka and P. R. Griffiths, *J. Colloid. Interfac. Sci.*, 1987, **120**, 370.
4. P. Angell, H. S. Campbell and A. H. L. Chamberlain, International Copper Association (ICA) Project No. 405, Interim Report, August 1990.
5. C. C. Gaylarde and I. B. Beech, in *Microbial Corrosion — 1* (Eds C. A. C. Sequeira and A. K. Tiller), Elsevier Applied Sciece, London and New York, 1988, pp. 20–28.
6. W. Fischer, I. Hänßel and H. H. Paradies, in *Microbial Corrosion —1* (Eds C. A. C. Sequeira and A. K. Tiller), Elsevier Applied Science, London and New York, 1988, pp. 300–327.
7. J. T. Walker, D. Wagner, W. Fischer and C. W. Keevil, *Biofouling*, 1993, **8**, 55–63.
8. D. Wagner, W. Fischer and H. H. Paradies, *Werkst. und Korr.os*, 1992, **43**, 496–502.
9. D. Wagner, W. Fischer and G. J. Tuschewitzki, International Copper Association (ICA), Project-No. 453, Final Report, 1992.
10. D. Wagner, H. Peinemann, H. Siedlarek, W. R. Fischer, P. Arens and G. J. Tuschewitzki, International Copper Association (ICA), Project-No. 453-A, Final Report, 1994.
11. W. Fischer, H. H. Paradies, D. Wagner and I. Hänßel, *Werkst. und Korr.os*, 1992, **43**, 56–62.
12. DVGW Working Sheet GW 392 "Copper tubes for capillary soldering in gas and water installations, rules and test determinations".

13. DIN 1988; "Potable water distribution systems on building sites, technical rules for building and running".
14. Deutsche Einheitsverfahren zur Wasser-, Abwasser- und Schlammuntersuchung, VCH Verlagsgesellschaft, Weinheim (FRG), 1988.
15. C. W. Keevil and J. T. Walker, *Binary*, 1992, **4**, 92–95.
16. A. H. L. Chamberlain, P. Angell and H. S. Campbell, *Brit. Corros. J.*, 1988, **23**, 197–98.
17. J. E. Hobbie, J. R. Daley and S. Jasper, *Appl. Environm. Microbiol.*, 1977, **33**, 1225–28.
18. W. A. Hamilton, *Trends in Biotechnology*, 1983, **1**, 36–40.
19. J. S. Colbourne, P. J. Dennis, R. M. Trew, C. Berry and G. Vesey, *Water Sci. Technol.*, 1988, **20**, 5–10.
20. D. E. Caldwell and J. R. Lawrence, *Microbial Ecology*, 1986, **12**, 299–312.
21. J. T. Walker, 'Investigation of biofilms in copper tube corrosion and the survival of *Legionella pneumophila* on alternative plumbing materials', PhD Thesis, Open University, 1994.

12

The Influence of Metal Ions on the Activity of Hydrogenase in Sulphate Reducing Bacteria

C. W. S. CHEUNG and I. B. BEECH

University of Portsmouth, Department of Chemistry, St Michael's Building, White Swan Road, Portsmouth PO1 2BL, UK

ABSTRACT

The effect of Cr, Ni and Mo and Fe ions on proliferation and activity of hydrogenase in the planktonic population of two marine isolates of sulphate reducing bacteria (SRB) grown in batch cultures was determined by using techniques of gas chromatography. The activities of hydrogenase in the planktonic population of SRB and population present in biofilms generated on mild steel surfaces were compared. Results showed that the presence of different metal ions had different effects on bacterial growth rates as well as on the activity of the hydrogenase enzyme. The activity of hydrogenase not only varied between SRB isolates but also between sessile and planktonic population of the same isolate.

1. Introduction

Different bacterial genera such as acid-producers, ammonium-producers, hydrogen-producers, slime-producers, iron-oxidisers, iron-reducers, sulphate-reducers and sulphur-oxidisers have all been implicated in the corrosion of metals and their alloys [1]. However, the group of microorganisms most frequently associated with corrosion failures are the bacteria that reduce sulphate to sulphide, collectively called sulphate-reducing bacteria (SRB). Anaerobic corrosion of steels due to SRB activities presents a very serious problem for oil and gas and shipping industries [2–5]. In the UK alone, estimated cost of damage from pipeline failures due to biocorrosion amounts to £150 to 250 million per annum [6].

Several mechanisms by which SRB can influence corrosion have been proposed [7, 8]. The theory most often referred to is that of cathodic depolarisation, first published by von Wolzogen-Kuhr and van der Vlugt in 1934 [9]. This theory postulates that the SRB utilise cathodic hydrogen for dissimilatory sulphate reduction, thereby increasing the rate of the cathodic reaction. Since the cathodic and anodic reactions are coupled, the dissolution of the metal would increase when the cathode is being depolarised. Alternative theories propose that both FeS and H_2S generated in the presence of SRB are involved in the cathodic depolarisation process [10–12]. The production of corrosive volatile phosphorus compounds by SRB has also been reported and it was concluded that these compounds could enhance the dissolution of metal under anaerobic conditions [13]. It has been demonstrated that under alternating aerobic and anaerobic conditions, which are likely to occur in marine environments, sulphide may react with oxygen and highly corrosive elemental sulphur will be formed [14–18]. It has also been suggested that the exopolymers secreted by SRB spp. may play a role in microbial corrosion, by facilitating bacterial attachment and subsequent formation of biofilms [19].

Several reports emphasised the role of hydrogenase enzymes in corrosion of mild steel due to SRB activity. These enzymes found in the cytoplasm, periplasmic space and attached to the membrane of some SRB *spp.*, play an important role in generation of energy within bacterial cells [8, 20, 21]. A number of pieces of evidence have demonstrated the activity of hydrogenase in catalysing removal of cathodically produced hydrogen from mild steel in the presence of the appropriate electron acceptors [22–27]. Significant increase in hydrogenase activity was measured in the presence of a glass surface [28]. Bryant *et al.* reported that a biofilm with active hydrogenase was associated with a significantly higher corrosion rate of mild steel comparing to a biofilm in which hydrogenase activity was not detected [29]. A study by Bovin *et al.* showed that the presence of cell-free extracts of hydrogenase enzyme could increase corrosion in mild steel [30]. Hydrogenase activity could still be detected in 6 month old SRB culture despite the absence of viable cells [31]. Bryant *et al.* also showed that the activity of hydrogenase enzymes was regulated by the concentration of iron present in the SRB growth media [32].

Several authors reported independently that under identical batch culture conditons very poor biofilms were formed by SRB on surfaces of stainless steel coupons comparing with prolifereous biofilms generated on the surfaces of mild steel surfaces [19, 33]. In each case, coupons had received the same surface finish.

The present study has been undertaken to investigate the reason for the differing dynamics of the biofilm development by SRB strains. The influence of Fe^{2+} ions as well as ions of metal present as alloying elements of steel such as Cr, Ni and Mo, on the growth of planktonic population of two marine isolates of SRB and on the activity of the enzyme hydrogenase, was determined. In addition the assessment of the activity of hydrogenase in biofilms formed by the two isolates on surfaces of mild steel coupons was also carried out to allow the comparison between the hydrogenase activity of attached and free cells.

2. Materials and Methods
2.1. Organisms

Two species of SRB, one recovered from a corroding pile in Portsmouth Harbour (Portsmouth strain) and another, isolated from a severely corroded hull of an oil storage vessel anchored off the Indonesian coast (Indonesia strain), were used in this study. Bacterial isolation was carried out in Postgate Medium B [34]. Purification of the species was performed on solid Postgate Medium E [34]. Cultures were screened for aerobic and anaerobic contaminants by spread plate technique on solid Nutrient and Anaerobic agar (Difco, UK). All subsequent experiments were carried out in Postgate Medium C at 37°C [34].

The selected SRB exhibited different behaviour towards mild steel under identical cultural conditions. Both initial attachment of cells to mild steel and corrosion rates due to the production of biofilms by these isolates on mild steel coupons significantly varied [35].

2.2. Influence of Cr, Ni, Mo and Fe on the growth of SRB

The growth of the two strains of SRB isolates in Postgate medium C containing Cr, Ni and Mo ions was monitored over a period of 14 days by haemocytometer counts. Aqueous solutions containing Cr^{3+} (added as $CrCl_3\ 6H_2O$), Ni^{2+} (added as $NiCl_2\ 6H_2O$), and Mo^{6+} (added as $Na_2MoO_4\ 2H_2O$) were filter sterilised (Millipore™ 0.45 ∝m membrane filter) and added to the bacterial media to achieve final concentration of 56.7 ppm (1.09 mM), 19.9 ppm (0.34 mM) and 3.1 ppm (0.03 mM) respectively. The above concentrations were chosen on the basis of the calculation of the loss of alloying elements from a stainless steel coupon (Cr: Ni: Mo = 18: 8: 2) with a total surface area of 40 cm^2, assuming that the corrosion rate is 1 μm/year in terms of the alloy thickness. The experiments were carried out in triplicate at 37°C. Controls consisted of SRB cultures grown in the absence of metal ions.

In order to determine the effect of Fe^{2+} on the growth and hydrogenase activity of SRB isolates, aqueous solutions containing Fe^{2+} at concentration 0 ppm (0 μM), 2 ppm (7.2 μM) and 5 ppm (18 μM) were filter sterilised and added as $FeSO_4\ 7H_2O$ to the culture medium. The growth rate of SRB was determined by the equation: $R = 3.3 \times (\log_{10}N - \log_{10}N_0)\ t^{-1}$ where R = growth rate (h^{-1}); N_0 = total population at the beginning of log phase; N = total population at the end of log phase and t = time (h).

Biofilms were generated on mild steel coupons (En2, C content < 0.2%) with total surface area of 22.4 cm^2. Coupons were degreased in alkaline cleaner for 5 min at 60°C, and rinsed in double distilled water. Corrosion products were removed by dipping coupons in HCl (12M) solution inhibited with hexamine (5 g L^{-1}) and then rinsed with double distilled water. Absolute alcohol was used to remove excess moisture. Prior to inoculation with bacteria, coupons were sterilised in 70% alcohol, flamed and placed inside sterile glass universal bottles containing Postgate Medium C. Aliquot of 1 mL of a 3 day old culture of SRB was inoculated into the bottles containing sterile culture media to achieve an initial inoculum concentration of 10^6 cells mL^{-1}, as estimated by haemocytometer count. The cultures were incubated at 37°C. On the seventh day of incubation coupons were removed from growth media and biofilms were dislodged from surfaces into phosphate buffer (0.1 M, pH 7) by sonication. The hydrogenase activity was quantified by gas chromatographic technique after disruption of cells by sonication. A11 experiments were performed in duplicate.

2.3. Measurement of Hydrogeanse Activity

Three day old SRB cultures were centrifuged at 10 000 × g at 4°C for 30 min. The pellet was further centrifuged at 13 000 × g for 30 min at room temperature. The pellet was then resuspended in 300 μL sodium phosphate buffer (0.1 M, pH 7) and sonicated to disruption. Hydrogenase activity was measured by monitoring hydrogen production. The assay of hydrogen evolution from dithionite-reduced methyl viologen was carried out in 10 mL gas tight vials with 3 mL of an anaerobic Tris HCl buffer (0.1 M, pH 7.6) containing 0.033% of bovine serum albumin, 1 mM methyl viologen, and 15 mM sodium dithionite. The samples were injected into the vial and hydrogen production was monitored by HP 5890 Series II gas chromatograph with a thermal conductivity detector. Gas sample (50 µL) were injected in duplicate into a chrompack carbPLOT P7 (megabore, 25 m long, 0.533 mm i.d.) capillary column. The column temperature was kept at 25°C. Nitrogen was used as the carrier gas at a flow rate of 3.5 ml min^{-1}. The temperature of the injection port and detector were controlled at 110 and 250°C respectively. Each run was completed in 2.5 min. The amount of hydrogen gas evolved was proportional to the activity of hydrogenase and reported as µmol H_2 min^{-1} mL^{-1}. Biofilm samples were treated in the same way as planktonic cells.

2.4. Statistical Analysis

Numerical data was evaluated by Student t-test and one way ANOVA with the aid of a statistical package Minitab (version 9.2).

3. Results
3.1. The Influence of Cr, Ni and Mo Ions on Bacterial Growth

The influence of Cr, Ni, and Mo ions on the growth of SRB isolates is illustrated in Figs 1 and 2. The presence of metal ions did not sigificantly ($p > 0.05$) affect the biomass of Portsmouth SRB isolate, comparing with the control (Fig. 1), irrespective of metal species. The growth rate of Portsmouth SRB isolate in the presence of Mo ions was however significantly lower than that of the control (Table 1). Mo and Ni were the two ions that significantly ($p < 0.05$) decreased the biomass and significantly ($p < 0.05$) increased the growth rate of Indonesia SRB isolate (Fig. 2 and Table 1). Although in the presence of Cr ions the production of biomass by Portsmouth SRB isolate was not affected, bacterial growth rate decreased significantly ($p < 0.05$). For both SRB isolates, Cr was the only ion that did not affect the biomass production.

Table 1.The effect of Cr, Ni and Mo ions on the growth rate of planktonic SRB in Postgate medium C

Isolate	Growth rates, h^{-1}			
	control	Cr^{3+}	Ni^{2+}	Mo^{6+}
Portsmouth SRB	0.0624 ± 0.0022	0.0633 ± 0.0026	0.0619 ± 0.0010	0.0513 ± 0.0013
Indonesia SRB	0.0419 ± 0.0015	0.0392 ± 0.0009	0.0962 ± 0.0028	0.0454 ± 0.0008

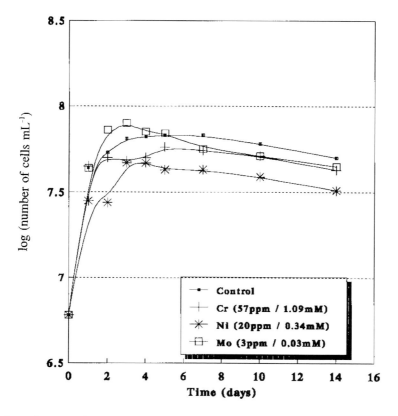

Fig. 1 Growth curves of Portsmouth SRB isolates in Postgate Medium C containing Cr, Ni and Mo ions.

The effect of Fe^{2+} on the growth rate of SRB isolates is presented in Table 2. The growth rate proved to be positively correlated with the concentration of Fe^{2+} added to the media. For all Fe^{2+} concentration tested, the growth rates of Indonesia SRB in planktonic phase were significantly higher than those of Portsmouth SRB ($p < 0.05$). The biomass produced by Portsmouth SRB was significantly lower ($p < 0.05$) at concentration of 0 ppm than at 2 ppm and 5 ppm of Fe^{2+} (Fig. 3). No such difference has been found in cultures of Indonesia SRB (Fig. 4).

3.2. Influence of Cr, Ni, Mo and Fe on Hydrogenase Activity

Table 3 shows the effect of Cr, Ni and Mo ions on the hydrogenase activity of the planktonic SRB. No hydrogenase activity could be detected in cells of Indonesia SRB regardless of cultural conditions. In the presence of Cr, Ni and Mo, the hydrogenase activity of Portsmouth SRB isolates decreased significantly ($p < 0.05$) compared with the controls.

The hydrogenase activity of both SRB isolates was found to be dependent on the

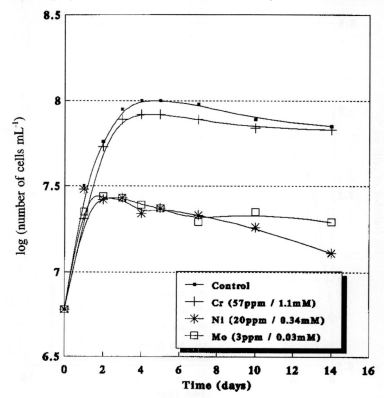

Fig. 2 Growth curves of Indonesia SRB isolate in Postgate Medium C containing Cr, Ni and Mo ions.

Table 2. The effect of Fe^{2+} on the growth rate of planktonic SRB in Postgate medium C

	Growth rates h^{-1}		
Isolate	0 ppm	2 ppm	5 ppm
Portsmouth SRB	0.0275 ± 0.011	0.0539 ± 0.009	0.0585 ± 0.009
Indonesian SRB	0.0749 ± 0.006	0.0763 ± 0.002	0.0798 ± 0.002

concentration of Fe^{2+} (Table 4). No activity was detected at Fe^{2+} concentration of 0 ppm. At 2 ppm the activity was only detected in cells of Portsmouth SRB. The activity in cells of Indonesia SRB was detected when the concentration of Fe^{2+} increased to 5 ppm. No significant difference was found between hydrogenase activity at 2 ppm and 5 ppm in cultures of Portsmouth isolate. At the same concentration of Fe^{2+}, the hydrogenase activity in culture of Indonesian isolate was always significantly lower ($p < 0.05$) than that of Portsmouth isolates. Although no hydrogenase activity was detected in planktonic population of Indonesia SRB, its activity was detected in biofilms (Table 5).

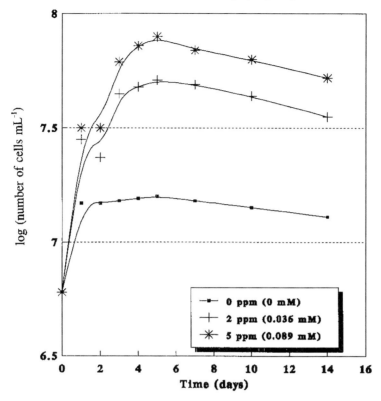

Fig. 3 Growth curves of Portsmouth SRB isolate in Postgate Medium C containing different concentration of Fe^{2+}.

Table 3. The effect of Cr, Ni & Mo ions on the activity ($\mu mol\ H_2\ min^{-1}\ mL^{-1}$) of hydrogenases in planktonic cells of SRB grown in Postgate medium C

Metal ions	Portsmouth	Indonesian
Medium C with < 4 ppm Fe^{2+}	2.31 ± 0.33	BD[a]
Medium C + Cr^{3+}	1.03 ± 0.24	BD[a]
Medium C + Ni^{2+}	1.61 ± 0.31	BD[a]
Medium C + Mo^{6+}	1.14 ± 0.28	BD[a]

a: below detection limit.

For both SRB isolates, activity of hydrogenase was higher in biofilms developed on surface of mild steel than in planktonic population (Table 5). Significantly higher activity ($p < 0.05$) was detected in Portsmouth SRB biofilm comparing with the biofilm formed in the presence of Indonesia SRB.

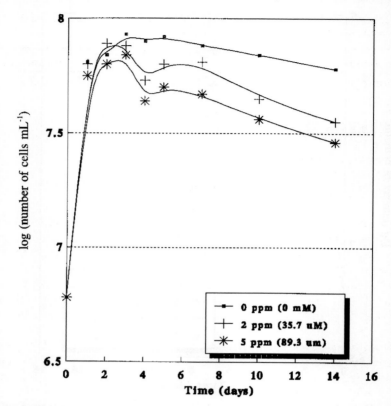

Fig. 4 Growth curves of Indonesia SRB isolate in Postgate medium C containing different concentration of Fe^{2+}.

Table 4. The effect of Fe^{2+} on the activity of hydrogenase (μmol H_2 min^{-1} mL^{-1}) in planktonic SRB cells

Fe/ppm	Portsmouth	Indonesia
0	BD[a]	BD[a]
2	2.69 ± 0.44	BD[a]
5	2.61 ± 0.31	1.37 ± 0.22

a: below detection limit.

4. Discussion and Conclusions

Although no hydrogenase activity was detected in planktonic populations of Indonesia SRB, the activity was detected in biofilms formed by these bacteria on surfaces of mild steel. Higher activity was found in biofilms of Portsmouth SRB than in planktonic populations of this isolate. It appears that the mild steel surface could trigger

Table 5. The activity of hydrogenase ($\mu mol\ H_2\ min^{-1}\ mL^{-1}$) in planktonic SRB cells and in 7 day old biofilms formed on mild steel surfaces in Postgate medium C

Isolate	Portsmouth	Indonesia
Planktonic population grown with < 4 ppm Fe^{2+}	2.31 ± 0.33	BD[a]
Biofilm population	4.75 ± 0.16	3.91 ± 0.08
planktonic population grown with 5 ppm of Fe^{2+}	2.61 ± 0.31	1.37 ± 0.22

a: below detection limit.

the activity of hydrogenase in both SRB strains. The activity in planktonic populations of both isolates was not as high as in their biofilms even when the growth media were supplemented with 5 ppm of Fe^{2+}. Therefore it is speculated that the concentration of Fe^{2+} ions at the surface/solution interface was higher than 5 ppm.

For SRB, Fe is an important cofactor for synthesis of enzymes such as hydrogenases and reductases (e.g. desulfoviridin) as well as for general processes of cell metabolism [34]. Postgate found that although the content of hydrogenase in *Desulfovibrio desulfuricans* was lower in iron-starved cells the decrease in the concentration of iron in the medium from 2 to 0.5 ppm did not affect the total amount of cells produced [36]. Czechowski *et al.* showed that the total amount of periplasmic hydrogenase in *Desulfovibrio vulgaris* in iron-limited media was less than that found in a non-iron limited media and that the concentration of Fe had no influence on the cell yield [23]. In our study, the hydrogenase activity in cultures of Portsmouth SRB isolate decreased when the concentration of Fe^{2+} in the medium was reduced from 2 ppm to 0 ppm and so did the yield of biomass.

SRB of the genus *Desulfovibrio* have an exceptionally high requirement for Fe [34]. Although the identification of the Portsmouth strain is not yet fully confirmed it is certain that the Indonesia isolate belongs to the genus *Desulfovibrio* [37]. The anaerobic environment around corroded metals is usually low in Fe since sulphide forms a strong bond with Fe^{2+}. Thus if SRB could only use unbound Fe^{2+} the activity of the enzyme in biofilms should be lower than that in planktonic cells. It has been suggested that SRB possess a mechanism of uptake of sulphide bound iron [38]. Indeed, transmission electron microscopy and atomic force microscopy studies showed that FeS particles were associated with the cell surface of SRB [39, 40]. Our studies showed an increase in hydrogenase activity within biofilms and are thus in support of the existence of such a mechanism. Bryant *et al.* demonstrated that the enzyme derepressed and the activity increased approximately fourfold when 5 ppm of Fe^{2+} were added to culture media, whereas the presence of 100 ppm of Fe^{2+} repressed the enzyme suggesting that hydrogenase acitivity was subject to an Fe induction-repression control mechanism [32]. In our study the hydrogenase activity was triggered on at 2 ppm for Portsmouth and 5 ppm for Indonesia isolates. The concentration of Fe^{2+} present in biofilms grown on steel coupons of the size used in our experiment (22.4 cm^2) could be as high as 7.33 ppm (based on the calculation of the dissolution rates of Fe). At this concentration the activity of the enzyme was high.

According to the theory of survival strategy it is advantageous for microbes in oligotrophic environments to spend their energy sparingly. This may account for the lack of detection of hydrogenase activity in SRB cultures grown without Fe^{2+}. No additional energy was spent to activate the enzyme. However when Fe^{2+} becomes available the production of hydrogenase is triggered.

Our study demonstrated that ions of metals present as alloying elements in stainless steels were able to inhibit the hydrogenase activity of Portsmouth SRB isolate. Failure of detection of the activity of hydrogenase in cultures of Indonesia isolate was most likely due to the fact that in order to activate the enzyme a concentration of Fe^{2+} higher than 2 ppm is required. The activity of hydrogenase not only varied between SRB isolates but also between sessile and planktonic populations of the same isolate.

It can be concluded that growth rates and biomass production of planktonic SRB population were adversely affected by Cr, Ni and Mo ions present in the growth medium. An increase in the concentration of Fe^{2+} ions resulted in higher bacterial growth rates and in greater activity of the enzyme hydrogenase. Study of the effect of metal ions on SRB metabolism and its importance for biocorrosion of both carbon and stainless steels are in progress.

5. Acknowledgements

The authors would like to thank University of Portsmouth for the financial support to Mr C. W. S. Cheung and Mr M. A. W. Hill for his technical assistance when using the gas chromatograph.

References

1. S. W. Borenstein, *Microbiologically Influenced Corrosion Handbook*, Woodhead Publishing Ltd, Cambridge, England, 1994.
2. J. L. Lynch and R. G. J. Edyvean, 'Biofouling in oilfield water systems — A review', *Biofouling*, 1988, **1**, 147–162.
3. R. Cord-Ruwish, W. Kleinitz and F. Widdel, 'Sulphate reducing bacteria and their activities in oil production', *J. Petroleum Technol.*, 1987, Jan, 97–106.
4. E. C. Hill, 'Microbial corrosion in ship engines', in *Microbial Corrosion*, The Metals Society, London, 1983, pp. 123–127.
5. E. S. Pankhurst, 'Significance of sulphate reducing bacteria to the gas industry: A review', *J. Appl. Bacteriol.*, 1968, **31**, 179–193.
6. W. A. Hamilton, 'Sulphate reducing bacteria and anaerobic corrosion', *Ann. Rev. Microbiol.*, 1985, **39**, 195–217.
7. T. E. Ford and R. Mitchell, 'The ecology of microbial corrosion', *Adv. Microb. Ecol.*, 1990, **11**, 231–262.
8. I. P. Pankhania, 'Hydrogen metabolism in sulphate reducing bacteria and its role in anaerobic corrosion', *Biofouling*, 1988, **1**, 27–47.

9. C. A. H. Von Wolzogen-Kuhr and L. S. Van der Vlugt, 'Graphitisation of cast iron as an electrochemical process in anaerobic soils',*Water*, 1934, **18**, 147–165.
10. R. A. King, J. D. A. Miller and J. S. Smith, 'Corrosion of mild steel by iron sulphides', *Brit. Corros. J.*, 1973, **8**, 137–141.
11. R. A. King and D. S. Wakerley, 'Corrosion of mild steel by ferrous sulphide', *Brit. Corros. J.*,1973, **8**, 41–45.
12. J. A. Costello, 'Cathodic depolarisation by sulphate reducing bacteria', *S. Afr. J. Sci.*, 1974, **70**, 202–204.
13. I. P. Iverson and G. J. Olson, 'Anaerobic corrosion by sulfate-reducing bacteria due to highly reactive volatile phosphorus compound', in *Microbial Corrosion*, The Metals Society, London, 1983, 46–53.
14. W.-C. Lee, Z. Lewandowski, S. Okabe, W. G. Characklis and R. Avci, 'Corrosion of mild steel underneath aerobic biofilms containing sulfate-reducing bacteria. Part I: at low dissolved oxygen concentration', *Biofouling*, 1993, **7**, 197–216.
15. W.-C. Lee, Z. Lewandowski, M. Morrison, W. G. Characklis, R. Avci and P. H. Nielsen, 'Corrosion of mild steel underneath aerobic biofilms containing sulfate-reducing bacteria. Part II: at high dissolved oxygen concentration', *Biofouling*, 1993, **7**, 217–239.
16. P. H. Nielsen, W.-C. Lee, Z. Lewandowski, M. Morrison and W. G. Characklis, 'Corrosion of mild steel in an alternating oxic and anoxic biofilm system', *Biofouling*, 1993, **7**, 267–284.
17. J. McKenzie and W. A. Hamilton, 'The assay of *in-situ* activities of sulfate reducing bacteria in a laboratory marine corrosion model', *Int. Biodeterior. Biodegrad.*, 1992, **29**, 285–297.
18. J. A. Hardy and J. L. Brown, 'The corrosion of mild steel by biogenic sulphide films exposed to air', *Corrosion*, 1984, **40**, 650–654.
19. I. B. Beech, C. C. Gaylarde, J. J. Smith and G. G. Geesey, 'Extracelluar polysaccharides from *Desulfovibrio desulfuricans* and *Pseudomonas flourescens* in the presence of mild and stainless steel', *Appl. Microbiol. Biotechnol.*, 1991, **25**, 65–71.
20. R. M. Fitz and H. Cypionka, 'Generation of a proton gradient in *Desulfovibrio vulgaris*', *Arch. Microbiol.*, 1991, **155**, 444–448.
21. J. M. Odum and H. D. Peck Jr, 'Hydrogen cycling as a general mechanism for energy coupling in the sulfate-reducing bacteria *Desulfovibrio* sp.', *FEMS Microbiol. Lett.*, 1981, **12**, 47–50.
22. R. D. Bryant and E. J. Laishley, 'The role of hydrogenase in anaerobic biocorrosion', *Can. J. Microbiol.*, 1990, **36**, 259–264.
23. M. H. Czechowski, C. Chatelus, G. Faque, M. F. Libert-Coquenpot, P. A. Lespinat, Y. Berlier and J. LeGall, 'Utilization of cathodically-produced hydrogen produced from mild steel by Desulfovibrio species with different types of hydrogenase', *J. Ind. Microbiol.*, 1990, **6**, 227–234.
24. B. S. Rajagopal and J. LeGall, 'Utilisation of cathodic hydrogen by hydrogen oxidizing bacteria', *Appl. Microbiol. Biotechnol.*, 1989, **31**, 406–412.
25. S. Daumas, 'Microbiological battery induced by sulfate reducing bacteria', *Corros. Sci.*, 1988, **28**(11),1041–1050.
26. I. P. Pankhania, A. N. Moosavi and W. A. Hamilton, 'Utilization of cathodic hydrogen by *Desulfovibrio vulgaris* (Hildenborough) on acetate', *J. Gen. Microbiol.*, 1986, **132**, 3357–3365.
27. J. A. Hardy, 'Utilisation of cathodic hydrogen by sulphate reducing bacteria', *Brit. Corros. J.*, 1983, **18**(4), 190–193.
28. R. E. Williams, E. Ziomek and W. G. Martin, 'Surface stimulated increase in hydrogenase production by sulfate reducing bacteria — consequences in corrosion', in *Biologically Induced Corrosion* (Ed. S. C. Dexter), NACE, Houston, TX, 1986, pp. 184–192.
29. R. D. Bryant, W. Jansen, J. Boivin, E. J. Laishley and J. W. Costerton, 'Effect of hydrogenase and mixed sulfate reducing bacterial populations on the corrosion of steel', *Appl. Environ. Microbiol.*, 1991, **57**(10), 2804–2809.

30. J. Bovin, E. J. Laishle, R. D. Bryant and J. LeGall, 'The influence of enzyme systems on MIC', *Corrosion '90*, Paper No. 128, NACE, Houston, TX.

31. C. Chatelus, P. Carrier, P. Saignes, M. F. Liber, Y. Berlier, P. A. Lespinat, G. Faque and J. LeGall, 'Hydrogenase activity in aged, nonviable *Desulfovibrio vulgaris* cultures and its significance in anaerobic biocorrosion', *Appl. Environ. Microbiol.*, 1987, **53**(7), 1708–1710.

32. R. D. Bryant, F. V. O. Kloeke and E. J. Laishley, 'Regulation of the periplasmic [Fe] hydrogenase by ferrous iron in *Desulfovibrio vulgaris* (Hildenborough)', *Appl. Environ. Microbiol.*, 1993, **59**(2), 491–495.

33. D. Feron, personal communication.

34. J. R. Postgate, *The Sulphate Reducing Bacteria*, Cambridge University Press, London, 1984.

35. I. B. Beech, C. W. S. Cheung, M. A. W. Hill, R. Franco and A. R. Lino, 'Study of parameters in biodeterioration of steel in the presence of different species of sulphate-reducing bacteria', *Int. Biodeterior. Biodegr.* 1995, in press.

36. J. R. Postgate, 'Iron and the synthesis of cytochrome C3', *J. Gen. Microbiol.*, 1956, **15**, 186–193.

37. A. R. Lino, personal communication.

38. E. J. Laishley, personal communication.

39. A. Steele, D. Goddard and I. B. Beech, 'The use of atomic force microscopy in study of the biodeterioration of stainless steel in the presence of bacterial biofilms', *Int. Biodeterior. Biodegr.* 1995, **34** (1), 35–46.

40. G. Bradley, 'An investigation into the structure and functional aspects of the cell wall of *Desulfovibrio*', PhD dessertation, City of London Polytechnic, 1985.

13

A Kinetic Model for Bactericidal Action in Biofilms

P. M. GAYLARDE and C. C. GAYLARDE*

Depto. de Pneumologia Pediatrica, Hospital de Clinicas de Porto Alegre, Rua Ramiro Barcelos, 2350, 90210 Porto Alegre, RS, Brazil
*Depto. de Solos, Fac. de Agronomia, UFRGS, Av. Bento Gonçalves, 7712, 91540-000 Porto Alegre, RS, Brazil

ABSTRACT

The standard methods for the determination of biocidal activity measure the rate of planktonic and/or sessile cell death. The role of diffusion within the biofilm on the kinetics of sessile cell death is generally not considered. It is known that, in the majority of cases, sessile cells are more resistant to biocides than are planktonic cells. The reasons for this increased resistance have been considered to be altered cell physiology or failure of the biocide to penetrate the biofilm. A model is presented which demonstrates how the rate of diffusion of biocides through a biofilm influences the rate of cell death. The calculations may be performed using a simple spreadsheet package. The results show a biofilm will decrease the rate of cell death even when microbial physiology in the biofilm is unaltered. Using this model, the increase in biocide concentration required to kill sessile, rather than planktonic, cells in the same physiological state in a given time may be estimated. The effects of altered cell metabolism may be considered only once the effect of diffusional time lag on the rate of cell death within the biofilm is known.

1. Introduction

Much recent work has focused on the importance of biofilms in microbial corrosion. It is well known that microorganisms present in biofilms are in some way protected from the action of biocides [1, 2] either by lack of penetration of the chemicals through the polysaccharide matrix [3] and/or because of an altered cell sensitivity [4]. An accepted method for measuring biocide activity in the laboratory is to determine a time-kill relationship with respect to biocide concentration. At any given temperature and biocide concentration, the rate of cell death is usually an approximately first order process [5, 6]. The physiological state of the microorganisms is important and factors such as nutrients, ionic strength, pH and the age of the culture modify the rate of cell death [7, 8].

Microorganisms present in biofilms often show marked differences in their metabolism and metabolic rate compared to planktonic organisms of the same strain and species grown in identical media [9] and this could affect their biocide sensitivity [10].

The most common methods used to assay biocides against sessile microorganisms determine minimum bactericidal concentrations [11], or reduction in cell numbers after a given time [12]. In this type of assay, the notion of innate microbial resistance in a system in which a diffusional time-lag must occur is nonsensical.

For the purpose of the development of the present model, it has been assumed that the rate of cell death is dependent solely on biocide concentration and is identical in both the planktonic phase and in the biofilm; that is, that the planktonic and sessile cells are, in themselves, equally sensitive to the biocide. Only when the interaction between the kinetics of diffusion and cell death are known can one attempt to assess the possible effects of microbial physiology in biofilms in a meaningful manner.

2. Description of the Kinetic Model

In the presence of many biocides, the number of viable cells declines exponentially with time if temperature and biocide concentration remain constant. Diffusion rates of chemicals through a membrane (or biofilm) are normally proportional to the concentration gradient in the direction of the measured flux. Although both of these are phenomenological relationships, they are readily treated mathematically as first order processes. The effect of concentration on the rate of cell death is complex and varies with the power of the concentration. The exponent varies between 0.5 and 19 for different biocides [13, 14].

In this model, it is assumed that cell death is a first order process determined by concentration and a concentration exponent. Diffusion into the biofilm is assumed to be from an ideally stirred system such that the diffusional resistance of the unstirred surface layer is both constant and small compared to the diffusional resistance of the biofilm. The rate of cell death is assumed to be proportional to concentration for cells both in the biofilm and in planktonic suspension. The half-life of the cells in the presence of biocide ($Bt_{1/2}$) is one time unit when the concentration of the biocide is unity. It is assumed that the activity coeff icient at any concentration is identical both in the biofilm and in solution and that it does not vary with concentration.

This system can easily be modelled on many standard spreadsheets; these graphs were prepared and the calculations performed on a Lotus-123 package. The memory limitation of personal computers using standard operating systems and spreadsheet packages does not allow the steps to be sufficiently small to avoid errors accumulating in iterative calculations. The cumulative errors may be held below 1% by avoiding complex models which generate multiple steps.

The spreadsheet calculations are as follows

1. Create a time series in column A $(0, 0 + dt, dt + dt...)$

2. Create a first order regression line $(6 - [A,n] \times (\log 2))$, where A is the column and n is the row number. This is the rate of cell death when the biocide kills 50% of the cells in 1 time unit ($Bt_{1/2} = 1$).

3. Calculate how the biocide concentration at any arbitrary point increases from $t = 0$ assuming a first order process. The diffusional half-life constant ($Dt_{1/2}$) may be any chosen value.

4. Using the value in (3), calculate a value using the chosen power exponent {(3)}.

5. Calculate the rate of cell death from the values in (4). Use the geometric mean of the values in line n and $n - 1$, except when $n - 1 = 0$, when $n/2$ should be used.

6. Calculate the remaining viable cell number at each time.

7. Take the logarithm of the number of cells remaining at each time and display it graphically against time.

These steps may be compressed into a smaller number of columns than is indicated.

3. Results and Discussion

Smith [5] discusses the effect of the diffusion of a biocide into cells on the rate of cell death when the concentration exponent is unity. The effect of deviations from unity is to prevent a precise solution. Since this is the most common case, the problem must be tackled by iterative calculation followed by successive approximations using smaller intervals until an approximately constant solution is obtained when it is clear that the increments are large.

Figures 1–3 show the theoretical time-kill curves for two biocides with very different kinetics. The phenolic biocide (Figs. 1 and 3) is assumed to have a concentration exponent of 6 and the heavy metal (Fig. 2) of 0.7. This signifies that, for a twofold increase in concentration of biocide, the rate of cell death increases by 2^6 and $2^{0.7}$ respectively. These figures are in accord with those reported in the literature.

For both biocides, line A represents the time-kill curve for planktonic cells at a nominal biocide concentration of 1 unit weight/unit volume and a $Bt_{1/2}$ value of 1 time unit. The other lines in Figs. 1 and 2 show the curves for sessile cells at various multiples of this concentration. The diffusion rate of the biocide into the biofilm is assumed to be such that the time taken to reach half maximum concentration ($Dt_{1/2}$) is 100 times that of the cell half-life ($Bt_{1/2}$). It is immediately obvious that for similar biocide concentrations the reduction in cell numbers in the biofilm will never equal that in the planktonic phase (line A will never cross the line representing sessile cell death for $C = 1$). Indeed, for the phenolic agent, the line for sessile cell death rate at $C = 1$ is almost horizontal, indicating that the cells within the biofilm are killed extremely slowly up to the time shown in Fig. 1. The slow initial rate of sessile cell death may be seen at all concentrations. Biocides with a high concentration exponent will tend to show this effect, as their reduced concentration in the biofilm will affect their biocidal action to a much greater extent than chemicals with a low concentration exponent, such as the heavy metal in this example.

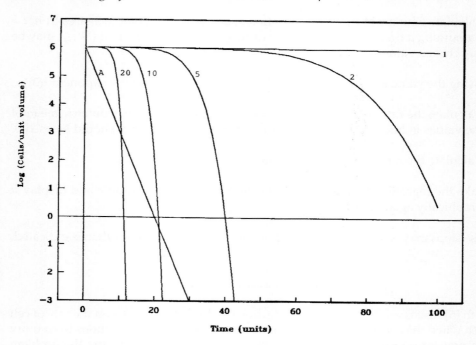

Fig. 1 Log of sessile cell numbers against time for a phenolic biocide with a concentration exponent of 6.
Cell half life ($Bt_{1/2}$) is 1 time unit when biocide concentration is 1 unit weight/unit volume and time taken for the biocide concentration in the biofilm to attain half the maximum value ($Dt_{1/2}$) is 100 time units. Line A represents death of planktonic cells ($Dt_{1/2} = 0$) at a concentration of unity. Numbers on the other lines indicate biocide concentrations external to the biofilm.

A 12-fold increase in biocide concentration is required to produce a 10^6-fold reduction in sessile cells in the same time interval as for planktonic cells when the $t_{1/2}$ value for diffusion within the biofilm ($Dt_{1/2}$) is 100 times the $t_{1/2}$ value for biocide-induced cell death ($Bt_{1/2}$) and the concentration exponent is 6 (the phenolic agent). When the concentration exponent is 0.7 (the heavy metal), the equivalent concentration is 17.

The effect of varying the ratio of $Dt_{1/2}$ to $Bt_{1/2}$ when the concentration coefficient = 6 is shown in Fig. 3. In this figure, biocide concentration and $Bt_{1/2}$ are unity, line A again represents the planktonic kill curve (i.e. $Dt_{1/2}:Bt_{1/2} = 0$) and the other lines represent values of $Dt_{1/2}:Bt_{1/2}$ of 5 to 100. Obviously, the shorter the diffusion time, the more efficient the biocide action. When the ratio is small, the sessile kill curve rapidly becomes parallel to the planktonic kill curve. The distance that the linear part of the curve is displaced to the right of the planktonic curve is the kill time lag (Lt); this is a function of the ratio $Dt_{1/2}:Bt_{1/2}$.

When the concentration exponent = 6, $Lt = [3.8 \, (Dt_{1/2}/Bt_{1/2})]^{0.97}$. When the concentration exponent = 0.7, $Lt = [5.4(Dt_{1/2})/(Bt_{1/2})]^{0.73}$.

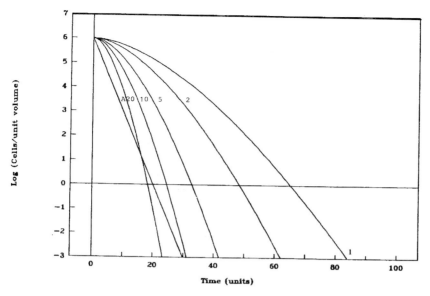

Fig. 2 Log of sessile cell numbers against time for a heavy metal biocide with a concentration exponent of 0.7. Conditions and notations are the same as in Fig. 1.

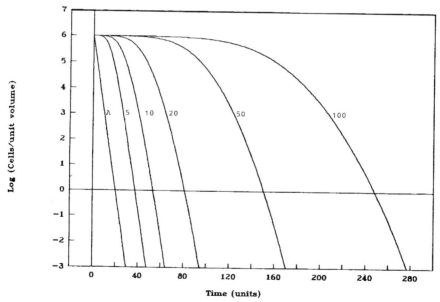

Fig. 3 Log of sessile cell numbers against time for a phenolic biocide with a concentration exponent of 6.
Cell half life ($Bt_{1/2}$) is 1 time unit. Biocide concentration is 1 unit weight/unit volume. Line A represents death of planktonic cells ($Dt_{1/2} = 0$). Numbers on the other lines indicate the ratio $Dt_{1/2}Bt_{1/2}$.

At all values of $Dt_{1/2}$ and of concentration there are striking departures from the normal linear semi-logarithmic time v. planktonic cell number plot (illustrated by the line A). The effect of diffusion rate on the rate of cell death is complex, since the factors determining the rate of increase of the free biocide in the biofilm are also complex. We have reported that the partition coefficient for the uptake of a simple biocide by a model biofilm is concentration dependent; we inferred that the biocide was held in the biofilm by ionic forces, as a complex and in solution [15]. With other biocides, hydrogen bonding, covalent bonding and Van der Waals forces may be important. When the biocide forms aggregates or complexes in solution or in the biofilm, the diffusion constants for each of the species will vary; formaldehyde and surfactants are just two examples of biocides that tend to have this property of auto-association. The rate of formation of these various complexes may be a simple first order process, or may be complex. Even when the concentration of a biocide in a biofilm may be determined empirically, much of it may be held in inactive forms. Most biocides are reactive and will bind to the known components of a biofilm. Simple diffusion (if it is simple) will act in concert with other rate constants to determine the 'free' concentration of biocide. These time-dependent processes will all ensure that the rate of cell death departs from a first order relationship. When diffusion or other factors cause the concentration of 'free' biocide to vary with time, a semi-logarithmic plot of cell number against time will not be linear. This departure from linearity shows that diffusional delays are an important factor in determining biocide activity. Until this aspect of biocide activity on sessile organisms is better understood, any suggestion about the role of physiological state of the cells is unfounded.

Application of biocides for short periods, even at high concentrations, may be ineffective if a sufficient biocidal level acting for sufficient time is not attained in the biofilm. As shown in Fig. 1, when the biocide has a large concentration exponent, no detectable biocidal death will occur for a long period even though the concentration in the medium outside the biofilm is many times that required to kill cells efficiently in the planktonic phase.

In practice, then, biocides that are unstable or which may penetrate biofilms slowly should be avoided, especially if they have a high concentration exponent. Prolonged exposure and/or high biocide concentrations are needed to treat organisms within biofilms.

References

1. C. W. Keevil, C. W. Mackerness, and J. S. Colbourne, 'Biocide treatment of biofilms', *Int. Biodeter.*, 1990, **26**, 169–180.
2. C. C. Gaylarde and J. M. Johnston, The effect of some environmental factors on biocide sensitivity in *Desulfovibrio* implications for biocide testing, in *Microbial Corrosion*, The Metals Society, London, 1983, pp.91–97.
3. J. W. Costerton, K. J. Cheng, G. G. Geesey, T. I. Ladd, J. C. Nickel, M. Dasgupta and T. J. Marrie, 'Bacterial biofilms in nature and disease', *Ann. Rev. Microbiol.*, 1987, **41**, 435–464.

4. P. Gilbert, P. J. Collier and M. R. W. Brown, 'Influence of growth rate on susceptibility to antimicrobial agents: biofilms, cell cycle, dormancy and stringent response', *Antimicrob. Aq. Chemoth.*, 1990, **34**, 1865–1868.

5. R. N. Smith, 'Kinetics of biocide kill', *Int. Biodeter.*, 1990, **26**, 111–126.

6. G. G. Meynell and E. Meynell, *Theory and Practice in Experimental Bacteriology*, Cambridge University Press, Cambridge, UK, 1970.

7. D. J. Evans, M. R. W. Brown, D. G. Alison and P. Gilbert, 'Susceptibility of bacterial biofilms to tobramycin: role of specific growth rate and phase in the division cycle', *J. Antimicrob. Chemoth.*, 1990, **25**, 585–591.

8. P. Gilbert and M. R. W. Brown, 'Cell-wall mediated changes in the sensitivity of Bacillus megaterium to chlorhexidine and 2-phenoxyethanol, associated with growth rate and nutrient limitation', *J. Appl. Bacteriol.*, 1980, **48**, 223–230.

9. H. Anwar, J. L. Strap and J. W. Costerton, 'Establishment of aging biofilms: possible mechanism of bacterial resistance to antimicrobial therapy', *Antimicrob. Ag. Chemoth.*, 1992, **36**, 1347–1351.

10. B. D. Hoyle, J. Jass and J. W. Costerton, 'The biofilm glycocalyx as a resistance factor', *J. Antimicrob. Chemoth.*, 1990, **26**, 1–6.

11. T. S. Whitham and P. D. Gilbert, 'Evaluation of a model biofilm for the ranking of biocide performance against sulphate-reducing bacteria', *J. Appl. Bacteriol.*, 1993, **75**, 529–535.

12. G. Lloyd, 'The development of "safer" compounds for biocidal use', *Int. Biodeterior.*, 1990, **26**, 245–250.

13. H. Chick, *The Theory of Disinfection. In A System of Bacteriology in Relation to Medicine*, Vol. 1, H. M. Stationery Office, London, 1930, p.179.

14. F. W. Tilley, 'An experimental study of the relation between concentration of disinfectants and the time required for disinfection', *J. Bacteriol.*, 1939, **38**, 499.

15. P. M. Gaylarde and C. C. Gaylarde, 'The kinetics of biocide uptake in a model biofilm', *Int. Biodeterior. Biodegr.*, 1992, **29**, 273–283.

14

Biocorrosion of Mild Steel by Sulphate Reducing Bacteria

I. T. E. FONSECA, A. R. LINO* and V. L. RAINHA

CECUL, Departamento de Quimica, Faculdade de Ciencias, Universidade de Lisboa, Rua da Escola Politecnica 58,1200 Lisboa, Portugal
*Instituto de Tecnologia Quimica e Biológica, Apartado 127, 2780 Oeiras, Portugal

ABSTRACT

Corrosion of mild steel was evaluated in the absence and in the presence of sulphate reducing bacteria, SRB, in a fresh and in an ageing culture. Parallel experiments were also performed in the nutrient medium in the presence and in the absence of organic nutrients.

Weight loss measurements have shown that the living bacterial medium is the most aggressive during the first day of exposure, with corrosion rate v_c (after 1 day) = 1.0376 mg cm^{-2}/day, but after 14 days v_c = 0.0195 mg cm^{-2}/day. The corresponding values in the nutrient medium without the SRB were 0.6033 and 0.0874 mg cm^{-2}/day.

Open circuit potential measurements have given E vs time curves showing regular oscillations, indicative of localised corrosion, in a fresh culture. In an ageing culture a steady-state plateau was obtained.

Potentiodynamic polarisation tests have shown that SRB play an important role on the kinetics and also in the mechanism of the anodic and the cathodic processes associated with the biocorrosion.

1. Introduction

The role of micro-organisms in the corrosion of metals has received increased attention in recent years. The presence and activities of microorganisms can cause serious problems, extending from industry to the military and wherever metals are present in our society, leading to severe economical consequences. These have been extensively reviewed by several authors[1–3]. The presence and metabolic activity of micro–organisms on metal surfaces lead to changes in H_2S concentration and pH level which influence the rates and also the type of corrosion.

Sulphate reducing bacteria, SRB, are an assemblage of bacteria that can grow in anaerobic medium by the oxidation of organic nutrients with sulphate being reduced to H_2S [2–4].

Lino et al. [5] have studied the growth of *Desulfovibrio desulfuricans* ATCC 27774 bacteria in a lactate/sulphate medium, at 37°C, in the presence of mild steel. From these studies growth curves such as those given in Fig. 1 were obtained. The curves in Fig. 1 show that there is a period in which the bacterial population increases almost exponentially, and then after about two days a steady-state of 60 × 10^6 cells/

cm² is reached. Great changes at the interfacial and in the bulk are to be expected during the period of exponential growth. During this period the sulphide concentration also increases at a high rate reaching a value of 900 ppm after 2 days of incubation. The increase of sulphide concentration is due to the growth of the bacterial population, as expected, according to the mechanism generally accepted [6, 7].

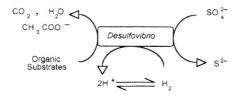

2. Experimental

Samples were made from a steel sheet with the following nominal composition: C 0.71, Mn 0.50, Si 0.30, S 0.015, P 0.020% Fe bal.

All the samples were first abraded with emery paper from No. 2 up to No. 0, polished with alumina from 1 µm up to 0.05 µm, then degreased with an electrolytic solution, FC-130® at 70°C, washed with distilled water and finally etched in HCl 7.4% for 30 min and dried with ethanol and cool air.

The micro-organism studied was the *Desulfovibrio desulfuricans* ATCC 27774.

Solutions with the composition given in Table 1 were prepared in Millipore–water by dissolving the compounds given in Table 1. The pH was adjusted to 7.6. All experiments were carried out at 37°C.

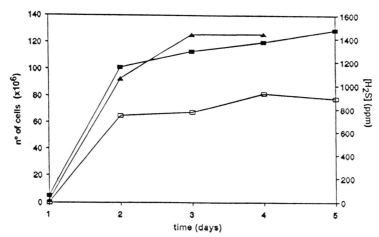

Fig. 1 Growth curve of Desulfovibrio desulfuricans *ATCC 27774 in a lactate/sulphate medium in the presence of mild steel. From Lino et al. [2, 3].* ■ *planktonic population (No. of cells* mL^{-1}*);* □ *sessile population (No. of cells* cm^{-2}*);* ▲ H_2S *concentration (ppm).*

Table 1. Composition of 1 L of X, Y and Z media

Medium	Compounds	Amounts	Conc. (mol. dm^{-3})
X	NH_4Cl	2 g	3.74×10^{-2}
	$MgSO_4.7H_2O$	2 g	8.0×10^{-3}
	K_2HPO_4	0.5 g	2.9×10^{-3}
	Na_2SO_4	4 g	2.8×10^{-2}
	$FeSO_4.9H_2O$	0.010 g	3.6×10^{-6}
	$CaCl_2$	0.2 g	1.8×10^{-3}
	$Na_2S.9H_2O$	0.25 g	1.1×10^{-3}
	Modified Wolfe's minerals	1 mL	–
Y	Medium X	987.5 mL	–
	Sodium lactate 60% (w/w)	12.5 mL	8.7×10^{-2}
	yeast extract	1 g	–
	Cysteine–HCl	0.25 g	1.6×10^{-3}
Z	Medium Y	900 mL	–
	Bacterial culture	100 mL	–

Prior to each experiment the solutions were purged with nitrogen to ensure anaerobic conditions. All the media were sterilised at 120°C for 30 min. Medium Z was freshly inoculated.

Potentiodynamic curves were run with a PARC waveform generator, mod. 173, coupled to a PARC potentiostat, mod. 178. The polarisation curves were run at 1 mV s^{-1} and recorded on a X–Y recorder from Philips, mod. PM 8120.

A conventional three-electrode two-compartment glass cell was used. The secondary electrode was a long helix of Pt wire. A commercial saturated calomel electrode (SCE) placed in a separate compartment was used as a reference electrode (E^0 = 0.242 V vs SHE).

3. Results and Conclusions
3.1. Weight Loss Measurements

A batch of steel samples treated as described in Section 2 was accurately weighed and then placed in a transversal position in bottles of about 30 mL volume and sterilised at 120°C for 30 min. Afterwards the bottles were completely filled with the solutions X or Y previously well deaerated and sterilised under the same conditions. Inoculation, to obtain solution Z, was made at this time. The bottles were then sealed with black screw caps (to maintain anaerobic conditions and to prevent unwanted contamination). Finally they were incubated at 37°C.

After exposure for periods of 1, 7, 14, 21 and 28 days the steel sample was taken from the respective bottle. Before the re-weighing the corrosion products were removed from the corroded samples by immersion in an ultrasonic bath during three minutes followed by the procedure used prior to exposure.

Corrosion rates evaluated from the difference in weights before and after exposure are given in Table 2. The variation of the corrosion rate as a function of time for each medium is also represented in Fig. 2.

The results in Table 2 and Fig. 2 show clearly that organic nutrients cause only a slight decrease in the corrosion rate of mild steel: 0.6033 against 0.7479 mg cm^{-2} d^{-1} after one day of exposure and 0.0819 against 0.1102 mg cm^{-2} d^{-1} after 28 days (a reduction of 20–25%).

Living SRB, in medium Z, causes a big increase in the corrosion rate during the first day; 1.0376 against 0.6033 mg cm^{-2} d^{-1} in medium Y; but quite soon this effect is reversed, i.e. after 7 days $v_c(Z) = 0.1100$ mg cm^{-2} d^{-1} and $v_c(Y) = 0.1515$ mg cm^{-2} d^{-1}. These data show that the corrosion rate of mild steel drops steeply to a very low value due to the presence of SRB (94% in medium Z compared with 7% in medium Y).

Table 2. V_c (mg cm^{-2}/d^{-1}) of steel samples immersed in X, Y and Z media

Time (days)	Medium X	Medium Y	Medium Z
1	0.7479	0.6033	1.0376
7	0.1261	0.1515	0.1100
14	0.1140	0.0874	0.0195
21	0.1352	0.0744	0.0169
28	0.1102	0.0819	0.0110

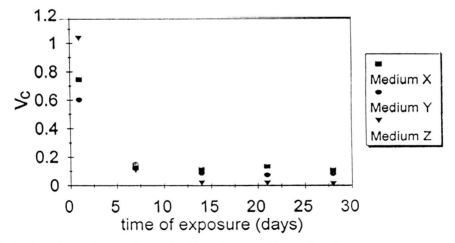

Fig. 2 Plots of corrosion rate V_c as a function of exposure time and medium.

According to the mechanism generally accepted an accumulation of sulphide as well as an increase of protons is brought about by the growth of the SRB, and this is the stimulus for the initial dissolution of iron at quite a high rate. But, on the other hand, a film of ferrous sulphide will be formed on the electrode surface, trapped by an organic biofilm, causing a partial inhibition in the anodic dissolution of iron.

In fact, visual observation of the steel samples after being exposed to a medium containing SRB show a black film over the electrode surface, while the equivalent medium free of bacteria produces a grey film and an orange–yellowish precipitate.

Apart from the initial value, corresponding to 1 day of exposure, corrosion rates in medium X are higher than in medium Y or Z. In fact, medium X is an aqueous solution with a quite high concentration of aggressive anions: 4×10^{-2} mol dm^{-3} of Cl$^-$, 3.6×10^{-2} mol dm^{-3} of SO$_4^{2-}$ and 1×10^{-3} of S^{2-} mol dm^{-3}. All these anions are known to promote the electrochemical dissolution of iron and steel.

Chloride is known to induce localised corrosion and this may be the reason why the corrosion of steel remains at quite a high value even after 28 days of exposure ($v_c(X) = 0.1102$ mg cm^{-2}/d^{-1}).

Medium Y also contains these anions but it also contains the organic nutrients, increasing the complexity of the system. The presence of yeast extract may explain the decrease in the corrosion rate observed over long periods of exposure, i.e. after two weeks. According to Tiller [3] yeast extract adsorbed on the electrode surface inhibits the corrosion of mild steel.

Medium Z suffers changes in the concentration of SO$_4^{2-}$, S^{2-} and protons as a consequence of bacterial growth.

3 2. Open Circuit Potential Measurements

Open circuit potentials, E_{ocp}, were measured against a saturated calomel electrode placed in a separate compartment.

The potentials of the steel electrode immersed in solutions X, Y, Z and Z^*, at 37°C, were measured by a high impedance voltmeter — a Hewlett Packard® 34401 A — interfaced to an IBM® 286 XT (see Fig. 3). This experimental set-up allowed readings to be made of the E_{ocp} at regular and short time intervals during long periods of exposure. Figure 4 shows plots of these readings taken during 2 days, at intervals of 2.5 min, for media X, Y, Z, and Z^*.

In the fresh bacterial culture, Z, E_{ocp} varied from –650 to –500 mV with regular oscillations, after an induction time of 5 h, while in an ageing culture (medium Z^*) E_{ocp} varied from –250 to –100 mV without showing any oscillations.

Oscillations are usually associated with localised corrosion. This process may be due to the metabolism of SRB, in which sulphide and protons are produced.

Plots of E_{ocp} as a function of immersion time given in Fig. 4. show large variations in the corrosion potentials.

Z^* is a four-day old culture, while Z is a fresh culture, at the beginning of the measurements. According to growth curves obtained by Lino et al. [5] in the presence of mild steel, Z is an exponentially growing culture, and Z^* is an ageing culture.

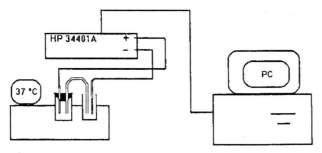

Fig. 3 *Diagram of the experimental set-up for the E_{ocp} measurements.*

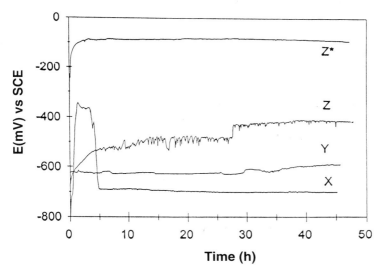

Fig. 4 *Plots of open circuit potential, E_{ocp}, as a function of immersion time.*

In medium X the E_{ocp} shifted initially to more positive values starting at –800 mV until it reaches a maximum of –350 mV after 2 h. The potential then dropped suddenly to a value of –700 mV. This value remained constant, at least during the next 40 h.

3.3. Anodic Polarisation Curves

The steel sample, a disc of 0.407 cm^2 was placed in a two compartment, three electrode cell filled with about 50 mL of the medium under study. A saturated calomel electrode was used as a reference electrode and a Pt foil as a secondary electrode. The cell was immersed in a bath at 37 °C.

The working electrode was left at open circuit for ten minutes, polarised at –1.0 V vs SCE for 3 min and then the potential was scanned from –1.0 to –0.65 V at a rate of

1 mV s^{-1}. The corresponding i vs E curves were recorded. Figure 5 shows a set of these curves.

The potentials at which these lines cross the potential axis are –0.85, –0.78 and –0.72 V vs SCE respectively for medium X, Y and Z. The anodic branches of the corresponding curves show clearly that corrosion in medium Z starts at less negative potentials but at a much higher rate.

Tafel analysis of the anodic and cathodic branches of the i vs E curves was performed. These plots are given in Fig. 6. Two linear sections can be drawn in each curve, respectively at low and high overpotentials. By low and high we mean overpotentials lower than 50–60 mV and higher than 60–70 mV, respectively. The second linear region extends from overpotentials of 60 to about 150 mV.

Corrosion potentials, and corrosion current densities, evaluated from the intersection of the first linear sections of the Tafel plots are given in Table 3.

E_{corr} values obtained either from the first or the second linear regions do not differ very much. The same is not true for the corrosion current densities: extrapolation of the low polarisation segment gives lower values. In any case the relationship for the various media is:

$$i_{corr}(Y) > i_{corr}(Z) > i_{corr}(X)$$

Tafel slopes of the anodic segments range between 0.040 and 0.100 V and for the cathodic segments between 0.060 and 0.400 V. Values of first and second linear regions are quite different: This difference is even higher for the cathodic side. Changes in Tafel slopes are, generally, indicative of changes in the mechanism [9]. Quantitative data, relevant to the establishment of the mechanism can be obtained from Tafel slopes, but only for less complex systems, i.e. for the dissolution of iron in a non-complexing anodic media [9].

Values of the Tafel slopes given in Table 3 demonstrate very well that SRB play an important role on the mechanism and also in the kinetics of the biocorrosion of mild

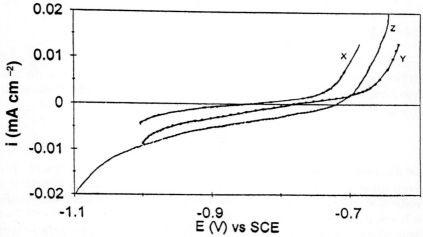

Fig. 5 Anodic polarisation curves of mild steel in media X, Y and Z. v = 1 mV s^{-1}.

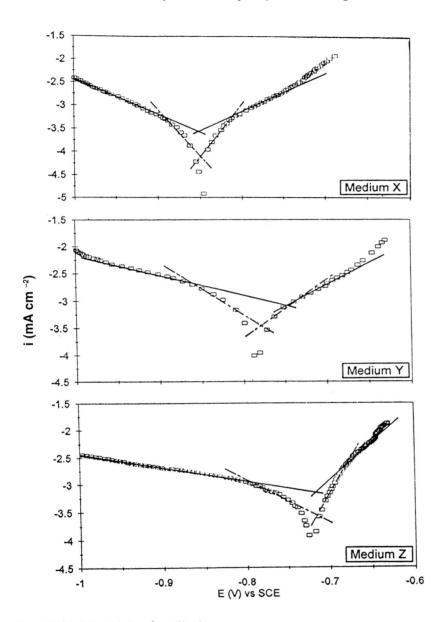

Fig. 6 Tafel plots with data from Fig. 5.

steel but the systems, particularly steel in medium Z, are too complex to allow well defined conditions under potentiodynamic polarisation. For these systems, whenever high polarisations and slow scan rates, i.e. 1 mV s^{-1}, are used, great changes in the environment and in the activity of the micro-organisms are to be expected. At

Table 3. Values of E_{corr}, i_{corr}, b_a and b_c from Tafel plots in Fig. 4

	Medium X	Medium Y	Medium Z
E_{corr} (V vs SCE)	–0.846	–0.765	–0.711
i_{corr} (A cm^{-2})	0.0993	0.4089	0.3036
b_{a1} (V)*	0.046	0.097	0.037
b_{a2} (V)*	0.108	0.090	0.074
b_{c1} (V)*	0.064	0.141	0.155
b_{c2} (V)*	0.126	0.261	0.405

*1 and 2 refers to low and high overpotentials, respectively.

high anodic polarisations, i.e. at η > +100 mV, corrosion products, such as iron sulphide, will be formed, which may or may not act as cathodes during cathodic polarisation; anodic and cathodic areas may vary and may be unknown, particularly in those media in which localised corrosion may take place; and the degree of colonisation of the electrode surface by the bacteria may change with polarisation. At high cathodic overpotentials H_2 evolution takes place. These and other factors will contribute to a certain degree of uncertainty.

In fact, potentiodynamic polarisation is a powerful technique for providing mechanistic information in corrosion studies particularly for systems undergoing uniform corrosion.

Additional potentiodynamic and potentiostatic experiments under various experimental conditions are still in progress, as well as surface and corrosion products analysis.

4. Acknowledgements

This work is a result of a cooperative research programme between Centro de Electroquímica e Cinética da Universidade de Lisboa (CECUL) and Instituto de Tecnologia Química e Biológica (ITQB). Financial support from Junta Nacional de Investigação Científica e Tecnológica (JNICT) is acknowledged.

References

1. A. K. Tiller, *Microbial Corrosion*, 1983, The Metals Society, London, 54–65.
2. R. Cord-Ruwisch, W. Kleinitz and F. Widdel, 'Sulphate-Reducing Bacteria and their Activities in Oil Production,' *J. Petroleum Technol.*, 1987, 98–105.
3. H. A. Videla, in *Corrosão Microbiológica*, Edgard Blücher Ltd, 3rd edn, São Paulo, 1993.
4. W. A. Hamilton, 'Sulphate reducing bacteria and anaerobic corrosion,' *Ann. Rev. Microbiol.*, 1985, **39**, 195.

5. A. R. Lino, M. J. Feio, C. Pinto, M. A. M. Reis, I. Beech and J. J. G. Moura, 'Biocide Action on Sessil and Planktonic Culture of *Desulfovibrio desulfuricans* ATCC 27774,' *Abstracts 9th Int. Biodeterioration and Biodegradation Symposium*, Leeds, U.K, 5–10 September 1993.

6. J. R. Postgate, *The Sulphate-Reducing Bacteria*, 2nd Edn, Cambridge University Press, Cambridge, UK, 1984.

7. T. Ford and R. Mitchel, 'The Ecology of Microbial Corrosion,' *Adv. Microb. Ecol.*, 1990, **11**, 231–262.

8. A. K. Tiller, private communication.

9. J. O'M. Bockris and A. K. Reddy, *Modern Electrochemistry*, Vol. 2, Plenum Press, New York, 1976.

15

Effect of Marine Biofilms on High Performance Stainless Steels Exposed in European Coastal Waters

J. P. AUDOUARD, C. COMPERE*, N. J. E. DOWLING, D. FERON[†],
D. FESTY*, A. MOLLICA**, T. ROGNE[††], V. SCOTTO**, U. STEINSMO[††],
K. TAXEN*** and D. THIERRY***

IRSID, BP 0142490 Fraisses, France
*IFREMER, BP 70, 29280 Plouzane, France
[†]CEA-LETC, Etab. de La Hague, 50444 Beaumont-Hague, France
**ICMM, Torre di Francia, Via de Marini, 6-IV p., 16149 Genova, Italy
[††]SINTEF, Rich. Birkelandsv. 3a, N-7034 Trondheim, Norway
***SCI, 101 Roslagsvaegen, Hus No. 25, Stockholm 10405, Sweden

ABSTRACT

It is increasingly apparent that marine biofilms play an important role in the extended lifetimes of ships, platforms and other offshore structures. The marine science and technology directorate in Brussels has initiated a collaboration between six research centres to examine the effects of biofilms which develop on high performance stainless steel alloys including UR SB8, SAF 2507 and 654SMO in the seas immediately surrounding the western European coastline. The exposure centres are located at Genova in the Mediterranean, Brest and Cherbourg in the eastern Atlantic, Kristineberg and Trondheim in the North sea. Recovery of biofilms at these sites have revealed a extremely complex relationship between the microfauna and their metal/biofilm interface. Open circuit potential measurements however, demonstrated a consistent tendency towards noble potentials around + 350 mV/SCE associated with the biofilm. Mediterranean biofilms appear to be consistently heavier than observed at the more northern sites. Stainless steels with intact passive films appear to behave similarly with respect to their biofilm interaction. A strong correlation between season and the cathodic reaction at the stainless steel surface was observed.

1. Introduction

The appearance of complex biofilms on engineering materials in marine environments has given much cause for concern in recent years [1]. The problems have been two fold (i) the mechanical loading of the substrate increasing drag and structure weight, and (ii) the actual deterioration of the substrate. This article is concerned with the latter aspect, and in particular the response of that class of alloys known as stainless steels. Perhaps the vast majority of the tonnage of steel presently arrayed in the sea is carbon ('structural') steel of various grades which were selected for their mechanical properties and cost ratio. Of necessity, carbon steel exposed to such corrosive environments has to be protected, usually involving a complicated arrangement of cathodic protection and anti-corrosion paints. The servicing of these anti-corrosion systems by cleaning, anode replacement, repainting etc. severely reduces the cost advantages of these steels and in consequence the use of inherently corrosion-resistant alloys becomes increasingly attractive. Other material selection criteria however can be almost indifferent of cost. Key parts of some offshore structures are of such crucial importance that fabrication in carbon steel is prohibited. These critical sections are usually involved in safety measures required for operational personnel, or required to reduce the risk of severe economic loss in the event of failure. Under such circumstances these sections are frequently fabricated from stainless steel.

Data published over the last decade [2–4] have shown that there are many uncertainties with respect to the use of stainless steels in the sea and in particular their response to biofilms. Not only is there a mechanical concern as mentioned previously, but the extraordinary effect of the greatly displaced open circuit potential in the so-called 'passive' region is often near the breakdown potentials for several of these materials, generating concern over their probability of failure. Although the displacement in potential has been attributed to several different electrochemical mechanisms, no conclusive information is currently available. It is known that functional biofilms are necessary for this event to occur since the application of sodium azide (a strong biocide) initiates a precipitous drop in the potential to levels which are associated with non-fouled surfaces.

1.1. Project Goals

The end result of this research project will provide insight into the relationship between the biochemical nature of the surface-attached microorganisms and the electrochemical state of stainless steel substrata in marine environments. The alloys selected are examples of the different strategies used to withstand the range of corrosion and mechanical requirements frequently encountered in the high-chloride marine environments. Thus, the microstructure and elemental composition are used variously to enhance characteristics of resistance to stress corrosion cracking, pitting, crevice corrosion, hydrogen-cracking as well as toughness and durability.

Seasonal exposure tests in European coastal waters were specifically designed to establish if the effects of biofilms upon metal substrates varied according to geographic and hydrological distribution. The measurements taken included the free corrosion potential, electrochemical impedance spectroscopy, galvanic coupling and biofilm collections executed when the potential reached particular prefixed values (which corresponded to biofilm interference at the metal/seawater interface). Subsequent biochemical analyses would allow the interpretation of the ennoblement of the potential in terms of microbiological events.

2. Materials and Methods
2.1. Materials

The results of initial tests reported in this article concern the behaviour of two metallurgical classes of stainless steels: (i) the gamma fully super-austenitics (UR SB8 of Creusot Loire Industrie and 654SMO of Avesta) and (ii) an example of the biphasic super-duplex containing approximately equal proportions of ferrite and austenite (SAF2507 of Sandvik Steel). Both these classes of stainless steels are recommended for seawater service under general conditions. The elemental distribution of the stainless steels used in this study (in both tube and plate form) are listed in Table 1.

Table 1. Compositions of alloys exposed in European coastal waters

Element Alloy	C	S	P	Si	Mn	Ni	Cr	Mo	Cu	N^2
SB8[a]	0.01	0.0008	0.01	0.24	0.93	25	25	4.7	1.4	0.21
645SMO[b]	0.01	0.001	0.02	0.16	3.6	21.8	24.5	7.3	0.43	0.48
SAF2507[c]	0.14	0.001	0.015	0.27	0.4	6.9	24.9	3.8	ND	0.28

These steels are produced by the following companies: **a** = Creusot-Loire Industrie, **b** = Avesta Sheffield and **c** = Sandvik Steel.
ND = Not determined.

The above materials were exposed in the coastal waters (with attendant variations in salinity, temperature, pollution etc.) adjacent to Trondheim (Norway), Kristineberg (Sweden), Cherbourg (France), Brest (France), and Genova (Italy). The exposure tests were executed from June to July 1993 for the summer season, from September to October for the autumn, and January to February for the winter season. In this first step no crevice-forming geometries were used. All participants tested 654SMO in the form of plates and tubes. The other alloys were only used in limited distribution. Biofilm growth was monitored using the evolution of the free corrosion potentials and galvanic currents arising from stainless steel coupons galvanically coupled with an iron anode, and secondly the biochemical analysis of the biofilm which accumulated over time as a function of specific potentials. Finally, the interfacial impedance of the biofilm/metal was monitored by electrochemical impedance spectroscopy.

2.2. Hydrological Parameters

The parameters of all seawater sites reflected the proximity of the respective coastlines, thus the salinity, for example, in each case was slightly more dilute than open ocean values due to run-off. These values are presented in Table 2.

Table 2. Average seawater parameters at the different station sites

Marine Station	Genova (Italy)	Cherbourg (France)	Brest (France)	Kristineberg (Sweden)	Trondheim (Norway)
Period	June/July Sept/Oct	June/July Sept/Oct	June/July Sept/Oct	June/July Sept/Oct	June/July Sept/Oct
ΔT (°C)	24.6 ± 0.7	15.9 ± 2	15 -- 16.5	8 -- 11	11 ± 1
	22.0 ± 1.2	15.5 ± 1.5	16.8 -- 15	12 -- 9.8	11 ± 2
pH	8.2 ± 0.1	8.1 ± 0.05	8.2 ± 0.5	7.8	8.0 ± 0.1
	8.3 ± 0.15	8.0 ± 0.1	8.1 ± 0.1	7.86	7.97 ± 0.08
S (%)	37.0 ± 0.17	29.0 ± 0.05	33	32.7	35.5 ± 4.5
	36.9 ± 0.4	30.0 ± 4	35.4	ND	32.5 ± 1
Chlorophyll a $\mu g\ L^{-1}$	0.34 ± 0.05	0.3	0.83 ± 0.35	0.1	0.23 ± 0.06
	0.6 ± 0.3	0.7	0.46 ± 0.09	ND	0.21 ± 0.03
O_2 ppm	5.2 ± 0.6	6.7 ± 0.3	8.4 ± 0.5	5.2	8.3 ± 0.3
	8.1 ± 0.6	6.9 ± 0.6	8.35 ± 0.05	ND	8.1
Org.matter mg L^{-1}	2.68 ± 0.35	1.4	0.68 ± 0.19	3	11.4 ± 1.01
	3.1 ± 0.7	1.7	0.6 ± 0.27	ND	5.1 --< 1
Ash mg L^{-1}	9.67 ± 0.52	ND	6.54 ± 0.47	10.5	ND
	11.3 ± 2.4	ND	6.54 ± 0.47	ND	ND

ND = Not determined.

2.3. Galvanic Coupling of Stainless Steels and Carbon Steel Anodes

The corrosion resistance of stainless steels is primarily controlled by the relative intactness of their passive films. The nature of this passive film is to some degree controlled by the elemental composition of the solid steel matrix, however it is principally composed of oxy-hydroxides of chromium. In order to test the cathodic properties of passive films of different stainless steels, plates and tubes of 654SMO and UR SB8 were exposed in seawater in connection with either an iron or carbon steel anode. The currents from these galvanic couples were measured as a function of time, and the increasing current density interpreted in terms of the upswing in free corrosion potential associated with the isolated coupons of the stainless steel.

2.4. Biofilm Collection and Treatment

The metal specimens (plates or tubes), withdrawn from the exposure system and leant on one edge to draw off excess water, were weighed to determine the total wet biofilm mass. The biofilm was then removed from the metal surfaces by ultrasonic washing in 20 mL of a buffer solution. The resulting suspension was centrifuged at 5000 rpm in order to separate the exopolymeric component (soluble in the supernatant) from the organisms and detritus (pellet). The two fractions were sent to ICMM (Genova for biochemical analysis. The samples were examined for total carbohydrate content using the antrone methods [5] after hydrolysis of the polymers with strong acids.

3. Results
3.1. Hydrological Characterisation of Exposure Tests

The five exposure sites selected are quite diverse in nature as well as geographical distribution. The sole Mediterranean site at Genova (Table 2) is distinguished by virtue of its relatively high salinity and temperature. All the eukaryotic and prokaryotic genera which were expected to appear in the surface biofilms, and to have an impact on this study, have biochemical mechanisms which would benefit from these conditions. In contrast, the northern exposure sites at Kristineberg and Trondheim are subject to considerably lower temperatures (Fig. 1) and a shorter period favourable for biological growth during the year. The two sites at Kristineberg and Trondheim are distinguished by their pristine conditions. The eastern seaboard of Europe is represented by two sites at Cherbourg and Brest located on two geographical promontories situated in Brittany and Normandy. While both sites are very close to the conjunction of the Eastern Atlantic and North seas they are still adjacent to the continental shelf, characterised by moderate temperatures and shallow seas quite rich in sea life. The two French sites are distinguished from the others by the aperiodic release of freshwater from large watersheds located along the coast. Under these conditions the salinity can vary drastically (Table 2). Other sites are, of course, subject to the usual minor fluctuations in salinity of any coastal zone resulting from rainwater runoff.

3.2. Evolution of the Open Circuit Potential

The evolution of the free corrosion potential of 654SMO over the summer exposure period for all the test sites involved is presented in Fig. 2. The trend confirms the information widely described in the literature that passive stainless steels in marine environments rapidly develop a high (and relatively stable) open circuit potential. This trend was observed at all sites although at varying rates depending upon whether tubes or plates were the subject of the study. The rates of increase reflected the seasonal variation i.e. temperature among other parameters. In Fig. 3 the effect of season is presented at the Brest and Genova exposure sites. Since there appeared to be no difference between the evolution of the potentials observed between the stainless steels Uranus SB8, SAF2507 and 654SMO a single representative plot is provided in Fig. 3. It was noted that while the coupons exposed at the Brest site gave strong evidence of a lag in the rise in potential associated with the winter exposure season, there was little evidence of this lag for the Genova site except for a considerably lower initial potential.

Effect of Marine Biofilms on High Performance Stainless Steels 203

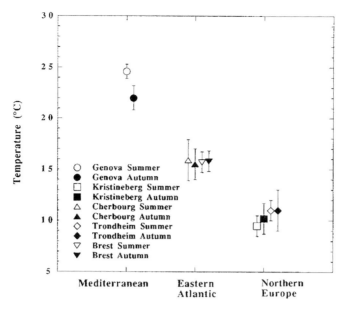

Fig. 1 Variation in temperature due to geographical position.

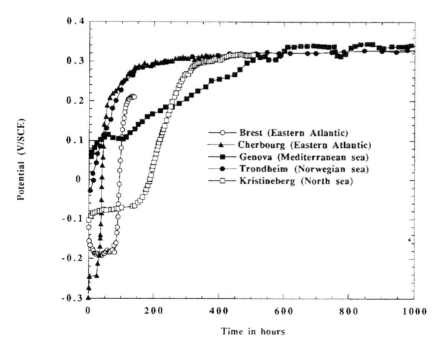

Fig. 2 Evolution of the free corrosion potential exhibited by 654SMO at five European marine stations during the summer exposure period.

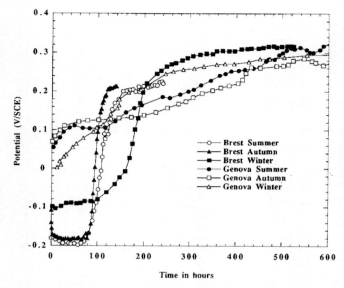

Fig. 3 Effect of season on the evolution of the open circuit potential as a function of time.

3.3. Biofilm Structure

The biofilm accumulation on the stainless steels surfaces was followed by scanning electron microscopy (SEM) at all the exposure sites. The classic evolution of the biofilm from initially few cells over several days to an easily discernible polymer film in which the microflora were embedded (7 days) was found. Figure 4 shows the complex distribution of organisms adhering to the surface after an immersion period of 3 days. Biofilms consisting of adherent microorganisms depend fundamentally, on the use of extra cellular polysaccharide (EPS) as a glue to hold the cells in a three dimensional matrix upon the surface of the substrate, in this case steel. To some extent therefore this class of compound can be used as a measure of biofilm biomass. The method of Dubois (Strickland and Parsons) breaks the polysaccharide into its monomers (sugars) and analyses them as the subunit.

3.4. Galvanic Coupling Tests

In order to test the cathodic characteristics of the various stainless steels the three main alloys under test were coupled with carbon steel or iron anodes. The open circuit potential of the couples was initially in the range of between -600 and -700 mV/SCE which indicates that the corrosion of the carbon steel/iron anode was principally controlled by the diffusion of oxygen to the stainless steel surface (thus the cathodic half reaction). Figure 5 shows that there is a relation between sites indicating at least a partial influence of temperature upon the lag time before the exponential increase of the galvanic current coupling the iron anode and the stainless steel. This current is proposed as a direct monitor for the accumulation of an active biofilm.

Effect of Marine Biofilms on High Performance Stainless Steels

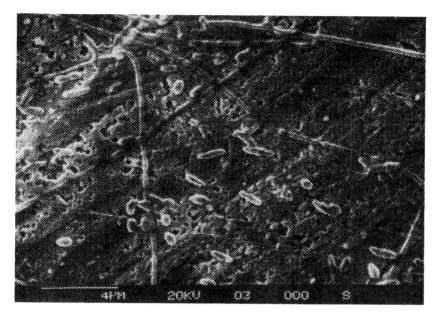

Fig. 4 SEM micrograph of a stainless steel after 3 days immersion at the Genova exposure site. (Photo kindly provided by G. Ventura.)

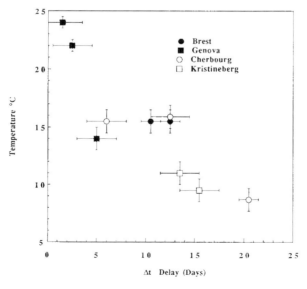

Fig. 5 Time lag before galvanic currrent increase following immersion of stainless steels as a function of temperature.

3.5. Accumulation of Biomass

The trend of biofilm accumulation with time is demonstrated in Fig. 6. No statistically significant difference was observed between the alloys SB8, SAF2507 and 654SMO at any site or during any season despite their variable concentrations of nickel and different microstructure. Despite this lack of difference in accumulated biomass at each site, there was a difference between sites, the interpretation of which was difficult. Data from the sole Mediterranean site at Genova not only exhibited-relatively high temperature and salinity, but also indicated that a flow rate of 1 m s^{-1} was optimum for maximum growth rate in water taken directly from this sea. The two moderate temperature stations at Cherbourg and Brest differed in their accumulation rates which were also influenced by the effect of retention of the seawater for an unknown period at the Cherbourg site before transmission to the seawater circuits. The Brest data however, confirm to a great extent the Mediterranean results where biofilm accumulation rates on tubes are greater than on plates. This effect is certainly due to the more rigorous initial attachment of the microflora under flow conditions and the more rapid passage of carbon and electron acceptors to the 'mature' biofilms. The differences in water introduction at the Atlantic sites and the northern European sites appear to be one principally of depth. The Trondheim fjord exposure station has its intake pipe entrance placed at 60 m depth while the Kristineberg (Gullmaren fjord) site intake pipe is at 40 m depth. While it can be argued that at this depth the temperature differential due to season at the sites is not large (Fig. 1) because only the surface waters in each case change temperature, the microflora will certainly be different and relatively constant throughout the year.

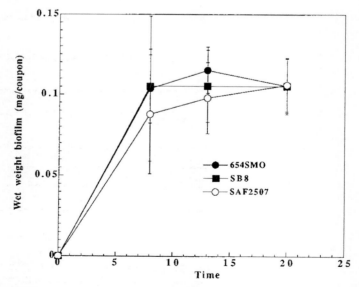

Fig. 6 Biofilm accumulation on three stainless steels at Kristineberg (North Sea) during the autumn exposure 1993.

3.6. Oxygen Reduction as a Function of Time

Coupling the iron anode and the stainless steels permitted a rough indication of the relative importance of the accelerated reduction of oxygen occurring on the stainless steel surface due to the biofilm. Subsequent studies at ICMM coupled these same electrodes directly via a series of resistances which enabled an estimation of the E/I curve associated with this phenomenon. Figure 7 shows the changing current associated with 5 decades of resistances which essentially demonstrate the evolution of the oxygen reduction curve as a function of exposure time.

4. Discussion

The use of stainless steel in marine environments is a choice of significant engineering and economic interest. Tuthill indicates that among all the potential materials available such as copper–nickel alloys, brasses, monel, austenitic and duplex alloys that are deployed in the sea, the AISI 316 and 316L grades are the most widely used [6]. It is unfortunate therefore, that these particular grades are subject to crevice corrosion within a very short space of time [7–8]. Incremental increase in the alloy content to match the service environment is logical under conditions where the environment is constant and known. Currently however, the contribution of microorganisms to marine corrosion is still an unknown quantity.

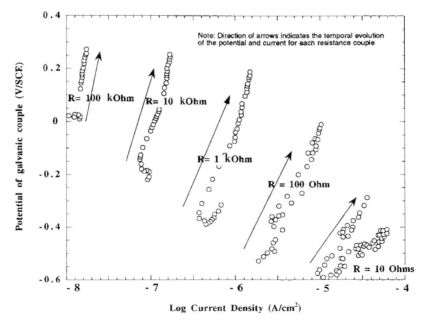

Fig. 7 Galvanic current between stainless steel and an iron anode as a function of the coupling resistance.

The MAST-2 project has currently examined the performance of the three stainless steels in Table 1. These have been some of the most highly alloyed grades (SB8, 654SMO and SAF2507) and are not expected to corrode in seawater under normal conditions (i.e. without supplemental oxidising chemicals, acids or stray currents). The data presented indicates that these three alloys support well the growth of superficial biofilms. The sites selected for the exposure stations offer a range of conditions which are found in the eastern European coastline. The sites differ not only in sources of water according to depth, temperature, and salinity etc. (Fig. 1, Table 2) but also to some degree in the rate at which the seawater is passed through the systems under test. Despite all the differences between the test sites and methods of electrochemical sampling, the growth of marine biofilms was observed to coincide with a displacement of several hundred millivolts in the noble direction on all the steels tested.

Figure 2 shows that the response between the test sites with respect to one of these stainless steels was reasonably uniform. Thus, in areas of diverse temperature, salt content, pollution and oxygen tension, the short term behaviour of the open circuit potential was quite similar. Equally, there appeared to be no significant difference in the quantity of biomass on the respective surfaces of the different stainless steels (Fig. 6).

Examination of the biofilm content with respect to extra cellular polymer and between sites did not reveal any significant differences from site to site. It was possible however to observe a broad relationship correlating the potential of the electrode to the quantity of EPS detected. The significance of this event is of course open to wide interpretation. It appears likely that the growth of biofilms at high Reynolds numbers will give more compact arrangements with more EPS present [9]. In the short term, as bacteria grow upon 'conditioned' surfaces the expectation is that early rapid protein manufacture reflects a high activity per cell. During this period EPS manufacture would not be favoured. Later on, as the cells approach the immediate nutrient limit they would produce more EPS to enable them to resist exfoliation and extend their ability to trap nutrients from the adjacent seawater (EPS may also act as an ion exchange medium). The production of EPS during the evolution of such biofilms cannot therefore, be considered to be linearly related to cell replication.

The interaction of the biofilms with the metal surface in electrochemical terms, is still quite ambiguous. While many authors have reported the important increase in potential and cathodic current density associated with living biofilms [2–4] the interpretation has eluded systematic testing. Impedance data produced at Kristineberg and Cherbourg have shown what appears to be a relatively high frequency electron transfer process. No supporting data have been obtained however. Several similar impedance investigations by Little and Mansfeld in Pacific ocean seawater with 254SMO have not demonstrated high frequency capacitive arcs. In the conventional sense, such a impedance spectrum would lend itself to interpretation by a porous, but largely inert coating with areas of enhanced conductivity. In terms of a bacterial biofilm it has not at all been established that some limited biofilm areas are specifically conductive to some molecular species that is electroactive at high positive potentials.

Seasonal variation of the biofilms has aptly been described by the variations in both potential and cathodic current (Figs. 3 and 5). While ocean seawater does not largely change in temperature during the year, this is not the case for coastal waters (with the notable exception of deep sea lochs such as the Norwegian and Swedish fjords). In comparative studies such as these where there is no *de facto* control, the separation of the effects of temperature and other seasonal variations such as salinity is impossible. Under these circumstances account must then be taken of a global increase in the rate of biochemical events due to more favourable temperatures and the fact that marine bacteria, for example, grow slightly better under brackish conditions than under full marine strength salinity (personal communication Fritz Widdel).

In addition, equinoctial tides (excepting the Mediterranean) and higher precipitation in the winter and spring months affect the corrosion pattern. Only in the northern European sites for reasons perhaps of depth and perhaps lower turnover rates was there less of an effect of the season.

4.1. Ennoblement of Stainless Steels

A considerable discrepancy has, however, been discovered between the effect of the biofilms on the potential of the stainless steels and the rise in current associated with the same biofilms when coupled to an iron or carbon steel anode. While the attachment of significant quantities of bacteria on these surfaces is clearly in the order of hours (Fig. 4 shows a biofilm of 3 days immersion), the potential increase occurs after perhaps 100 h. In contrast, the rise in current of the coupled electrodes occurs several days afterwards with the sole exception of Genova where the increase in current was very rapid (Fig. 5). These observations imply that there is a separate event to accelerate the cathodic half-reaction on the stainless steel from that process which elevates the free corrosion potential. Given these observations, the supposition becomes more credible that the biofilm electrochemically isolates the metal surface from the seawater and permits a very weak current to maintain the elevated potential.

While the current data suggest that the effect of marine biofilms upon passive stainless steels appear to be equal, this is perhaps not the case under active corrosion conditions. The next series of tests in the MAST-2 project will try to establish the relationship between the nature and quantity of the biofilm and the initiation and propagation of each alloy in crevice corrosion conditions. Highly alloyed materials appear to be subject, under some conditions [9], to microbially influenced corrosion.

5. Conclusions

Despite the errors implicit in all round robin-type tests the obtained data support the following conclusions:

1. The ennoblement of passive stainless steels was observed at all exposure sites and

without exception. This phenomenon was apparently independent of the thickness of the biofilm.

2. The dispersion of EPS results among other biochemical tests between sites was probably due to differences in sampling and exposure techniques and not due to any real biochemical difference in microflora between sites.

3. No differences in alloy performance was observed, whether by biofilm accumulation or potential fluctuations (corrosion was never observed).

4. A strong correlation of the variation in potential and galvanic coupling current was observed with respect to season. This effect was, however, limited at the Mediterranean site.

References

1. E. Hill and Shennan, *Microbial Problems in the Offshore Oil Industry*, Watkinson Press, 1987.
2. A. Mollica, A. Trevis, E. Traverso, G. Ventura, V. Scotto, G. Alabiso, C. Marcenaro, U. Montini, G. DeCarolis and R. Dellopiano, 'Interaction between biofouling and oxygen reduction rate on stainless steel in seawater', in *Proc. 6th Int. Congr. on Marine Corrosion and Fouling*, pp. 269–281, Athens, 1984.
3. S. C. Dexter and G. Y. Gao, 'Effect of seawater biofilms on corrosion potential and oxygen reduction of stainless steel', Article 377, *Corrosion '87*. Proc. Conf., NACE, Houston, TX, 1987.
4. R. Johnsen and E. Bardal, 'The effect of a microbiological slime layer on stainless steel in natural seawater', *Corrosion '86*, NACE, Houston, TX, 1986.
5. J. D. H. Strickland and P. R. Parsons, *A practical handbook of seawater analysis*. Fisheries research board of Canada, Bulletin 167. Ottawa, 1972.
6. A. Tuthill, 'Usage and performance of nickel stainless steels in natural waters and brines', Paper No. 408, *Corrosion '88*. Proc. Conf., NACE, Houston, TX, 21–25 March, 1988.
7. E. B. Shone, R. E. Malpas and P. Gallagher, 'Stainless steels as replacement materials for copper alloys in seawater handling systems', *Trans. I. Mar. E.*, **100**, 193–206.
8. J. Charles, J-P. Audouard, F. Dupoiron, J. M. Lardon, P. Soulignac and D. Catelin, Paper No. 116, *Corrosion '89*, Proc. Conf., NACE, Houston, TX, 17–21 April, 1989.
9. M. W. Mittelman, D. E. Nivens, C. Low and D. C. White, Differential adhesion, activity and carbohydrate: protein ratios of *Pseudomonas atlantica* monocultures attaching to stainless steel in a laminar shear gradient, *Microb. Ecol.*, 1990, **19**, 269–278.
10. P. J. B. Scott, J. Goldie and M. Davies, 'Ranking alloys for susceptibility to MIC — a preliminary report on high-Mo alloys', *Mat. Performs.*, January, 1991.

16

Detection and Characterisation of Biofilms in Natural Seawater by Analysing Oxygen Diffusion under Controlled Hydrodynamic Conditions

A. AMBARI*, B. TRIBOLLET*, C. COMPERE,
D. FESTY and E. L'HOSTIS

IFREMER, Centre de Brest, BP 70, 29280 Plouzane, France
*UPR 15 CNRS, Physique des Liquides et Electrochimie, 4 Place Jussieu 75252 Paris Cedex 05, France

ABSTRACT

The laboratories involved in this work have conducted experimental work on the analysis of oxygen mass transport for characterising the biofilm formed on passive metal electrodes. This technique is based on the analysis of the oxygen reduction current when a rotating disk electrode is polarised to a potential corresponding to the diffusion plateau of oxygen. The variation of the limiting current with time (in days) is due to the biofilm formation and hence to the modification of the mass transport properties through it. For the steady flow condition, a mathematical model is proposed and the agreement between experimental data and fitted curves is excellent. This fitting procedure gives a parameter that is proportional to the biofilm thickness over oxygen diffusion coefficient ratio (δ_f/D_f). This experimental procedure conducted for different immersion times shows that it is possible to follow, *in situ*, the biofilm development on a gold electrode as well as on a stainless steel electrode immersed in natural seawater. The detection threshold is 24 h. After developing the basic theory, this paper presents the results for gold electrodes. The thickness of a biofilm developed in natural seawater has been calculated as well as the oxygen diffusion coefficient in the biofilm. The results are in agreement with the current knowledge on biofilm properties.

1. Introduction

Our first results [1] established that the method of using the dissolved oxygen in seawater to characterise the biofilm is efficient. We have shown that the measured steady state current due to oxygen reduction ($O_2 + 2\,H_2O + 4e^- \rightarrow 4OH^-$) on a fouled stainless steel rotating disk electrode polarised at -750 mV in natural seawater can be fitted by the following law. The potentials are expressed vs Ag/AgCl/seawater reference electrode.

$$I = I_0 + \frac{1}{I_{\Omega \rightarrow \infty}^{-1} + I_L^{-1}} \tag{1}$$

$$I_{\Omega \to \infty} = \frac{nFSD_f C_\infty}{\delta_f} \quad (2)$$

$$I_L = K\Omega^{0.5} \quad (3)$$

$$K = 0.62nFS\sqrt{\frac{2\pi}{60}} D^{\frac{2}{3}} v^{-\frac{1}{6}} C_\infty \quad (4)$$

where I_0 is the hydrogen reduction current, $I_{\Omega \to \infty}$ is a parameter characterising the biofilm, proportional to the layer permeability D_f/δ_f, I_L is the Levich current, δ_f is the biofilm thickness, D_f is the coefficient of diffusion through the biofilm, C_∞ is the oxygen content of seawater, K is the Levich coefficient, Ω is the disk rotation speed, S is the electrode area ($S = 0.2$ cm^2), F is the Faraday constant, n is the number of electrons, D is the coefficient of diffusion in the electrolyte and v is the kinematic viscosity.

The biofilm is taken as a porous layer and the system under investigation is schematised in Fig. 1. The oxygen concentration gradient is distributed between the fluid and the porous layer (biofilm). Equation (1) expresses the mass balance equations in the porous layer where the concentration distribution is only determined by the molecular diffusion and in the electrolytic solution where the concentration distribution is governed by the convective diffusion.

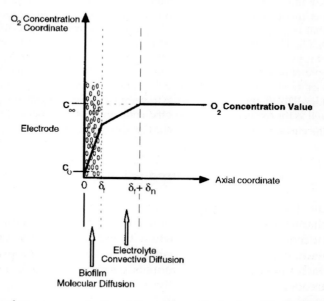

Fig. 1 *Variation of oxygen concentration vs the axial coordinate where δ_f is the biofilm thickness, δ_n is the diffusion layer thickness, c_∞ is the concentration of the electroactive species in the bulk solution and c_o the concentration of the electroactive species in the biofilm.*

To simplify the discussion we will write the expression as follows:

$$I = A + \frac{1}{B + C\Omega^{-0.5}} \tag{5}$$

where $A = I_0$, $B = I_{\Omega \to \infty}^{-1} = cst. \delta_f / D_f \ S^{-1}$

These three different parameters A, B and C can be fitted from the experimental results.

The first experimental results showed scattered values. Nevertheless, the amount of measurements performed indicated that the parameters A and C seem to be independent of time exposure (e.g. the biofilm formation) and B seems to increase with exposure time.

The electrohydrodynamical impedance (EHD) diagrams ($Z_{EHD} = \frac{\Delta I}{\Delta \Omega}$) have been drawn in the same conditions. This technique enables calculation of the diffusion time through the biofilm [2] (δ_f^2 / D_f) and with the previous technique, the calculation of the two parameters: biofilm thickness δ_f and diffusion coefficient D_f. The impedance diagrams are slightly shifted as a function of time and rotation speed indicating a small value for the diffusion time.

These results, which were obtained on stainless steel rotating disk electrodes during the summer, showed the feasibility of the steady state measurement technique to detect and monitor biofilm development *in situ*. Thus, we continued our experimental work with an inert electrode. A gold electrode was used.

2. Experimental Techniques

The disk electrode was fixed to a rotating mercury contactor driven by a direct current motor (Fig. 2). The motor speed control was by means of a tachometer. When steady state experiments were performed, the control unit input was connected to a ramp generator (Fig. 2). The rotating speed and oxygen reduction current measurements were performed by means of a digital voltmeter and a computer. For EHD impedance measurements, the speed modulation was by means of the generator of the transfer function analyser unit (Fig. 3), the constant rotation speed was fixed by the control unit. The oxygen reduction current, the rotation speed measurements and the impedance calculation were performed by the transfer function analyser unit.

The biofilm was developed on eighteen-carat gold electrodes immersed in natural seawater from the Atlantic Ocean. The water was continuously renewed, salinity was 35 g L^{-1} temperature was 15°C, oxygen content 8.1 ppm (i.e. 2.5 10^{-7} mol.cm^{-3} at equilibrium with air at the test temperature and atmospheric pressure), pH was 8.2. The water surface was exposed to daylight. Before immersion, the electrodes were polished using alumina powder (3 μm).

To reduce the contribution of kinetic reaction, the electrode was polarised to a potential within the limits of the oxygen limiting diffusion plateau. Figure 4 presents

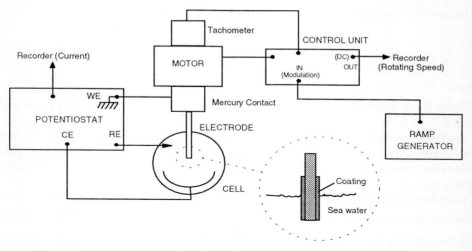

Fig. 2 *Steady state measurement. Experimental technique principle.*

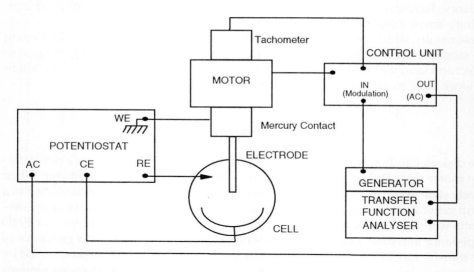

Fig. 3 *Electrohydrodynamic impedance. Experimental technique principle.*

the polarisation curves obtained with bare gold electrodes in testing conditions at different rotation speeds. For each rotation speed, potentiodynamic polarisation curves were recorded in the cathodic direction from the corrosion potential up to –1200 mV and then in the reverse direction. It can be seen that the oxygen limiting diffusion plateau is well defined and that there is effectively no difference between the forward and reverse sweeps. The hydrogen current clearly appears below –1000 mV. On the basis of these curves, the experiments were performed at –900 mV. At this potential, a current corresponding to hydrogen reduction is added to the oxy-

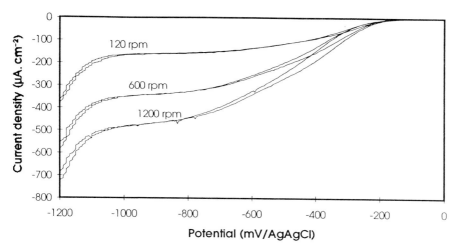

Fig. 4 Polarisation curves on bare gold electrode in natural seawater.

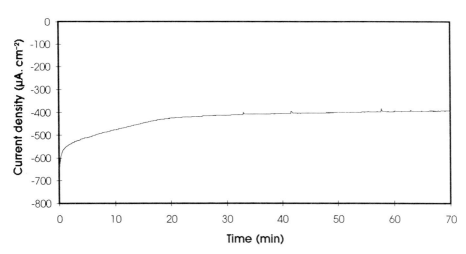

Fig. 5 Reduction current measurement vs exposure time in natural seawater at 1200 rpm. Potential: −900 mV vs AgAgCl/seawater reference electrode.

gen reduction current. This non-diffusional component corresponds to the parameter A in expression (5).

The current measurement vs time for a bare electrode polarised at this potential value is plotted in Fig. 5 for a constant rotating speed of 1200 rpm. After 20 min, the oxygen reduction current was stable and an experiment could be run by modulating the rotation speed.

3. Results
3.1. Steady State Current Measurement

After polarising the electrode at –900 mV for 20 min, the steady state current was recorded vs the square-root of the angular speed. The record presented in Fig. 6 is obtained on a gold electrode without and with a 14 days biofilm. The curve obtained without biofilm is a straight line with an intercept at $\Omega = 0$ near the zero value. This means that the contribution of the hydrogen reaction is very low. The curve obtained with a 14 days biofilm is no longer a straight line but has a clear curvature in agreement with expression (5). A fitting procedure was used to determine the parameters A, B and C for each set of experiment. In Fig. 7 $(I-A)^{-1}$ is plotted vs $\Omega^{-1/2}$. The results correspond to those shown in Fig. 6 and the fitted parameters are:

	$A (\mu A)$	$B (\mu A^{-1})$	$C (\mu A^{-1} \cdot rpm^{1/2})$
Without biofilm	1.3	0	0.2
14 days biofilm	4.1	$1.4 \cdot 10^{-3}$	0.2

The difference between calculated currents from fitted parameters using expression (5) and experimental values are plotted in Fig. 8. We can see that the agreement is excellent, so that the proposed model can be used to analyse the steady state current.

The next step of the study consisted in immersing three electrodes in natural seawater and performing steady state measurements every day. The three parameters were calculated and were plotted versue time in Figs 9–11.

In Fig. 9, the hydrogen reduction current (parameter A) is rather scattered. This

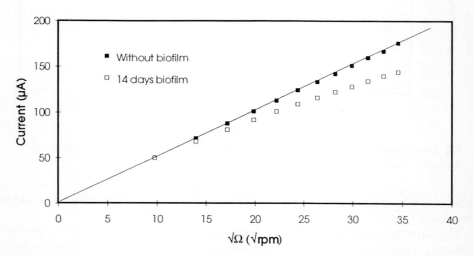

Fig. 6 Steady state measurement on gold electrode in natural seawater, without biofilm and with a 14 days biofilm. The solid line is the fitted Levich law.

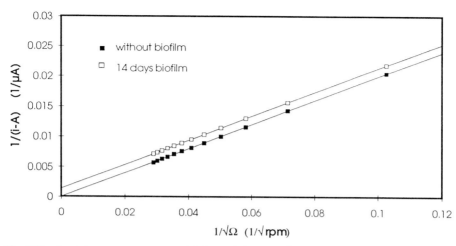

Fig. 7 Steady state measurement on gold electrode in natural seawater, without biofilm and with a 14 days biofilm. The solid lines are the fitted values, the marks are the experimental results.

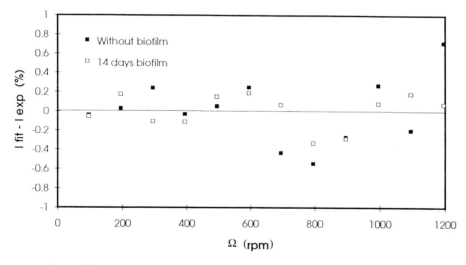

Fig. 8 Difference between experimental and fitted values.

can be due to the fact that the hydrogen reduction current is an exponential function of the potential, and thus, a small variation of the potential results in a large variation of current. On the other hand, the pH value at the interface may fluctuate resulting in a variation of hydrogen overpotential. We can note that within 5 days the average value is 2.6 ± 1.0 µA and after 8 days, 4.7 ± 1.0 µA.

In Fig. 10, we can see that the parameter B is small and almost constant around $6 \pm 3 \times 10^{-4}$ µA^{-1} within 5 days and gradually increases up to $23 \pm 6 \times 10^{-4}$ µA^{-1} around 20 days.

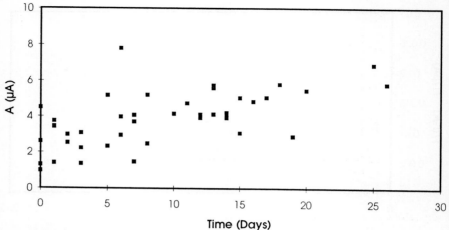

Fig. 9 Parameter A *value vs time exposure.*

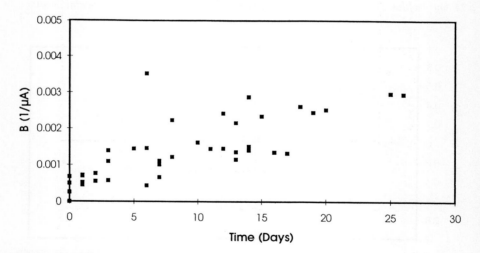

Fig. 10 Parameter B *value vs time exposure.*

In Fig. 11, the parameter C is constant for almost the totality of the exposure period (0.20 ± 0.01 μA^{-1}. $rpm^{1/2}$). The scatter is low except between 3–8 days.

3.2. Electrohydrodynamical Impedance

We first checked the reducibility of data by the dimensionless frequency $p = \omega/\Omega$ (where ω is the pulsation) on bare electrodes. The results show the validity of the hypothesis of diffusion limitation of oxygen reduction by the diffusion through the electrolyte (Fig. 12).

Detection and Characterisation of Biofilms in Natural Seawater

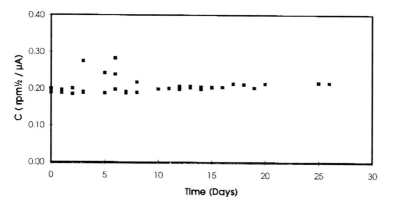

Fig. 11 Parameter C value vs time exposure.

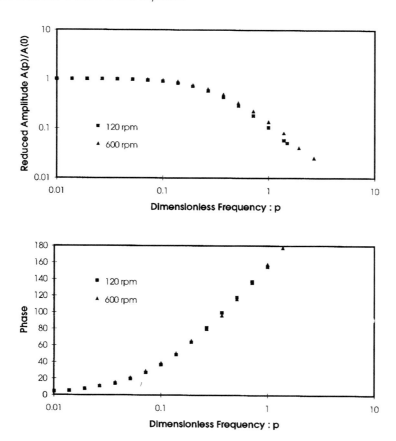

Fig. 12 Electrohydrodynamic impedance Bode diagrams of a bare gold electrode in natural seawater at different rotation speeds. $p = \dfrac{\omega}{\Omega}$.

In Fig. 13, the EHD diagrams obtained on an electrode immersed 10 days show that the data are not perfectly reducible by dimensionless frequency p. There is a slight but significant shift towards low frequency when the rotation speed increases from 120 to 1200 rpm. This systematic effect was confirmed by plotting the EHD diagrams recorded at 600 rpm on a bare electrode and after 10 days exposure (Fig. 14). This shift is due to the presence of a porous layer (i.e. a biofilm).

4. Discussion

From the steady state measurements obtained on gold electrodes it is possible to detect the biofilm formation in natural seawater.

From the Levich equation (4), C^{-1} is directly proportional to the oxygen content (C_∞) and reactive area. The results showed that C is constant with immersion time,

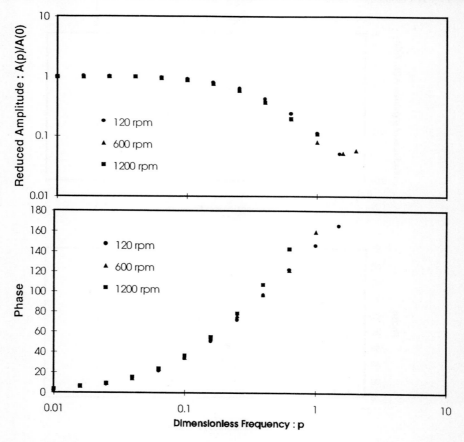

Fig. 13 Electrohydrodynamic impedance Bode diagrams of a gold electrode exposed 10 days in natural seawater at different rotating speeds. $p = \dfrac{\omega}{\Omega}$.

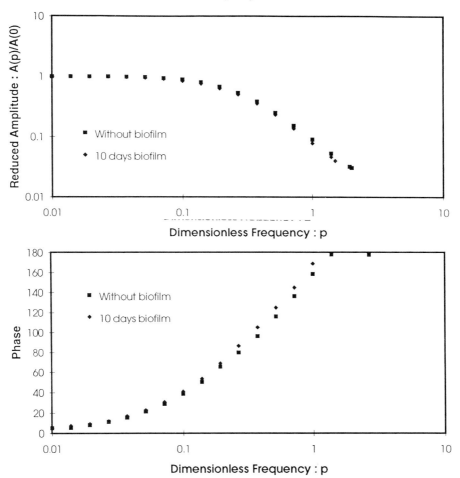

Fig. 14 Electrohydro dynamic impedence Bode diagrams of a gold electrode in natural seawater, without biofilm and with a 10 days biofilm, at 600 rpm. $p = \dfrac{\omega}{\Omega}$.

meaning that the reactive area is not significantly affected by biofilm growing. Taking C equal to 0.2 µA^{-1} rpm$^{1/2}$, the coefficient of diffusion [3] D in the electroyte is equal to 1.65 10^{-5}cm^2 s^{-1} and v is equal to 10^{-2} cm^2.s^{-1}, C_∞ calculated from the parameter C is equal to 2.4 10^{-7} mol cm^{-3}. This value is quite comparable to the value experimentally determined and is used to calculate the ratio δ_f/D_f from the equation of parameter B:

$$B = \frac{\delta_f}{nFD_f C_\infty S} \qquad (6)$$

Taking $B = 2.10^{-3}$ µA^{-1} (10 days), we obtained $\delta_f/D_f = 35$ cm^{-1} s.

The slight shift shown on EHD diagrams means that the diffusion time is smaller than unity: $2\pi f \cdot (\delta_f^2 / D_f) < 1$. Taking f equal to the higher frequency (f = 10 Hz) and the previously calculated value of δ_f/D_f (35 cm^{-1} s), we obtained a maximum biofilm thickness of 4 µm and the order of magnitude of oxygen diffusion coefficient is the same in the biofilm as in seawater ($D_f \approx 1.0\ 10^{-5}$ cm^2 s^{-1}). These values are in agreement with the current knowledge on biofilm properties [4] and allow one to calculate, with the expression [5], (7):

$$D_f = D\varepsilon^{1.5} \tag{7}$$

where ε is the porosity — a porosity value around 0.7 corresponds to a very porous film.

5. Conclusion

From the results of the experiments carried out with a gold electrode, it appears that the use of flow modulation techniques to detect *in situ* a biofilm development is acceptable. These techniques are based on oxygen diffusion analysis through the biofilm, considering it as a non reactive and solid porous layer. The proposed mathematical model describes with a very good accuracy the steady state measurement results that can be used to calculate the contribution of hydrogen reduction current, biofilm thickness and oxygen diffusion coefficient values. The results are in good agreement with the current knowledge on biofilm properties. Nevertheless, the proposed mathematical model may need to be correlated to a hypothesis that is less simple than a simple porous and rigid layer. We may have to consider the possibility of chemical reaction of oxygen with the biofilm as well as the presence of a non solid film having the mechanical properties of a gel.

6. Acknowledgements

Our thanks go to Lyonnaise Des Eaux CIRSEE for financial help.

References

1. D. Festy, F. Mazeas, M. El Rhazi and B. Tribollet, 'Characterization of the biofilm formed on a steel electrode in seawater by analysing the mass transport of oxygen', *Proc. 12th Int. Corros. Congr.*, NACE, Houston, TX, USA, 19–24 September, 1994, **5B**, 3717–3725.
2. C. Deslouis, B. Tribollet, M. Duprat and F. Moran, 'Transient mass transport at a coated rotating disk electrode', *J. Electrochem. Soc.*, 1987, **134**, 2496–2501.
3. C. Deslouis, O. Gil, B. Tribollet and G. Vlachos, 'Oxygen as tracer for management of studies and turbulent flows', *J. Appl. Electrochem.*, 1992, **22**, 835–842.
4. J. V. Matson and W. G. Characklis, 'Diffusion into microbial aggregates', *Water Res.*, 1976, **10**, 877–885.
5. A. C. West, 'Comparison of Modelling Approaches of Porous Salt Film', *J. Electrochem. Soc.*, 1993, **140**, 403–408.

17

Determination of Biofilm on Stainless Steel in Seawater in Relation to the Season by Analysing the Mass Transport of Oxygen

B. TRIBOLLET*, C. COMPERE, F. DARRIEUX and D. FESTY

IFREMER, Centre de Brest, BP 70, 29280 Plouzane, France
*UPR 15 CNRS, Physique des Liquides et Electrochimie, 4 Place Jussieu 75252 Paris Cedex 05, France

ABSTRACT

Steady state measurements of the oxygen reduction current vs the rotation speed of a stainless steel disk electrode have been carried out as a function of immersion time in natural flowing seawater. They yield information on the active electrode area and on transport properties from the oxygen diffusion rate through the biofilm. This method allowed the biofilm growth on a stainless steel electrode immersed in seawater to be followed. The effect of the season on biofilm growth is discussed.

1. Introduction

Very promising results were obtained on gold electrodes [1] concerning the detection and the characterisation of biofilms in natural seawater by analysing oxygen diffusion under controlled hydrodynamic conditions. Earlier results [2] with stainless steel electrodes are now revised taking into account the effect of the season. With stainless steel the interpretation is more complicated than with an inert material and will necessitate complementary experiments with impedance techniques in order to separate, ata fixed cathodic potential, the different components of kinetic and limiting flux of the overall current.

2. Experimental Techniques

The basic experimental conditions and devices are similar to those described in Ref. [1]. The stainless steel was an Uranus SB6 (Cr 23, Ni 25, Mo 3%, bal. Fe). SB6 is a trade name of Creusot Loire Industrie. The active area was 0.5 cm^2. The experiments were performed in a continuously renewed natural seawater and for two exposure periods, in winter and summer seasons. The physico–chemical parameters were as follows: temperatures of 11°C in winter and 15°C in summer, a relatively constant pH around 8.2, a salinity around 35 g L^{-1} and oxygen contents of 3.1 10^{-7} mol cm^{-3} in winter and 2.6 10^{-7} mol cm^{-3} in summer. The samples are polished with emery paper down to 1200 grit, cleaned in alcohol using an ultrasonic bath and air dried.

The cathodic reduction current values of four stainless steel electrodes were recorded after 30 min of polarisation at –750 mV(Ag/AgCl/seawater). Only the fourth cycle of current vs angular speed measurements is considered. The angular speed varied from 120 up to 1600 rpm. The evolution of the cathodic reduction current was followed for sixteen days for both exposure periods.

3. Results

Figure 1 confirms the generally described evolution of the stainless steels free corrosion potential in natural seawater [3]. This figure presents the behaviour of three different electrodes in continuously renewed natural seawater during the summer exposure period. The reasons and mechanisms of the free corrosion potential ennoblement associated with the biofilm are not well understood [4, 5].

For the experiments performed on gold electrodes, the limiting cathodic current measurements were analysed using the following expression (Ref. [1]):

$$I = I_0 + \frac{1}{I_{\Omega \to \infty}^{-1} + I_L^{-1}} \tag{1}$$

$$I_{\Omega \to \infty} = \frac{nFSD_f C_\infty}{\delta_f} \tag{2}$$

$$I_L = K\Omega^{0.5} \tag{3}$$

$$K = 0.62 nFS \sqrt{\frac{2\pi}{60}} D^{\frac{2}{3}} v^{-\frac{1}{6}} C_\infty \tag{4}$$

where I_0 is the hydrogen reduction current, $I_{\Omega \to \infty}$ is a parameter characterising the biofilm, proportional to the layer permeability D_f/δ_f, I_L is the Levich current, δ_f is the biofilm thickness, D_f is the coefficient of diffusion through the biofilm, C_∞ is the oxygen content of seawater, K is the Levich coefficient, Ω is the disk rotation speed, S is the electrode area, F is the Faraday constant, n is the number of electrons, D is the coefficient of diffusion in the electrolyte and v is the kinematic viscosity.

For stainless steel, the analysis of steady state measurements cannot be made in the same manner as for the gold electrodes. In fact, as shown on the typical polarisation curve obtained at 600 rpm with a scanning rate of 0.083 mV s^{-1} (Fig. 2), it appears that the reduction current measurements vs the angular speed do not take place on a diffusion plateau for a polarisation value of –750 mV(AgAgCl/seawater). The diffusion plateau is not as well defined as in the case of gold electrodes. A slight diffusion plateau only occurs on the reverse scan from –1100 mV up to the free corrosion potential after cleaning of the active surface brought about by hydrogen evolution.

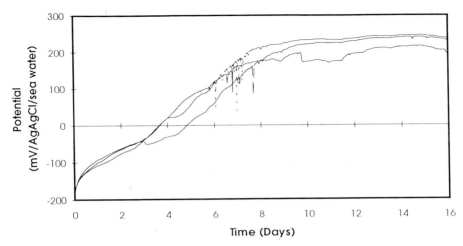

Fig. 1 Time dependence of the free corrosion potential of a stainless steel electrode in natural seawater.

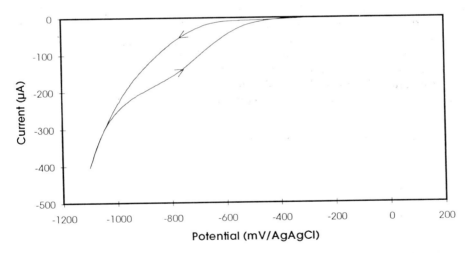

Fig. 2 Polarisation curves for a stainless steel electrode free from biofilm in natural seawater. Disk rotation speed: 600 rpm; scan rate 0.083 mV s^{-1}; active area: 0.5 cm^2.

The reduction current has been plotted vs time for a bare electrode polarised at −750 (mV/AgAgCl/seawater) and at a constant rotating speed of 1200 rpm (Fig. 3). After 45 min, the reduction current is not stable and continues to increase. However, the current measurement vs angular speed lasted about 4 min, a time during which the current may increase by about 6%. This value is at a maximum at the beginning of the immersion.

Hence the kinetic flux of current I_K, given by the charge transfer reaction assuming that the mass transfer is infinitely fast, cannot be neglected. The expression (1) becomes:

$$I = I_0 + \frac{1}{I_K^{-1} + I_{\Omega \to \infty}^{-1} + I_L^{-1}} \quad (5)$$

$$I_K = kC_\infty \exp b\eta \quad (6)$$

where I_K = kinetic flux, b = Tafel coefficient, k = kinetic parameter, and η = overpotential.

To simplify the discussion we will write the expression as follows:

$$I = A + \frac{1}{B' + C\Omega^{-0.5}} \quad (7)$$

where $A = I_o$, $B' = I_{\Omega \to \infty}^{-1} + I_K^{-1}$, $C = K^{-1}$ and B' is not directly proportional to the diffusion rate of oxygen through the biofilm as in the case of gold electrodes.

The proposed mathematical model is based on the growth of a non-reactive and solid porous layer. The agreement between experimental values of the cathodic current and these calculated with eqn (7) is very good as shown in Fig. 4 using the fitted and experimental values of $(I-A)^{-1}$ vs $\Omega^{-0.5}$ obtained during the summer period. However, Fig. 5 shows the curves of the cathodic reduction currents vs the inverse square root of angular speed obtained during summer for one particular electrode. It appears that after 12 days of immersion the mathematical model does not apply any more and that the plot of $(I-A)^{-1}$ vs $\Omega^{-0.5}$ is no longer a straight line parallel to the

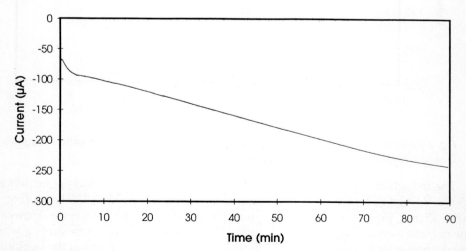

Fig. 3 *Reduction current measurement vs exposure time on a stainless steel electrode free from biofilm in natural seawater at 1200 rpm. Potential: –750 mV vs Ag/AgCl/seawater reference electrode; active area: 0.5 cm².*

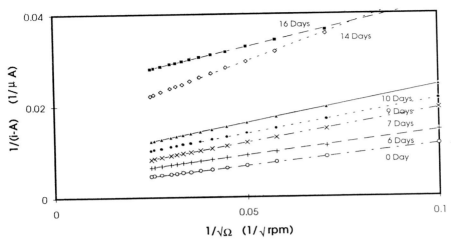

Fig. 4 Steady state measurement vs exposure time on a stainless steel electrode in natural seawater during summer period. The solid lines are fitted values, the ciphers are the experimental results.

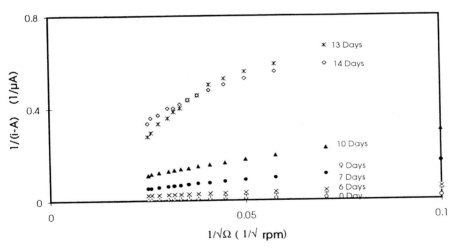

Fig. 5 Cathodic reduction currents vs the inverse square root of angular speed during summer exposure period for one specific electrode.

Levich linear variation. This could be due to the non-homogeneous and random growing of the biofilm. A biological study of the biofilm is in progress. The curves of the cathodic reduction currents following eqn (7) are of course only considered. Then the quadratic mean deviation between experimental and calculated values was in all cases below 0.5, which is very good.

After a fitting procedure with the experimental results, following the considerations described above, the three parameters A, B' and C were evaluated as a function of time for both exposure periods. Figures 6–8 give the evolution of these parameters, the plots (a) characterise the winter period and the plots (b) the summer period.

The hydrogen reduction current values (values A, Fig. 6) show scatter, with a scattering range five times more significant than on gold electrodes. Since four different stainless steel electrodes are used for each set of experiments, the difference in active area could partly explain this scattering. Nevertheless, these values seem to be independent of exposure time and seasons with a mean value around 50 µA cm^{-2}, which is considerably higher than the mean value of 12.5–15 µA cm^{-2} obtained on gold.

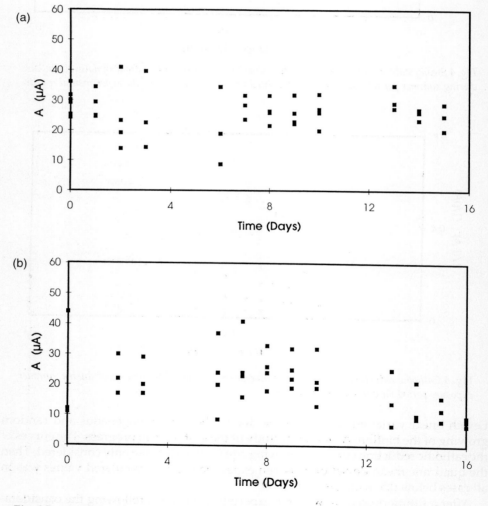

Fig. 6 Parameter A value vs time during (a) winter and (b) summer period exposure.

A gradual increase in the B' values is observed as a function of time for both exposure periods (Fig. 7). However, the summer values are higher than the winter values by a factor of ten. By considering the B' values obtained on stainless steel electrodes free from biofilm (i.e. at $t = 0$) it becomes possible to determine the kinetic flux of current I_K. If we assume that the kinetic flux of current does not vary significantly with immersion time, eqn (2) and (7) permit the development of $\dfrac{\delta_f}{D_f}$ with time to be followed directly.

The C values are relatively constant and similar in winter periods and during the first eight days of immersion for the summer period. However, the values and the scattering both increased after 8 days of immersion during summer (Fig. 8).

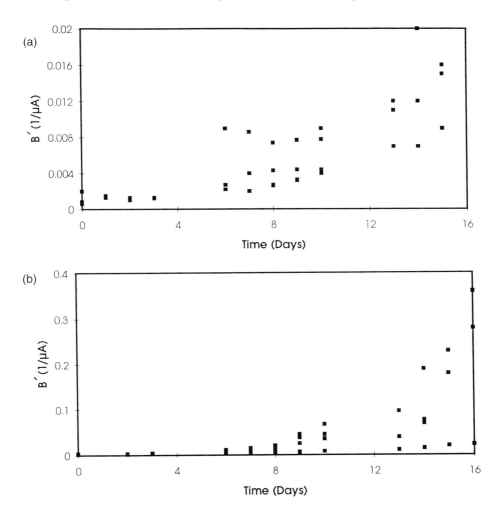

Fig. 7 Parameter B' value vs time during (a) winter and (b) summer period exposure.

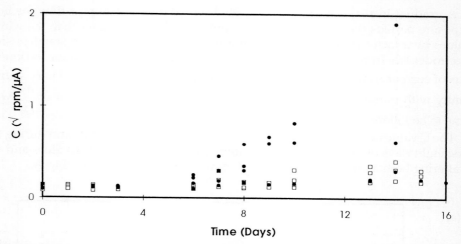

Fig. 8 *Parameter C value vs time during winter (□) and summer (•) period exposure.*

From the Levich equation, the expression of parameter C is:

$$C = (0.62 nFS\sqrt{\frac{2\pi}{60}} D^{2/3} \nu^{-1/6} C_\infty)^{-1} \qquad (8)$$

where D = diffusion coefficient of the electroactive species in the electrolyte, ν = kinematic viscosity, F = Faraday number, C_∞ = oxygen content in seawater, n = number of electrons and S = electrode area.

C^{-1} is directly proportional to the oxygen content and reactive area. The mean values of C are 0.17 and 0.19 rpm$^{0.5}$ (µA)$^{-1}$ for the winter and for the first eight days of exposure in the summer period respectively. From these results and from the experimental values of C_∞ = 3.1 10^{-7} mol cm^{-3} in winter and C_∞ = 2.6 10^{-7} mol cm^{-3} in the summer periods, and from D = 1.65 10^{-5} cm^2 s^{-1}, ν = 10^{-2} cm^2 s^{-1}, it appears that only half of the surface is reactive.

Combining eqns (2), (7) and (8), we can write:

$$\frac{B' - I_K^{-1}}{C} = \frac{0.62\sqrt{\frac{2\pi}{60}} D^{2/3} \nu^{-1/6} \delta_f}{D_f} = 2.8 \times 10^{-4} \frac{\delta_f}{D_f} \qquad (9)$$

and plot the value $\dfrac{\delta_f}{D_f}$ as a function of time for both exposure periods (Fig. 9). We observe that these values are of the same order of magnitude in winter and summer periods. This allows us to envisage the growing of a dense and relatively thin biofilm during winter compared with a thicker and looser biofilm during summer.

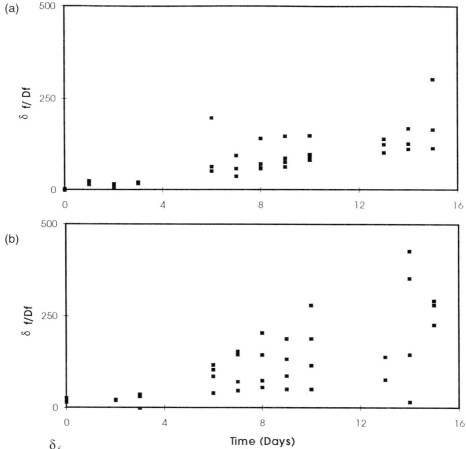

Fig. 9 $\frac{\delta_f}{D_f}$ value vs time during (a) winter and (b) summer period exposure.

4. Conclusion

Steady state measurements of the oxygen reduction current vs the rotation speed of a stainless steel disk electrode allowed the evolution of the oxygen diffusion through a biofilm as a function of immersion in natural seawater to be followed. It appears that the oxygen diffusion is of the same order of magnitude irrespective of the seasons. This technique also allows us to determine that only half of the surface is reactive during the winter and during the first days of exposure in the summer period. However, results performed on stainless steel electrodes presented a more important scattering range than on gold electrodes and no longer followed the proposed mathematical model for the case of a thick biofilm, i.e. as obtained after ten days of immersion during a summer period. In these particular conditions, the model has to be reviewed in terms of a non-solid and reactive layer.

References

1. A. Ambari, B. Tribollet, C. Compère, D. Festy and E. L'Hostis, 'Detection and Characterisation of Biofilms in Natural Seawater by Analysing Oxygen Diffusion under Controlled Hydrodynamic Conditions', *Proc. 3rd European Federation of Corrosion Workshop on Microbial Corrosion,* Estoril, Portugal, 13–16 March 1994, this volume, pp. 211–222.
2. D. Festy, F. Mazeas, M. El Rhazi and B. Tribollet, 'Characterization of the Biofilm Formed on a Steel Electrode in Seawater by Analyzing the Mass Transport of Oxygen', *Proc. 12th Int. Cong.,* NACE, Houston, TX, USA, 19–24 September, 1994, **5B**, 3717–3725.
3. J. P. Audouard, C. Compère, N. J. E. Dowling, D. Feron, D. Festy, A. Mollica, T. Rogne, V. Scotto, U. Steinsmo, K. Taxen and D. Thierry, 'Effect of Marine Biofilms on High Performance Stainless Steels exposed in European Coastal Waters', *Proc. 3rd European Federation of Corrosion Workshop on Microbial Corrosion,* Estoril, Portugal, 13–16 March, 1994, this volume, pp. 198–210.
4. P. Chandrasekaran and S. C. Dexter, 'Factor Contributing to Ennoblement of Passive Metals due to Biofilms in Seawater', *Proc. 12th Int. Congr.,* NACE, Houston, TX, USA, 19–24 September, 1993, **5B**, 3696–3707.
5. A. Mollica, G. Ventura and E. Traverso, 'On the Mechanism of Corrosion Induced by Biofilm Growth on the Active–Passive Alloys in Seawater', *Int. Congr. on Microbially Influenced Corrosion* (Eds N. J. E. Dowling, M. W. Mittleman and J. C. Danka), Knoxville, TN, 7–12 October, 1990, Ch. 2, 25–31.

18

The Importance of Bacterially Generated Hydrogen Permeation Through Metals

J. BENSON and R. G. J. EDYVEAN*

Company Research Laboratory, BNFL, Springfields Works, Preston, PR4 0XJ, UK
*Department of Chemical Engineering, The University of Leeds, Leeds, LS2 9JT, UK

ABSTRACT

Biological fouling can occur in any system where seawater is present, even when in very small quantities. It is therefore, of great importance to any industry where cooling water and heat exchange facilities are required and especially in the oil and gas industry where oil/water/gas mixtures, providing both a suitable environment and a nutrient supply, are common. Microbiological induced corrosion of metals has been well documented.

However, the products of some bacteria, particularly in anaerobic environments, can influence failure of metals by promoting hydrogen damage. This study compares the effect of both biogenic and abiotic hydrogen sulphide on hydrogen permeation through BS4360 50D steel. Differences between these two environments have been noted.

1. Introduction

Biological fouling can occur in any system where seawater is present. Even when only very small quantities of seawater are available bacterial biofilms will develop. It is therefore, of great importance to any industry where cooling water and heat exchange facilities are required and especially in the Oil and Gas industry where oil/water/gas mixtures, provide both a suitable environment and nutrient supply for biofilm development. Biologically enhanced corrosion has been well documented [1, 2]. However, the effect of marine fouling on the permeation of hydrogen through steel has largely been overlooked [3].

Hydrogen atoms produced electrochemically may enter a metallic lattice and permeate through the metal. Such permeation is known to have an extremely detrimental effect upon the mechanical properties of metal as interatomic bonds become damaged causing loss of ductility and tensile strength [4], a phenomenon referred to as hydrogen embrittlement. Hydrogen embrittlement is known to enhance crack propagation and thus, the likelihood of catastrophic failure is increased. Any external supply of monatomic hydrogen is likely to have an effect.

Biotic hydrogen sulphide (H_2S) is produced by sulphate reducing bacteria (SRB) when conditions within a biofilm become favourable for their anaerobic growth. Sulphide ions not only enhance corrosion by stimulating anodic dissolution but also poison the recombination of hydrogen atoms to the gaseous molecule. Thus, more hydrogen is made available for entry into the steel. The damage that results from hydrogen entry and permeation is referred to as hydrogen embrittlement.

To date research involving the hydrogen damage of metals has been performed largely using non-biotic hydrogen sulphide test solutions, for example NACE solution [5]. The susceptibility of materials to hydrogen entry and failure is then assessed under these 'worst' case conditions. However, this is not thought to be particularly representative of conditions offshore, since the levels of H_2S produced are likely to be much lower. In order to assess the effect of biologically produced hydrogen sulphide on the permeation of hydrogen through BS4360 50D steel and on hydrogen damage, the following tests were performed:

2. Materials and Methods
2.1. Preparation of Biologically Active H_2S Environments

Biologically active H_2S environments were created by the natural decomposition of marine algae in artificial seawater (Instant Ocean, Underworld). *Enteromorpha* spp. and *Ulva* spp., were collected from the north-east and south-west coast of England. 100 g (wet weight) of the algal material, together with 100 mL of a previous active SRB culture were placed in 1 L capacity plastic containers, filled with artificial seawater, sealed and allowed to decompose at room temperature. Access was provided for sampling and a water trap prevented any build up of gas.

The H_2S content of each culture was determined at weekly intervals by the iodimetric method [6].

The corrosivity of these biologically active environments was assessed by performing polarisation measurements on working electrodes removed from the cultures [7].

2.2. Preparation of Abiotic H_2S Environments

The abiotic H_2S environments were produced on the day of testing. 0.6 g $Na_2S.9H_2O$ was dissolved in 250 mL of previously deaerated 3% NaCl solution. The pH of this stock solution was then adjusted to pH 7.4. The H_2S content of the solution was then determined (see above) and depending on the test requirements was diluted until a suitable H_2S content was achieved.

2.3. Measurement of Hydrogen Permeation

In order to measure the permeation of hydrogen generated from a biological culture through the steel specimens the following electrochemical technique was used. The method has been developed from a technique first pioneered by Devanathan and Stachurski [7], which enables the instantaneous rate of hydrogen permeation through a metal to be measured.

A hydrogen permeation cell that enables the effect of biogenic hydrogen sulphide in solution to be studied has been constructed (Fig. 1). From a deaerated reservoir, 0.1M sodium hydroxide was pumped through the measuring chamber at a rate of 15 mL/min. An anodic potential of +200 mV was applied to the working electrode and the cell was allowed to stabilise. At this point the charging chamber was filled with the test solution and a cathodic potential of −1000 mV was applied to the working electrode. The resultant ionisation current was recorded with time.

The working electrodes (50 mm × 20 mm) were made from BS4360 50D steel plate, ground to a final thickness of 0.5 mm. Thinner specimens were avoided to ensure that the structural properties of the steel were retained. Prior to insertion the working electrodes were finished to P1200 metallographic grinding paper (Buehler-Met), degreased with Decon 90 surfactant and dried with acetone.

Tests were performed using both biotic and abiotic cultures with a range of H_2S concentrations. Control tests were performed using artificial seawater. Due to the effect of pH on hydrogen permeation [8] the pH of the test solutions used was maintained around pH 7–8.

3. Results

Artificial seawater with no hydrogen sulphide or sulphate reducing bacterial activity provides little hydrogen for permeation and Fig. 2 gives a typical 'base line' curve. This typical permeation curve shows a small, but marked delay before any hydrogen is recorded in the measuring chamber. The hydrogen level than climbs to a low steady state that is sustained by the cathodic reaction taking place on the charging surface.

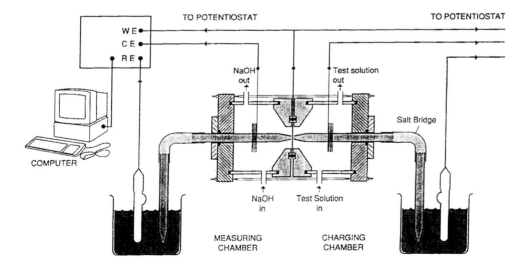

Fig. 1 *Schematic diagram of the hydrogen permeation cell. WE = working electrode; CE = counter electrode; RE = reference electrode.*

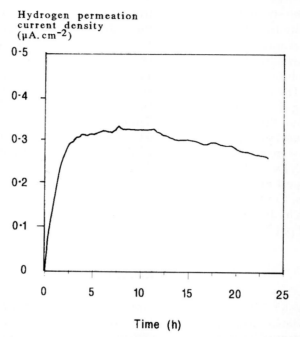

Fig. 2 Hydrogen permeation through BS4360 grade 50D steel in artificial seawater. Test conditions: Specimen thickness = 0.5 mm, charging potential = –1000 mV, pH = 8.

In contrast, when the test solutions contain H_2S the hydrogen permeation through the steel membrane is markedly increased, even at low levels of H_2S. Results for hydrogen permeation from solutions containing around 100 ppm H_2S are given in Fig. 3. Two results are shown, in one (biotic culture) the H_2S has been generated by the action of sulphate-reducing bacteria on decaying seaweed, in the other (abiotic culture) a similar level of H_2S was generated chemically in sterile conditions with no bacterial or seaweed present. Both curves show higher levels of hydrogen permeation than the seawater control. However, the biotic culture shows considerably greater hydrogen permeation and a reduction of lag time than the abiotic conditions. Neither experiment produces the steady state conditions found in the control situation. This may indicate changing conditions at the charging metal surface.

Results for hydrogen permeation maxima are given in Fig. 4 for a range of H_2S levels produced both by biological activity and artificially. In all cases the biologically produced H_2S had greater effect that the equivalent level of H_2S produced artificially.

4. Discussion

Hydrogen sulphide is both toxic and highly corrosive [1]. It can occur naturally but is usually produced *in situ* by bacterial sulphate reduction. Levels of biologically

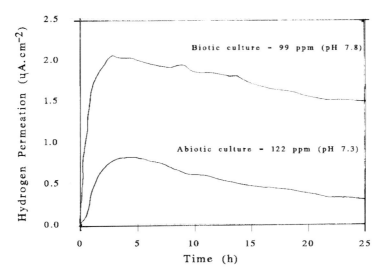

Fig. 3 Hydrogen permeation through BS4360 grade 50D steel at −1000 mV in both abiotic and biotic sulphide cultures.

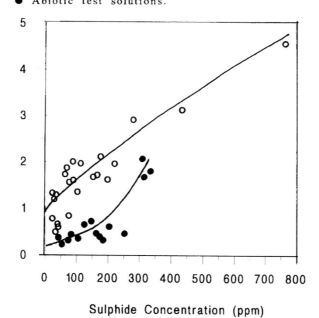

Fig. 4 Relationship between maximum hydrogen permeation and sulphide concentration in both biotic and abiotic sulphide solutions.

generated hydrogen sulphide vary considerably, depending on the conditions. In pure culture studies Miller (1949) found that, given ideal conditions, SRB could produce more than 2000 ppm H_2S. However, realistic levels in anaerobic environments on offshore production platforms range from 15 ppm to possibly 800 ppm, depending on nutrient availability (Wilkinson, 1983.; Efrid & Lee, 1979.; Thomas et al., 1988). However, it is generally thought that H_2S levels of between 50 and 200 ppm may be representative of SRB active conditions at a metal surface.

The results in Figs 2 and 3 clearly show that the test solutions containing H_2S cause a significant increase in the levels of hydrogen permeating through the steel. Even at the lower H_2S concentrations hydrogen permeation through the steel is greater than when the cell was charged with artificial seawater alone. These results support the findings of Thomas et al. (1988), who found that any amount of H_2S would enhance corrosion fatigue. Corrosion fatigue is due to the combined effects of an aggressive environment and a cyclic stress (for example, wave action) and is a considerable problem offshore. The rate of crack growth under corrosion fatigue conditions is controlled by dissolution of the metal at the crack tip and by hydrogen embrittlement (Bristoll and Roeleveld, 1979). This would suggest therefore, that the levels of hydrogen permeation caused by the presence of the H_2S enhances the likelihood of hydrogen embrittlement, which under favourable conditions may in turn promote corrosion fatigue.

The shapes of the permeation curves for the tests performed using H_2S at pH 7–8 are very similar. Following an initial period of rapid hydrogen flux an optimum level of permeation current is usually achieved within the first 3 hours of the test. After this period the level of the permeation current tends to gradually decrease. This fall in the level of hydrogen permeating through the steel is thought to be largely due to the formation of surface deposits on the charging side of the working electrode. These deposits are thought to in some way to interrupt the passage of hydrogen into the steel.

Throughout this study it has been shown that when the steel is charged with biotic H_2S the resultant levels and rate of hydrogen permeation are greater than when abiotic H_2S is used. In contrast, Thomas et al. (1988) found that corrosion fatigue was greater with abiotic H_2S solutions. They argued that the presence of organics and bacterial metabolic products provided a barrier to hydrogen embrittlement. In this study the biogenic test solution was removed from the original culture with a pipette. Thus, little of the organic material and only planktonic bacteria would have been transferred to the charging chamber of the cell, too little to apparently inhibit hydrogen entry.

One possible explanation of the these results (Figs. 2 and 3), is the nature of the surface sulphide deposits formed on the entry side of the working electrodes in each test solution may be different. It has been shown that the corrosive effects of hydrogen sulphide are highly dependent on the chemical and physical nature of the surface film corrosion product (Edyvean, 1990). The formation of iron sulphides from sulphide ions will in certain conditions not protect the metal beneath and that some may have a strong influence on the corrosion rates and corrosion products of corrosion (Smith and Miller, 1975). It is possible therefore, that the sulphide deposits formed from the biogenic test solution are of such a nature as to increase the rate of hydrogen entry into the steel. The abiotic sulphide films may in turn be more protective, which would explain the differences observed.

Thus, biotic H_2S does promote hydrogen entry into steel which in turn may lead to enhanced levels of hydrogen embrittlement.

5. Conclusion

Only relatively low levels of hydrogen are required to cause hydrogen embrittlement. It would seem likely therefore, that the levels of hydrogen sulphide produced within a biofilm would be sufficient to have a considerable effect on a susceptible steel.

6. Acknowledgements

The authors would like to thank Mr R Greenwood and Miss E Lake for their help, British Steel Swinden Laboratories for their assistance and British Gas Plc for the funding of this project.

References

1. R. G. J. Edyvean, *Int. Biodeterior.*, 1987, **23**, 199–231.
2. E. C. Hill, J. L. Shennan and R. J. Watkinson, (eds), in *Microbial Problems in the Off-shore Oil and Gas Industry*, Wiley & sons (pub), Chichester, New York, 1986.
3. T. Ford and R. Mitchell, *MTS J.*, 1990, **24** (3): 29–35.
4. J. P. Hirth and H. H. Johnson, *Corrosion*, 1976, **32**, 3.
5. *NACE Standard TM10177/90 (Section 5)*. Test method for laboratory testing of metals for resistance to SSC in hydrogen sulphide environments, 1990.
6. A. I. Vogel, *A Text-Book of Quantitative Inorganic Analysis including Elementary Instrumental Analysis*. 3rd edition, 1961, Longmans, UK.
7. J. J. Devanathan and Z. Stachurski, *Proc. Roy. Soc.*, 1962, **270**, 90–102.
8. C. J. Thomas, R. G. J. Edyvean and R. Brook, 'Biologically enhanced corrosion fatigue', *Biofouling*, 1988, **1**, 65–77.

Part 3

Microbial Corrosion: Case Studies

19

Influence of Metal–Biofilm Interface pH on Aluminium Brass Corrosion in Seawater

P. CRISTIANI, F. MAZZA* and G. ROCCHINI

ENEL Spa, Environment and Materials Research Centre, v. Rubattino, 54, 20134, Milan, Italy
*University of Milan, Dept of Physical Chemistry and Electrochemistry, v. Golgi, 19, 20133, Milan, Italy

ABSTRACT

Aluminium brass is largely used for tubes of seawater cooled steam condensers of power plants, and may be susceptible to microbiological corrosion.

Comparing the results of several experiments in natural seawater and artificial seawater acidified to pH 6.3, it may be hypothesised that, in the case of aluminium brass, one of the mechanisms of the action of biofilm on corrosion is the occurrence of acidification of the metal–biofilm interface. In fact, noticeable differences have been found in the corrosion current density values (up to 2 μA cm^{-2}) and in the surface morphology of specimens exposed to natural seawater — which proved more aggressive — compared with the results observed in artificial seawater at the same pH (8.3). On the other hand, there were significant similarities with the results obtained from tests using acidified artificial seawater.

Potentiodynamic cathodic curves, free corrosion potential measurements and electrochemical measurements for evaluating the polarisation resistance, were carried out on all the samples. Chemical analyses and analysis of the surface morphology were also conducted.

1. Introduction

Aluminium brass is largely used in marine applications, particularly for the tubes of seawater cooled steam condensers in power stations [1].

The characteristics of resistance to corrosion and the good thermal conductivity of this alloy are well known [2], as well as is its resistance to the growth of marine organisms — copper being toxic towards most of them [3].

In spite of these characteristics, the surfaces of aluminium brass tubes immersed in seawater are quickly covered by colonies of bacteria which, unlike macrofouling organisms, are not affected by the toxicity of copper, because they are protected by a mucopolysaccharide matrix which they secrete around themselves [4]. The formation of bacterial colonies is one of the first stages in the development of the biofilm which, along with other organic matter dispersed and adsorbed, corrosion products, algae (in the presence of light) and other microorganisms, results in a complex microfouling that adheres to the metallic surface [5].

This microfouling introduces an undesirable resistance to the heat exchange, modifies the chemical and physical characteristics of the surface, and facilitates the development of corrosion processes [6].

Most of the aspects of corrosion induced by microorganisms are not yet very clear [7, 8] and the modifications of chemical and physical quantities, occurring below the biofilm, cannot be detected easily without the introduction of disturbances caused by the measuring system and, it is only recently that more reliable microsensors have been developed [9].

As a result, it is useful to simulate in the laboratory, the conditions which significantly modify the nature of the passive film at the metal–biofilm interface, using properly conditioned artificial seawater. It is important to determine the influence of pH, which is one of the main parameters controlling corrosion processes.

2. Experimental Conditions

The effect of the water pH change on the electrochemical behaviour of the aluminium brass and the surface morphology has been studied during several laboratory tests carried out using sterile artificial seawater, at pH 8.3 and pH 6.3. The results were compared with those obtained under analogous experimental conditions, using sea water drawn from the cooling water canal of the thermoelectric power station at Vado Ligure (Tyrrhenian sea), during course of a wide research programme, conducted by ENEL, on microbiological corrosion and on the impact of microfouling on materials [10].

2.1. Testing Using Artificial Seawater

The tests were conducted in a circulating loop [11] with three parallel branches each containing three electrochemical cells (Fig. 1). The branches were working at water velocities of 0.5, 1 and 2 ms^{-1} and at a temperature of 30°C. 75 L of artificial seawater, prepared according to ASTM D 1141-52 specification was circulated in the loop.

The concentration of dissolved oxygen in the solution was maintained at about 8 ppm by means of an air intake in the expansion tank.

In each measuring cell were mounted five specimens, the central one being used as a counter electrode for electrochemical measurements (Fig. 2).

A total of 15 samples of aluminium brass, CDA 687, 55 mm long and with an i. d. of 20 mm (area of the exposed surface = 34.6 cm^2), were used for each branch.

The chemical composition of CDA 687 alloy was:

Zn 21.6, Al 2.02, As 0.026, Fe 0.006, Pb 0.003, C < 0.001%, Cu bal.

The specimens were cut from an unused condenser tube and no internal surface finishing was made. Polarisation resistance evaluations were performed on all samples each week and their free corrosion potential was measured each day. Furthermore, potentiodynamic cathodic curves were performed on some specimens every two weeks.

Influence of Metal–Biofilm Interface pH on Al-Brass Corrosion

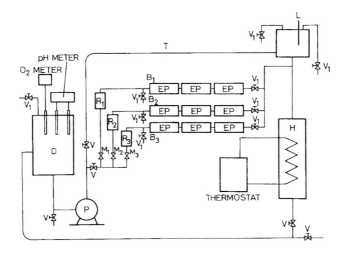

Fig. 1 Laboratory loop. EP: electrochemical probe; R1-3: flow meters; B1-3: branches at different flow rate; M: membrane valves for flow control; V: butterfly valves; L: expansion tank; D: detectors tank.

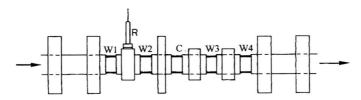

Fig. 2 Electrochemical cell. W1-4: Working electrodes, C: Counter electrode, R: Ag/AgCl reference electrode.

Ingold type 363 Ag/AgCl electrodes were used as reference electrodes.

Chemical and morphological analyses of the sample surfaces, covered with gold or graphite, were also carried out, by using a Scanning Electron Microscopy (SEM) with microprobe. SEM observations were carried out using a voltage of 20 kV. EDS (Energy Dispersive Spectrometry) microanalyses were carried out with a resolution of 20 eV per channel using a voltage of 20 kV and acquisition time of 100 s. Some preliminary analyses by X-ray photoemission spectroscopy (XPS) and X-ray diffraction (XRD) were also done.

Electrochemical measurements were carried out by using an EG & G mod. 'VERSASTAT' computerised potentiostat.

The polarisation resistance values were computed by analysing polarisation curves traced from −20 to +20 mV with respect to the corrosion potential, with a sweep rate of 1 V h^{-1}. For calculation of the polarisation resistance, only the linear zone of the curves was taken into consideration, as shown in Fig. 3. From the values of R_p, an estimate of the corrosion current density I_{corr} was made, according to the formula [12]:

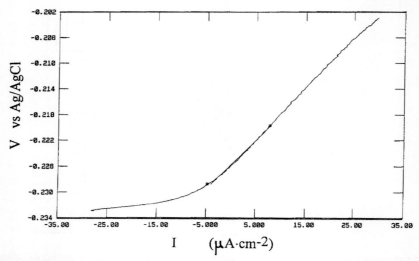

Fig. 3 *Polarisation curve 13 days after the start of the test in artificial seawater. T = 30 °C, pH: 8.3, solution flow rate: 1 ms^{-1}, R_p = 751.86 $k\Omega$ cm^2, I_{corr} = 0.0288 μA cm^{-2}.*

$$I_{corr} = \frac{\beta_a \beta_c}{2.303(\beta_a + \beta_c) R_p} \quad (1)$$

where the value of 100 mV was arbitrarily given to the Tafel slopes β_a and β_c; therefore:

$$I_{corr} = 21.71 \cdot R_p^{-1} \quad (2)$$

Some cathodic branches of the polarisation curves, using the potentiodynamic technique, with the previous sweep rate and starting from the free corrosion potential were also traced.

At fixed intervals, specimens were taken out for SEM observation, interrupting the flow only long enough to replace the samples removed.

At the end of every test, the samples were washed in distilled water and dried to determine the weight loss; then they were cut and prepared to observe the internal morphology.

The tests were carried out at pH 8.3 and pH 6.3 and can be summarised as follows:

(i) 26, 55, 81 days at pH 8.3;

(ii) 55, 81 days at pH 8.3, then, 22 days at pH 6.3.

2.2. Tests Using Natural Seawater

Some tests were performed under conditions similar to those described above, using samples cut from the same tube and in seawater drawn from the cooling canal of the thermoelectric power station at Vado Ligure.

The following are some of the characteristic parameters of the seawater: seasonal changes in temperature of about 10°C from a winter minimum of 14°C (January–February) to a summer maximum of 24°C (July–August), oxygen content of about 8 ppm, TOC of about 5–8 ppm with occasional peaks of about 20 ppm in the case of sea storms and algal blooms in spring.

The water circulated at a velocity of 0.5–0.6 ms^{-1} inside the tubes of the experimental loop. Changes in weight were not determined, because the samples were used for microbiological analysis.

3. Results

The electrochemical techniques used during the test proved to be very useful for following qualitatively the development of aluminium brass corrosion.

Figures 4 and 5 show some trends of the polarisation resistance, and Fig. 6 shows the behaviour of the corrosion current density relating to the three conditions.

3.1. Artificial Seawater at pH 8.3

During the course of tests at pH 8.3 over a few days, the polarisation resistance rose from an initial value of a few kΩ cm^2 to values of about two order of magnitude higher (400–1000 kΩ cm^2) and kept these high values throughout the test period (Fig. 5). The corrosion current densities, computed from the expressions (2), had values of about 0.02–0.05 µA cm^{-2} (Fig. 6) and the samples did not suffer any significant weight loss but showed an increase in weight due to the formation of oxides and surface precipitates.

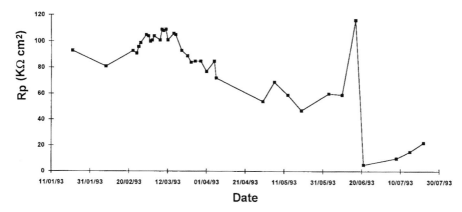

Fig. 4 Time behaviour of the polarisation resistances. Specimens exposed to natural seawater at Vado Ligure.

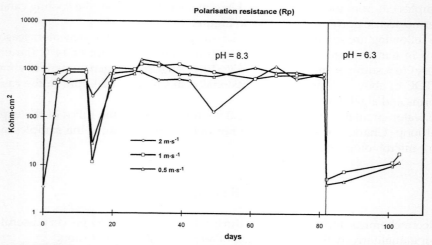

Fig. 5 Time behaviour of polarisation resistances. Specimens exposed to artificial seawater at pH 8.3 and pH 6.3, at different flow rate.

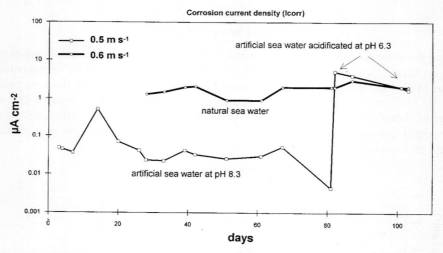

Fig. 6 Time behaviour of corrosion current density. Specimens exposed to natural seawater at Vado Ligure and specimens exposed to artificial seawater: before at pH 8.3, after at pH 6.3.

The passive state was attained rapidly during the first days and maintained throughout the test period, except for some samples that reverted to the initial values of the polarisation resistance (Fig. 5, 6) for short periods, which coincided with the stopping of the loop for solution or sample changes.

The limiting current density of the cathodic process, obtained from the potentiodynamic curves, reached values of about 50–100 µA cm^{-2}, during the early hours of the test, and then values of 2–3 order of magnitude lower (0.05–0.5 µA cm^{-2}) during the rest of the test period. The free corrosion potential values shifted with time from –265 to –200 ± 20 mV vs Ag/AgCl.

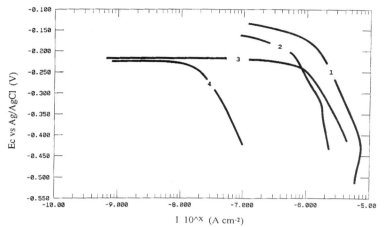

Fig. 7 *Potentiodynamic cathodic polarisation curve scarried with scanning speed of 1 V h^{-1}; **1**: natural seawater, summer; **2**: natural seawater, winter; **3**: artificial seawater, pH 6.3; **4**: artificial seawater, pH 8.3.*

No significant differences were observed between the tests at different solution velocities. Electron microscope observations revealed the presence of crystalline precipitates (Fig. 9), which, by X-ray diffraction analysis, were found to be made up mostly of aragonite ($CaCO_3$). The layer of precipitates was found to be more compact with increase in solution velocity in the range 0.5–2 ms^{-1} (Fig. 8). XPS analysis also revealed the presence of a high amount of Cu(I). Cu(II) was found only in the first molecular layers, in contact with seawater.

3.2. ASTM Acidified Seawater at pH 6.3

Artificial acidified seawater at pH 6.3 showed initially very small polarisation resistance values, about 1 kΩ cm^2, as at the beginning of the test at pH 8.3, which could correspond to corrosion of bare alloy. In the course of fifteen days, the value of R_p increased to about 40 kΩ cm^2 (Fig. 5). Then, the values of the corrosion current density, that were in some cases greater than 10 µA cm^{-2}, decreased to 0.5–3 µA cm^{-2} during the course of the tests, as shown in Fig. 6. A steady value for the corrosion current density was not found because of the short duration of the tests.

The surface analyses revealed the presence of crystalline compounds of zinc, aluminium and copper in the corrosion products. These compounds are distributed in different areas on the sample surface, as shown in Fig. 10, especially in the tests with a flow rate of 0.5 ms^{-1}. Copper compounds, green in colour, in the lower region of the specimens were found. These compounds were distributed in agglomerates lightly attached to the surface (Fig. 11), and correspond to carbonates and/or hydroxychlorides of copper.

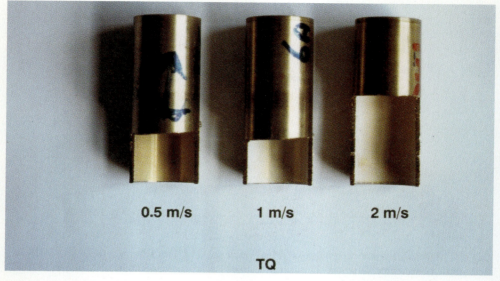

Fig. 8 Aluminium brass exposed to artificial seawater at 30°C, pH: 8.3, for about 1 month.

Fig. 9 SEM micrograph of the deposits of aragonite. Central specimens of Fig. 8. (Solution flow rate = $1\ ms^{-1}$.)

On the upper part of the specimens exposed at a solution velocity of 0.5 ms^{-1}, deposits of zinc and aluminium were predominant (Fig. 12); no deposits were found on the lateral sides. An explanation for this phenomenon could be the distribution of the bubbles of air which can accumulate on the upper part of the sample, determining the division between an upper cathodic region rich in oxygen and a lower anodic region, on the surface of the specimens exposed. In the section of loop with a solution velocity of 1 ms^{-1}, a very small amount of copper were found over the compounds of zinc and aluminium.

In this case, it may be assumed that most of the copper entered the solution, as Cu(II), as shown by the greenish colour of the circulating water.

Chloride was found on all samples, in association with copper (Fig. 11).

It must be noted that, in tests conducted at pH 6.3, the free corrosion potential did not show values significantly different from those observed during the tests at pH 8.3 (–220 ± 20 mV vs Ag/AgCl).

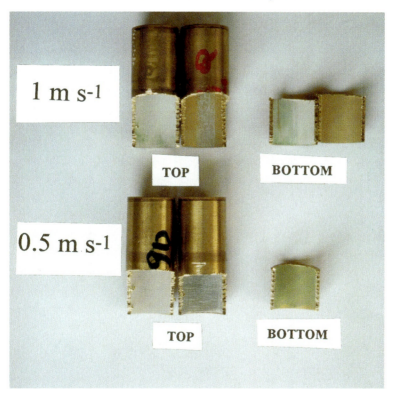

Fig. 10 *Aluminium brass samples exposed to artificial seawater acidified to pH 6.3, T = 30°C. Note the different types of deposits and the varied distribution of these deposits in the upper and lower regions of the samples.*

Fig. 11 *Corrosion products, green coloured, of aluminium brass in seawater acidified at pH 6.3, T = 30°C, flow rate = 0.5 ms^{-1}. SEM micrograph.*

Fig. 12 *Zinc and aluminium compounds, white coloured on aluminium brass in seawater acidified at pH 6.3, T = 30°C, flow = 0.5 ms^{-1}. SEM micrograph.*

3.3. Natural Seawater

During the course of a year's experience with the loop at Vado Ligure [13], the corrosion current density of aluminium brass attained values that varied seasonally; winter they were found to be around 0.5 µA cm^{-2} (R_p of about 50 kΩ cm^2) and during the summer around 1.5–3 µA cm^{-2} (R_p of about 1–10 kΩ cm^2), as shown in the example in Fig. 4.

When the samples were observed after one month exposure, the surface appeared to be uniformly covered with spongy spherical bodies. EDS microanalysis revealed the presence of copper, calcium, magnesium, phosphorus, silicon and chlorine (Fig. 13), underneath a thin biological film.

A year later, corrosion was observed on surfaces of the samples. (Fig. 14). Corrosion products were found to be distributed irregularly and often in layers on the surface, under the biofilm. The chemical nature and morphology of these compounds were found to be similar to those obtained in tests using artificial seawater acidified to pH 6.3 (Figs 11, 12). XPS analysis of some zones of the internal surface of the tubes showed the presence of Cu(II), in the superficial layers, in the form of carbonates and hydroxychlorides, green in colour, which may be both malachite as well as

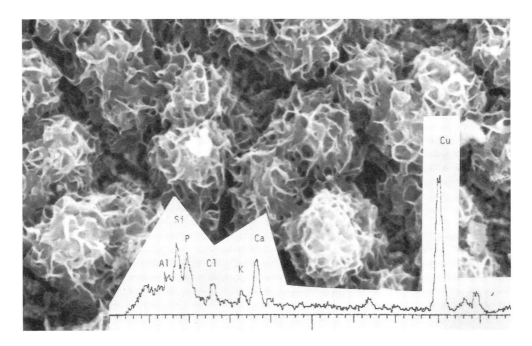

Fig. 13 *Aluminium brass exposed to natural seawater at Vado Ligure for 2.5 months. SEM micrograph and EDS emission spectrum.*

Fig. 14 *Aluminium brass exposed to natural seawater for more than a year.*

atacamite, which were already found both in seawater [14] and in fresh water [15]. Preliminary X-ray diffraction analysis of the crystalline phases revealed the presence of paratacamite and forms of carbonate different from malachite (Fig. 15). Cu(I) was found only in the innermost layers, adhering to the metal. In other parts of the surface of the tubes, in areas of localised corrosion, considerable quantities of compounds of zinc and aluminium were found (Fig. 16). The XRD technique revealed the presence of hydroxy-zinc compound under the green deposit.

4. Discussion

Natural seawater was found to be more aggressive than artificial seawater at pH 8.3, especially during spring and summer, in concomitance with the maximum development of biofouling. Assuming that this higher aggressivity was because of microorganisms and organic substances present in the sea, these data then confirm the experimental finding [16], that the corrosion induced by microorganisms in the biofilm on copper is of the order of 1 $\mu A\ cm^{-2}$. In fact, values of corrosion current density of 0.5–3 $\mu A\ cm^{-2}$ in natural seawater, and values of about 0.05 $\mu A\ cm^{-2}$ in artificial seawater at the same pH of 8.3 were observed (Fig. 6).

On the other hand, the behaviour of aluminium brass in artificial seawater acidified at pH 6.3 and in natural seawater was found to have many similarities. The corrosion current density assumed values to the order of magnitude of some $\mu A\ cm^{-2}$ (Fig. 6); the cathodic characteristics were found to be similar (Fig. 7) and the same types of chemical compounds were found in the corrosion products.

These facts suggest that one of the main mechanisms of microbially induced corrosion may be the acidification of the metal–biofilm interface.

The results from other experiments on aluminium brass [17, 18] may be explained

NATURAL SEA WATER

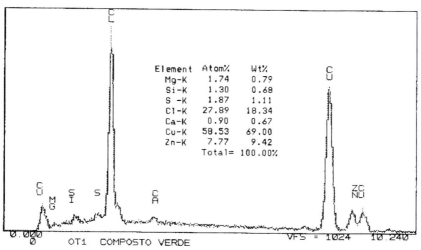

Fig. 15 Copper(II) corrosion products, green coloured, of aluminium brass exposed to natural seawater at Vado Ligure for more than a year. SEM micrograph and EDS emission spectrum.

on this basis. In fact, where the samples exposed to natural seawater acidified at pH 6.3 for more than three months, showed values of R_p similar to, and even lower than, those observed on samples exposed to non-acidified seawater, may be explained on this basis.

The capacity of microorganisms to modify physico–chemical parameters, espe-

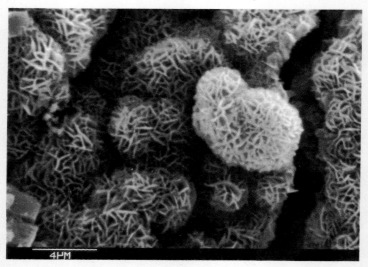

Fig. 16 Zinc and aluminium compounds, on aluminium brass exposed to natural seawater at Vado Ligure for more than a year. SEM micrograph and EDS emission spectrum.

cially the pH, is well known [19]. However, only recent experimental data concerning stainless steels [20], have shown the close relationship between local variations of pH and the microbial corrosion processes.

Some consideration should be given to the results of chemical and morphological analysis.

The presence of calcium carbonate above a layer of copper oxides, found on the surface of the samples exposed to the artificial seawater at pH 8.3, may result from the conditions of alkalinity at the surface, i.e. as caused by the buffer effect of the carbonate–bicarbonate equilibrium. In these conditions, the oxides of copper are stable and highly protective, in accordance with the experimental results given in [21] and [22].

At pH 8.3 the overall cathodic reaction is:

$$O_2 + 2H_2O + 4e \rightarrow 4OH^- \qquad (3)$$

whereas the anodic reaction may be represented by

$$2Cu + H_2O \rightarrow Cu_2O + 2H^+ + 2e \qquad (4)$$

Cu_2O species appears to be more stable, together with CuO which forms later through the reactions:

$$Cu_2O + H_2O \rightarrow 2CuO + 2H^+ \qquad (5)$$

in accordance with the potential–pH Pourbaix diagram [23] for copper.

Since there is no macroscopic separation between the anodic and cathodic regions, there is no local variation in pH.

The stability of oxides on the metallic surface makes the material passive and explains the very low corrosion rate. Furthermore, under these conditions, the buffer effect of the seawater is sufficient to neutralise any subsequent alkalinity or acidity produced by the dissolution of oxides and the other corrosion products present on the surface.

At pH lower than 7, the Pourbaix equilibrium lines for copper [23] show that the most stable solids are mixed salts (instead of oxides); these include atacamite ($Cu_2(OH)_3Cl$) and malachite ($CuCO_3 \cdot Cu(OH)_2$):

$$2CuO + H^+ + Cl^- + H_2O \rightarrow Cu_2(OH)_3Cl \qquad (6)$$

$$2CuO + H^+ + HCO_3^- \rightarrow CuCO_3 \cdot Cu(OH)_2 \qquad (7)$$

The predominance of malachite or atacamite in the reaction:

$$Cu_2(OH)_3Cl + HCO_3^- + 2H_2O \rightarrow CuCO_3 \cdot Cu(OH)_2 + Cl^- + H_2O \qquad (8)$$

depends upon the concentration of the total inorganic carbon [23], as shown in Fig. 17. The presence of these compounds (Figs 11, 13) in the corrosion products on samples surface exposed to seawater provides indirect proof of the conditions of local acidity of the solution.

Furthermore, substantial deposits of carbonates were not found on the specimens exposed to natural seawater. On the same specimens, calcium was found only in association with organic substances, biological organisms and high amounts of corrosion products.

Further proof of acidity at the metal–biofilm interface is provided by SEM observations and XPS analysis, which give results similar to those of laboratory tests at

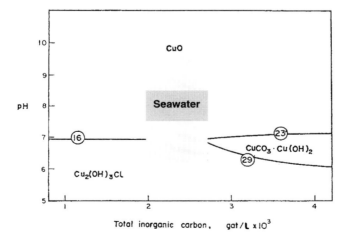

Fig. 17 Domains of stability of solid phases $CuCO_3 \cdot Cu(OH)_2$, $Cu_2(OH)_3Cl$, CuO in seawater at 25°C, after Ref. [22].

pH 6.3, especially as regards the presence of copper, zinc, aluminium and chlorine compounds.

As regards chlorine compounds, it must be stated that they were always present in the surface layer of samples exposed to natural seawater, but were never found on specimens exposed to artificial seawater at pH 8.3.

The stabilisation of copper by compounds other than oxides plays a determining role in the corrosion process as its salts, being more soluble and less bonded to the surface than oxides, especially the cuprous, entering the solution, do not form a protective layer.

As regards zinc and aluminium compounds (Fig. 12), their protective capacity and their presence in the corrosion products may be closely related to the solution pH. Furthermore, the reduction of corrosion rate during the exposure of samples to acidified seawater may be explained by the good protective characteristics attributed to aluminium compounds.

There may be various mechanisms for the production of acidity at the metal–biofilm–water interface. In the case of aerobic bacteria, acidity may be produced by the release of carbon dioxide during respiratory metabolism, and by the secretion of organic and inorganic acids, the nature of these depending upon the species present. Lowering of the local pH and the onset of putrefaction in fouling, in the absence of oxygen, also favour the growth of anaerobic bacteria, especially sulphate reducers, which contribute drastically to the acidification of the environment by producing H_2S.

Finally, the entire biofilm acts as a barrier between the seawater and the metallic surface, restraining the diffusion of species formed at anodic sites and hence the buffer effect of the seawater.

In apparent contradiction with the previous observation, the lowering of the pH of natural seawater, by acidification, eliminates fouling already present on the tubes, as has been verified by recent studies [17, 18]. This fact may be explained by assuming that aerobic microfouling can survive only when the acidification is localised. As a result, the acidity produced by metabolic processes of bacteria, in the absence of a buffering effect of the seawater or with sufficient change of water in the acidified region, could have a self inhibition effect on the growth of aerobic fouling.

Furthermore, in natural seawater, the phenomenon of acidification alone cannot justify all the events that occur on the aluminium brass surface in the presence of organic microfouling. In fact, it must be stated, that during the course of laboratory tests carried out at pH 8.3, the corrosion potential did not undergo significant shifts compared with those observed at pH 6.3. In both cases, the values shifted in time from −265 mV to −200 ± 20 mV vs Ag/AgCl, unlike the case in natural seawater, where the potential underwent an increase of about 150 mV or more, and passed from −265 mV to ±−100 mV vs Ag/AgCl, as is clearly recognised in the case of other alloys with active–passive transitions, and especially in the case of stainless steel [24].

A possible explanation of the phenomenon may be linked to the detailed study of the mechanism of cathodic reactions, which may lead to the formation of hydrogen peroxide on the metal–solution interface, by means of an intermediate reaction, strongly influenced by the pH of the solution [25, 26], in which the microbial metabolism could play an active role. In fact, in a recent study [19] on stainless steels, a certain amount of hydrogen peroxide was found on the metal–biofilm interface. However, in the case of copper alloys, conditions may differ considerably, as copper and its compounds catalyse the reduction of hydrogen peroxide [27].

5. Conclusions

1. The electrochemical measurements associated with the study of surface analysis techniques provide results that are very useful for the evaluation of the corrosion behaviour of aluminium brass exposed to seawater.

2. The results obtained highlight the prominent role played by variations, however small, of the pH of the metal–solution interface on the development of a protective film. The presence of calcium carbonate as aragonite covering a thin film of copper oxides on aluminium brass exposed to artificial seawater at pH 8.3, provides a proof of the alkaline conditions present close to the specimen surface and explains the negligible values of the corrosion current. The presence of Cu(II) in the form of carbonates and hydroxychlorides, and of aluminium and zinc compounds on the surface of samples exposed to natural seawater, may be due to the development of local acidity on the metal–biofilm interface. This observation may explain the fact that natural seawater is more aggressive then artificial seawater at the same pH.

3. On the basis of the similarities found in the tests, it may be assumed that the production of local acidity on the metal–biofilm interface is one of the mechanisms that promote corrosion induced by both aerobic and anaerobic microorganisms. However, because of the short duration of the experimentation carried out, it is advisable to plan other studies, in order to verify the validity of the suggested hypotheses and to quantify the effects of acidification of seawater in long-term tests.

References

1. G. A. Bianchi, A. Cerquetti, P. Longhi, F. Mazza and S. Torchio, *Problemi di corrosione nei condensatori di vapore delle centrali termiche e nucleari*, Associazione Italiana di Metallurgia, Milan, Italy, 1973.
2. V. Calcut, 'Copper Alloys in Marine Environments', *Metallurgia*, 1988, **55** (1), 38–40.
3. U. R. Evans, *The Corrosion and Oxidation of Metals, Scientific Principals and Practical Applications*, C. Arnold, London, 1960, p. 585.
4. G. G. Geesey, M. Mittelman, T. Iwaoka and P. R. Griffiths, Role of bacterial exopolymers in the deterioration of metallic copper surfaces, *Mat. Perform.*, 1986, **2**, February, 37–40.
5. R. Mitchell, 'Mechanisms of biofilm formation in seawater', in *Proc. Ocean Thermal Energy Conversion (OTEC) Biofouling and Corrosion Symposium*, Seattle, Washington, USA, 1977, pp. 45–49.
6. A. K. Tiller, 'Metallic Corrosion and Microbes', in *Proc. 2nd EFC Workshop on Microbial Corrosion* (Ed. C.A.C. Sequeira), EFC Publication No. 8, The Institute of Materials, 1992, pp. 1–8.

7. S. C. Dexter (Ed.), *Biologically Induced Corrosion,* NACE, Houston, TX, 1986.
8. *Microbial Corrosion: 1988, Workshop Proceedings* (Ed. G. J. Licina), EPRI ER– 6345, Palo Alto, CA, USA,1989.
9. B. Little, 'An Overview of Microbiologically Influenced Corrosion', *Electrochim. Acta,* 1992, **37** (12), 2185–2194.
10. P. Cristiani, 'Attivita di ricerca sulla corrosione microbiologica e monitoraggio elettrochimico del fouling', ENEL-DSR/CRTN internal report n. G6/92/04/MI, 1992, Milan, Italy
11. A. Colombo, P. Cristiani, G. Rocchini, F. Mazza and S. Torchio, 'Comportamento di leghe di rame per scambio termico in acqua di mare sintetica in condizioni fluidodinamiche', ENEL-DSR/CRAM internal report n. G6104186/MI, Milan, Italy, 1987.
12. M. Stern and J. A. Geary, *J. Electrochem. Soc.* **104**, 56, 1957.
13. G. Salvago, G. Fumagalli, P. Cristiani and G. Rocchini, 'Fouling on aluminium brass tubes in seawater and its electrochemical monitoring', *8th Int. Congr. on marine Corrosion and Fouling,* Taranto, Obelalia, vol. 19, p. 331–341, 1993.
14. E. D. Mor and A.M. Beccaria, *Proc. 3rd Int.Congr. on Marine Corrosion and Fouling.* 1972, Gaithersury, MD. National Bureau of Standards, Special Publication.
15. D. Wagner, W. Fisher and H. H. Paradies, 'First Results of a Field Experiment in a Country Hospital in Germany Concerning the Copper Deterioration by Microbially Induced Corrosion', *Proc. 2nd EFC Workshop on Microbial Corrosion,* (Ed. C. A. C. Sequeira), EFC Publication No. 8, The Institute of materials, 1992, 1–8.
16. S. C. Dexter, *Biologically Induced Corrosion,* (Ed. S. C. Dexter), NACE, Houston, TX, p. 144, 1986.
17. P. Cristiani, work in progress.
18. G. Salvago, G. Fumagalli and G. Taccani, 'Fouling formation on al-brass tubes in acidified flowing seawater', *Corrosion '93,* Paper No. 310, NACE, Houston, TX, 1993.
19. H. A. Videla, *Int. Biodeterior. Biodegr.*, 1992, **29**, (3–4), 195–212.
20. P. Chandrasekaram and S. C. Dexter, 'Factors Contributing to Enobblement of Passive Metals Due to Biofilms in Seawater', *Corrosion Control for Low-Cost Reliability, Proc. 12th Int. Corrosion Congr.,* pp. 3696, 3707, NACE, 1993.
21. S. B. Adeloju and Y. Y. Duan, 'Influence of Bicarbonate Ions on the Stability of Copper Oxides and Copper Pitting Corrosion', *Progress in the understanding and prevention of corrosion,* **3** (Eds J. Costa and A. D. Mercer), (10th European Corrosion Congress, Barcelona, 1993.) The Institute of Materials, London, 1993.
22. H. H. Strehblow, U. Collisi and P. Druska, 'Control of Copper and Copper Alloys Oxidation', Symposium Proceedings, *Metallurgie,* **6**, Rouen, France, 1992.
23. G. Bianchi and P. Longhi, 'Copper in sea-water, potential-pH diagrams', *Corros. Sci.,* 1973, **13**, 853–864.
24. A. Mollica, G. Alabiso, V. Scotto, E. Traverso and G.Ventura, 'Influenza della componente biologica sulla corrosione di leghe attivo–passive in acqua di mare', atti del convegno *Corrosione Marina,* Genova, 21–23 Sept. 1988.
25. G. Bianchi, F. Mazza and T. Mussini, *Proc. 2nd Int. Congr. on Metallic Corrosion,* NACE, Houston, TX, 1966.
26. G. Salvago, G. Fumagalli, G. Taccani, P. Cristiani and G. Rocchini, 'Electrochemical and Corrosion Behaviour of Passive and Fouled metallic Materials in Seawater', *Proc. 2nd EFC Workshop on Microbial Corrosion,* EFC Publication No. 8, The Institute of Materials, London, 1992.
27. W. Machu, *Des Wasserstoffpeoxyd und der Perverbindunger,* J. Springer ed., Vienna, 1937.

20

Effect of Seasonal Changes in Water Quality on Biofouling and Corrosion in Fresh Water Systems

R. P. GEORGE, P. MURALEEDHARAN, J. B. GNANAMOORTHY,
T. S. RAO* and K. V. K. NAIR*

Metallurgy Division, Indira Gandhi Centre for Atomic Research, Kalpakkam-603102, India
* Water and Steam Chemistry Laboratory, IGCAR Campus, Kalpakkam-603102, India

ABSTRACT

This paper describes the seasonal variations in the water quality of a fresh water reservoir at Kalpakkam and its influence on the corrosion rates of carbon steel and on the biofouling rates of stainless steel, brass, titanium and admiralty brass. The study conducted during May 1992 to August 1993 showed that the quality of inflow water changes on storage in the open reservoir due to excessive growth of macro and micro algae in the reservoir. There was an increase in the nutrient content and conductivity during the months of May to September 1992 followed by a decrease in all the water quality parameters except oxygen and silica contents during October to December 1992. Corrosion rates of carbon steel determined by short-term exposures showed an increase from 7.6 mpy (May 1992) to 10.6 mpy (August 1992) to 15.1 mpy (November 1992) and then a decrease during subsequent months. The Langelier index value in November indicated that the reservoir water should not be corrosive. However, the total viable count of bacteria (3×10^7 cfu cm^{-2}) on the coupons was found to be at a maximum in November '92. Thus, the higher corrosion rate in November could be attributed to the possibility of microbially induced corrosion (MIC). Another possible contributing factor could be the relatively high dissolved oxygen observed in the reservoir during that period. Studies on the extent of biofouling by algae on various materials showed short-term fouling rates to be relatively high on titanium and long-term fouling rate to be the highest on stainless steel.

1. Introduction

The fresh water requirements of both the Madras Atomic Power Station (MAPS) and the Fast Breeder Test Reactor (FBTR) at Kalpakkam are met by the water stored in an open reservoir at the site. The reservoir which has an area of 130 m × 130 m and a depth of 2.13 m at overflow level can hold up to 7.5 million gallons (34.1 ML) of water. Water to this reservoir comes from bore wells in the Palar river bed (located about 20 km away from the site) where the ground water is collected through infiltration galleries and is carried by 60 cm dia. cast iron pipes. The vigorous rainfall during the north-east monsoon period (October to January) at Kalpakkam recharges the ground water at the Palar river-bed.

The FBTR has an open recirculating type of service water system which uses water drawn from the open reservoir and collected in a sump near the reactor plant. The major material of construction of the service water system is carbon steel (CS). Within five years of commissioning of the system, problems such as pin-hole leaks and blockage of carbon steel pipes with corrosion products were experienced [1]. This led to an investigation which suggested that long periods of stagnation of the cooling water in the pipe lines and absence of regular chlorination was the primary cause for the severe corrosion, since, recent studies on CS corrosion in FBTR service water by Rao et al have indicated the possibility of MIC.

Based on the pattern of rainfall and its influence on hydrobiological characteristics [3], the seasons at Kalpakkam fall into three distinct periods, viz., summer (February to June), south-west monsoon (July to September) and north-east monsoon (October to January). To understand the seasonal variations in water quality and their influence on the corrosion rates of carbon steel, the present study was carried out during the period May 1992 to August 1993. Biofouling on stainless steel, titanium, brass and admiralty brass were also studied during this period. This paper reports the results of these studies and discusses it in the general context of water quality and carbon steel corrosion.

2. Experimental Procedure
2.1. Water Quality Analysis

Water samples were collected from the inlet to the reservoir (Palar water) as well as from the open reservoir (reservoir water) at fortnightly intervals and analysed for pH, conductivity, dissolved oxygen, total alkalinity, total hardness, total dissolved and suspended solids, chloride, sulphate and nutrients like nitrite, nitrate, phosphate and silicate by standard methods. Estimation of total viable and total heterotrophic counts of bacteria was done as per the procedures given by Pelczar *et al*. [5] Detection of iron-bacteria and sulphate-reducing bacteria (SRB) was done using procedures given in ASTM methods (D 932-72 and D 993-58). Chlorophyll content of the water was estimated by methods given by Parsons *et al*. [6].

2.2. Studies on Corrosion and Biofouling

Carbon steel coupons for the corrosion studies were made from a pipe material from FBTR service water system (ASTM-A-53, Grade-B steel). For long-term and monthly corrosion studies CS coupons of $50 \times 30 \times 2$ mm size were used. Seasonal short-term corrosion and biofouling studies were carried out using carbon steel and type 304 stainless steel (SS) coupons of size $20 \times 20 \times 1.6$ mm. Long-term biofouling studies were carried out using araldite mounted coupons of type 304 SS, titanium, brass and admiralty brass having approximately 5 cm^2 surface area. All the coupons were polished with successive grades of abrasive paper up to a final 600 grit finish, cleaned and degreased with acetone, prior to exposure in the reservoir water. The coupons were suspended in the open reservoir from stainless steel frames using nylon threads, so that the coupons were at a depth of about one metre from the water level in the reservoir.

To evaluate the long-term corrosion rates, 24 coupons of CS were suspended at the start of the experiment (May 1992). Two coupons were removed at monthly intervals. In order to assess any variation in the corrosion rate of carbon steel during different months, coupons (in duplicate) were suspended at the beginning of each month and were withdrawn at the end of the month. The corrosion products were scraped off and the coupons were descaled along with a control coupon using Clarke's solution [7] and weighed to calculate the corrosion rate from the weight loss measurements.

The scraped out corrosion products from the CS coupons exposed for various periods in the open reservoir were powdered, pelleted and analysed using X-ray diffraction. From the d values obtained, various phases of iron oxide present in these corrosion products were identified using X-ray diffraction data file.

With a view to finding out the influence of seasonal variations in water quality on corrosion rates and biofouling, studies were conducted in the months of May, August and November for a period of 15 days. Twelve coupons each of type 304 SS and CS were exposed in the open reservoir and coupons in duplicate were withdrawn on the 2nd, 4th, 8th, 10th and 15th day. To study the rate of build up and composition of the biofilm in the open reservoir, Perspex slides were exposed in the reservoir water for various durations. The biofilm thickness was estimated microscopically. Chlorophyll and nutrient contents in the film were also analysed by the procedures described earlier.

To determine the long-term biofouling rates in the open reservoir, coupons of type 304 SS, titanium, brass and admiralty brass were suspended and withdrawn at monthly intervals. After carefully removing these coupons from the araldite mould, one set was rinsed in distilled water and the organisms were fixed in 2% glutaraldehyde overnight at 6°C dehydrated in 30%, 50%, 80% absolute alcohol and observed in a scanning electron microscope. From the other set, the biofilm was scraped into distilled water and then filtered through a Millipore filter for chlorophyll estimation.

3. Results and Discussion
3.1. Water Quality of Open Reservoir *vis-a-vis* Palar River Water

Analysis of the water quality of the Palar water and the reservoir water over a year showed that the quality of Palar water changes on storage in the open reservoir. The values of water quality parameters of the reservoir water such as dissolved oxygen content, pH, temperature, microbial density and chlorophyll content (Fig. 1) were always higher than those of Palar water while the amount of dissolved solids, conductivity and nutrient content were lower in the reservoir water (Fig. 2) as compared to Palar water.

Extensive growth of macrophytes like *Vallisneria*, *Najas* and *Ceratophyllum* were observed in the open reservoir during this period. Investigations in our laboratory [8] have shown that extensive growth of macro and micro algae is one of the factors responsible for the change in Palar water quality on storage in the open reservoir. These studies showed that the pH and P-alkalinity increase while hardness and cal-

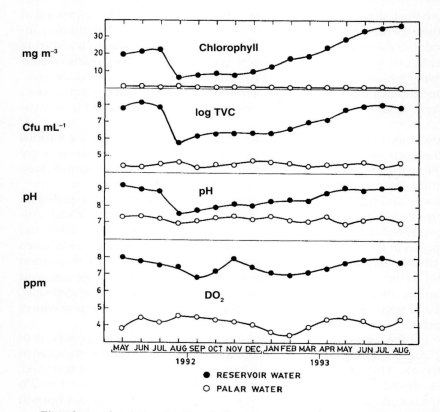

Fig. 1 Seasonal variations in chlorophyll content, total viable count of bacteria, pH, and dissolved oxygen content in reservoir water and Palar water.

cium content decrease due to the presence of algae in the reservoir water. It appears possible that the utilisation of CO_2 for algal photosynthesis shifts the pH to higher values leading to calcium carbonate precipitation. It has been suggested by other workers also that an increase in the pH of natural waters often indicates intense photosynthetic activity [9]. According to Atkins and Harris [10] a pH greater than 8.1 shows that the rate of photosynthesis exceeds that of respiration in that system.

The Langelier index (LI) of Palar water was negative (–0.1 to –1.1) throughout the period of study whereas the LI of reservoir water was negative only during August and September '92 and positive during the remaining part of the study. This could be due to the lowering of the pH of the reservoir water following manual cleaning to remove weeds during August to October.

3.2. Seasonal Changes in the Open Reservoir Water Quality

The seasonal variations in water temperature and the average rainfall during the various months of this study are shown in Fig. 3. The water temperature was found

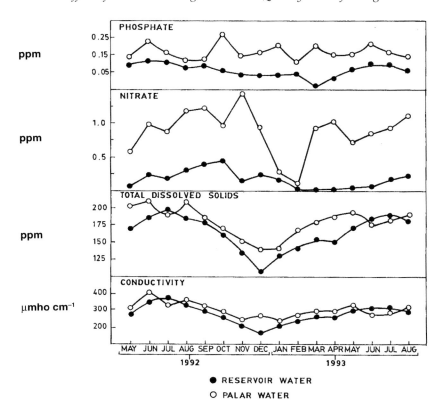

Fig. 2 Seasonal variations in dissolved solids, conductivity, and nutrient content in reservoir water and Palar water.

to be maximun in May and minimum in December. The pH of the reservoir water varied from 7.5 to 9.2 (Fig. 1). Following excessive weed growth during the period May to July '92, weed cleaning was undertaken in August '92. Removal of weeds drastically lowered the pH of the reservoir water to a range of 7.5 to 7.8, confirming the suggestion that photosynthesis is responsible for the higher pH observed in the reservoir water as compared to Palar water. The pH values of the reservoir water showed distinct maxima in May '92 and May '93 and a minimum in September '92. Total alkalinity values of the reservoir water ranged from 60 to 110 mg L^{-1}. The most important feature was a lowering of the total alkalinity values with the onset of north-east monsoon rains in October '92 (Fig. 4). Other water quality parameters such as conductivity (Fig. 2), hardness, chlorides and sulphate (Fig. 4) in the reservoir water showed maxima during the summer and minima during the north-east monsoon period. The dissolved oxygen content of the reservoir water varied from 6.5 to 8.5 mg L^{-1} during the year showing two maxima, one during May to July '92 and the other during November '92. During the period May to July '92 when the reservoir water exhibited relatively high chlorophyll content, the nitrate content of

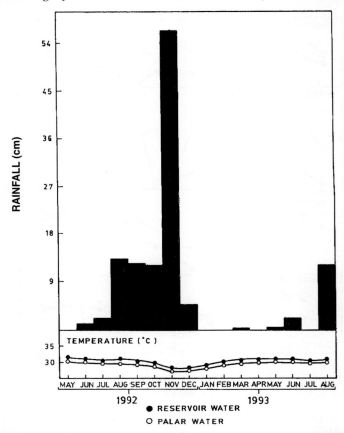

Fig. 3 Seasonal variations in the temperature of reservoir water and average rainfall at Kalpakkam.

the water also showed a significant decrease from 0.48 to 0.09 mg L^{-1}. The phosphate content was relatively high during the above period (0.05–0.3 mg L^{-1}).

During May to July '92 the pH, temperature and oxygen content exhibited relatively high values. High pH and oxygen content reflect an increased photosynthetic activity [9]. This is also confirmed by the maxima observed in chlorophyll content and microbial density during this period. Due to extensive consumption of nitrates by photosynthesis the normal N:P ratio of 16:1 in natural waters decreased to a value of 0.6:1 during May to June. Low N:P ratio (< 16:1) throughout the year in the reservoir water shows that photosynthetic activity in the reservoir is nitrogen-limited [11].

3.3. Effect Of Changes In Water Quality on Corrosion and Biofouling of Materials

Figure 5 illustrates the variations in monthly as well as long-term corrosion rates of CS for the period May '92 to May '93. Total viable count of bacteria on the CS cou-

Effect of Seasonal Changes in Water Quality on Biofouling and Corrosion 267

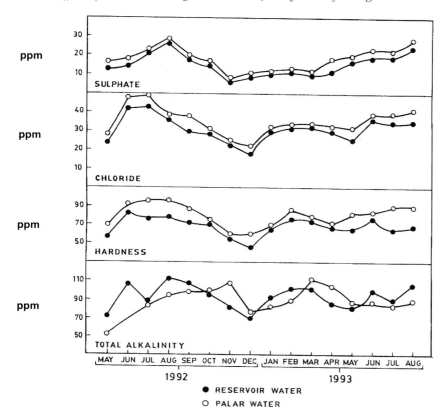

Fig. 4 *Seasonal variations in total alkalinity, hardness, chloride and sulphate in reservoir water and Palar water.*

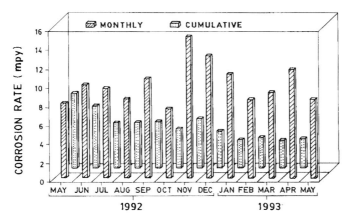

Fig. 5 *Monthly and long term corrosion rates of carbon steel coupons exposed in the reservoir water.*

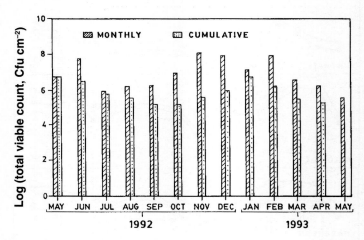

Fig. 6 Total microbial density on the carbon steel coupons exposed for monthly and long term intervals in reservoir.

pons for various months are presented in Fig. 6. The monthly corrosion rates of carbon steel showed a peak value of 15 mpy in November and a value of 8 mpy in May '92 and May '93 (Fig. 5). Total viable bacterial counts (TVC) on the coupons were also at a maximum in November (Fig. 6).

Long-term corrosion studies of carbon steel showed that the corrosion rate of 8 mpy observed in the first month decreased to a value of 3.2 mpy after 12 months (Fig. 5). The total viable count of bacteria on the CS coupons showed a decrease after 3 months (from 5.4×10^6 cfu cm^{-2} on 1-month coupon to 6.2×10^5 cfu cm^{-2} on 3-month coupon) possibly due to the spalling of corrosion products. However, the differential count of specific bacteria like iron oxidising bacteria (IOB), *Pseudomonas* sp., and sulphate reducing bacteria (SRB) on the coupons exposed for 3 months were significantly higher than those observed on the 1-month coupon. The IOB count was found to increase from a negligible value on the 1-month coupon to a value of 1.5×10^6 cfu cm^{-2} on the 3-month coupon and found to decrease on further exposure (54 cfu cm^{-2} on the coupon exposed for 11 months). Yellowish brown tubercles were observed on all the coupons after three months of exposure (Fig. 7). Descaling of these coupons revealed shallow pits beneath the tubercles. However, the tubercles did not grow significantly with further exposure and coupons exposed for one year showed only moderate tuberculation. This is in contrast to the very large tubercles formed on the CS coupons exposed to the tap water in the laboratory [8]. One reason for the absence of large tubercles on the coupons exposed to reservoir water is the formation of algal mats on these coupons. More detailed experiments are necessary to get a clearer picture of the role of algae in the inhibition of tuberculation by IOB.

XRD analysis of the corrosion products on CS coupons exposed in the reservoir water for various durations showed major peaks of maghemite (α–Fe$_2$O$_3$) and goethite (γ-FeOOH). Other workers [12, 13] have reported that goethite and maghemite were

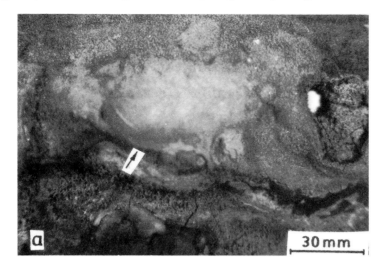

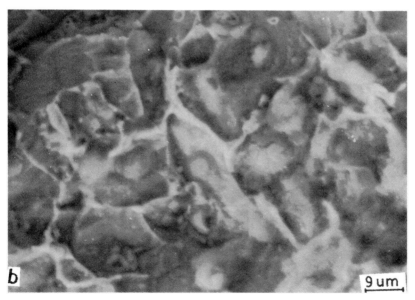

Fig. 7 (a) Tubercles initiated on the carbon steel coupons after 3 months' exposure in the reservoir water, and (b) cleaned tuberculated CS coupon showing the mode of attack.

the major iron oxide phases in the presence of bacterial activity whereas lepidocrocite and magnetite were the predominant phases in the oxide scales formed on the cast iron in water in the absence of bacterial activity [14]. Goethite has also been isolated from mineralised *Gallionella* (IOB) colonies [15]. The prevalence of goethite and maghemite in the corrosion products observed in the present study possibily indicates IOB mediated corrosion (MIC) of CS in the reservoir water.

3.3.1. Biofouling and corrosion rates during different seasons

The results of short-term corrosion studies on CS coupons, conducted in the months of May, August and November 1992, covering all the three seasons, showed an increase in corrosion rate from 7.6 mpy in May to 10.6 mpy in August, to 15.1 mpy in November. Although, there were significant increases in the values of water quality parameters such as nutrient content and hardness during the south-west monsoon season, the possible reason for the increase in the corrosion rate in August appears to be the lowering of pH of the water caused by the removal of the extensive weeds. The decrease in pH shifted the LI of the water to negative values increasing the chemical corrosivity of the water. This tendency seems to have undergone a reversal following the onset of the north-east monsoon season when an increase in pH; and a shift in LI to a positive value (+0.5) was observed making the water less corrosive. However, the oxygen content of the water and the microbial population on the coupon were significantly high in this season.

XRD studies showed dominant peaks of $CaCO_3$, in addition to other iron oxide phases on the coupons exposed in May '92 when the pH and photosynthetic activity were relatively high in the reservoir water. The lower corrosion rates of CS in May can be possibily due to the algal mediated $CaCO_3$ precipitation on the coupons. Reports in the literature [16] also support the view that a rust coating containing $CaCO_3$ is sufficiently dense and continuous to offer protection from corrosion.

Studies on the nutrient status of reservoir water and biofilm showed that there was a nutrient enrichment in the biofilm as compared to the surrounding waters during the north-east monsoon season [17]. In the present study the microbial density on the SS coupon showed a two-fold increase and the thickness of the biofilm formed on Perspex was also higher than in the other seasons (Table 1). There are reports in the literature [18] that in nutrient-rich water, the bacteria will show a very

Table 1. Seasonal variations in the microbial density and the biofilm thickness on the coupons exposed in the reservoir water

	Microbial density of water (cfu mL^{-1})	Microbial density on SS coupon of 15 day exposure (cfu cm^{-2})	Biofilm thickness on Perspex of 15 day exposure (μm)
November '92 (N.E. monsoon)	3×10^6	1.3×10^8	160
August '92 (S.W. monsoon)	3×10^5	6.7×10^6	120
May '92 (summer)	3×10^7	3.0×10^6	—

feeble tendency to form a biofilm whereas in a nutrient-limited medium the biofilm thickness increases and the cells constituting the film produce copious quantities of extracellular polysaccharides [19]. This suggests that high corrosion rates observed in November could be attributed to the possibility of MIC.

3.3.2. Long-term biofouling studies

To study the rate of biofilm formation and biofouling, coupons of type 304 SS, titanium, brass and admiralty brass were exposed in reservoir water for various durations. The coupons were withdrawn after exposure for 2, 8 and 11 months and the bacterial densities were estimated. The results are tabulated in Table 2. It was found that the bacterial densities were at a maximum on titanium coupons. Although the initial bacterial density on brass and admiralty brass coupons was relatively low, it showed an increase with time of exposure.

Scanning electron microscopic observations of these coupons showed a diverse flora of algae on type 304 SS and titanium but only a single type of filamentous alga on brass (Fig. 8). Firmly attached diatoms were observed on SS coupons, whereas only frustules of diatoms were observed on brass and admiralty brass coupons (Fig. 9) possibily indicating copper toxicity.

In order to study the extent of biofouling by algae on these materials, the chlorophyll content per unit area on these coupons was estimated. The short-term biofouling was found to be higher on the titanium coupons (0.84 µg chlorophyll cm^{-2}) as compared to that on type 304 SS (0.31 µg cm^{-2}) and brass (0.15 µg cm^{-2}). But after one year of exposure, maximum algal fouling was found on type 304 SS (3.6 µg cm^{-2}) followed by titanium (2.2 µg cm^{-2}). Brass and admiralty brass showed relatively low fouling (1.51 µg cm^{-2} and 1.33 µg cm^{-2} respectively).

Table 2. Variations in the microbial density on various materials exposed for different durations in the reservoir water

	Materials Microbial density (cfu cm^{-2})			
	2 months' exposure May '92 to July '92	4 months' exposure May '92 to September '92	8 months' exposure May '92 to January '93	11 months' exposure May '92 to April '93
Stainless Steel	1.3×10^9	2.03×10^5	3.8×10^9	4.9×10^9
Titanium	4.3×10^9	—	1.3×10^9	7.05×10^9
Brass	4.3×10^3	1.06×10^4	3.1×10^5	8.7×10^7
Admiralty Brass	—	4.5×10^3	—	7.8×10^7

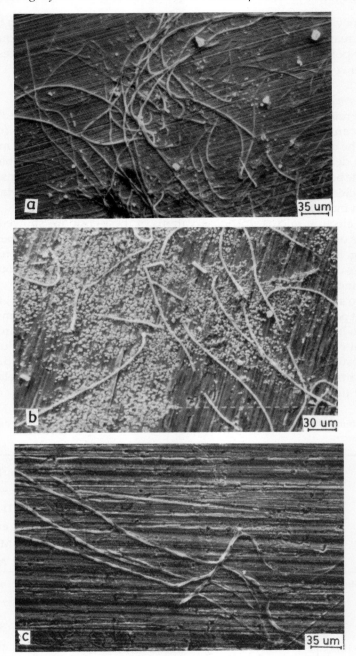

Fig. 8 SEM pictures of algal biofilms on the coupons exposed in the reservoir water: (a) SS coupon with diverse flora of alga; (b) Titanium coupon with diverse flora of alga; (c) brass coupon with single type of filamentous alga.

Effect of Seasonal Changes in Water Quality on Biofouling and Corrosion 273

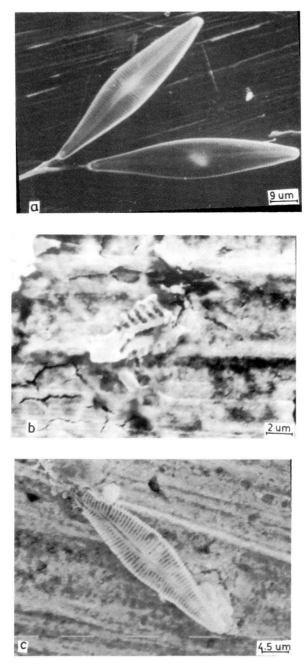

Fig. 9 SEM pictures showing (a) firmly attached viable diatoms on the SS coupon exposed in the reservoir water, (b) and (c), frustules of diatoms on brass and admiralty brass coupon respectively.

4. Summary and Conclusions

A study was conducted to understand the seasonal changes in the quality of Palar water and the open reservoir water and its influence on the corrosion rates of carbon steel and biofouling rates of stainless steel, brass, titanium and admiralty brass. Salient conclusions of the above study are the following:

1. The quality of inflow water was found to change on storage in the open reservoir.

2. The reservoir water quality also showed seasonal changes with an increase in the nutrient content and conductivity during the months of May to September '92 and a decrease in all the values of water quality parameters except oxygen and silica contents during the north east monsoon in October '92.

3. Corrosion rates of carbon steel determined by short-term exposure in the reservoir water showed an increase from 7.6 mpy (May '92) to 10.6 mpy (August '92) to 15.1 mpy (November '92). The corrosion rates during the months more or less reflected the strong influence of the seasonal (monthly) changes in the microbial density on the coupons. The seasonal changes in the reservoir water (such as L. I., DO_2, pH, TDS and conductivity) seem to have no influence on, or correlate with, the changes in corrosion rates.

4. Tubercles with shallow pitting were observed on all the carbon steel coupons after an exposure for three months in the reservoir water. However, the tubercles did not grow significantly with further exposure time.

5. Studies on the extent of biofouling due to algae on various materials showed a higher biofouling on titanium coupons under short-term exposures, whereas on long-term exposure stainless steel showed maximum biofouling. Brass and admiralty brass coupons showed relatively low fouling.

6. Scanning electron microscopic examination revealed a diverse flora of algae on stainless steel and titanium coupons but only a single type of filamentous alga on brass coupons. Firmly attached diatoms were observed on SS coupons, whereas only frustules of diatoms were observed on brass and admiralty brass coupons.

5. Acknowledgement

The authors are thankful to Mr. E. Chandrasekharan, Micrometerological Laboratory at Kalpakkam, for providing the rainfall data in this study. They are very grateful to Dr. V. P. Venugopalan and Mr. K. K. Satpathy, WSCL, for various discussions and many useful suggestions during the course of this investigation. They are also grateful to Dr. P. K. Mathur, Head, WSCL, IGCAR for providing facilities for microbiological studies.

6. References

1. R. P. Kapoor, D. Jambunathan and R. Selvam, *Proc. Nat. Symp. Water Treatment*, 1987, Madras, India.
2. T. S. Rao, M. S. Eswaran, V. P. Venugopalan, K. V. K. Nair and P. K. Mathur, *Biofouling*, 1993, **6**, 245–259.
3. K. V. K. Nair and S. Ganapathy, *Mahasagar*, 1983, **16**(2), 143.
4. APHA *Standard Methods for the examination of water and waste water*, 14th edition, American Public Health Association, USA, 1989.
5. M. J. Pelczar (Jr), E. C. S. Chan and N. R. Kreig, *Microbiology*, 5th edition, McGraw Hill, New York, 99–147, 1986.
6. T. R. Parsons, Y. Maita and C. M. Lalli, *A Manual of Chemical and Biological Methods of Sea Water Analysis*, Pergamon Press, Oxford, p.173.
7. *Annual book of ASTM standards*, Section 3, **03.02**, G1-90, 39–42, 1986.
8. R. P. George *et al.*, unpublished results.
9. K. V. K. Nair, Ph.D Thesis, University of Bombay, 1979.
10. W. R. G. Atkins and G. T. Harris, *Sci. Proc. Roy. Dublin Soc.*, 1924, **18**, 2–21.
11. L. A. Codispoti, *Productivity of Ocean: Present and Past* (Eds W. H. Berger *et al.*), Dahlem Conf., Wiley, 1989.
12. E. Ramous, M. Magrini, P. Matteazzi and G. Ripetto, *Microbial Corrosion 1*, (Eds C. A. C. Sequeira *et al.*), Elsevier, London, 460, 1988.
13. B. Badan, M. Magrini and E. Rramous, *J. Mat. Sci.*, 1991, **26**, 1951–1954.
14. B. McEnaney and D. C. Smith, *Corros. Sci.* 1980, **20**, 873–886.
15. E. C. Harder, *Iron depositing bacteria and their geologic relations*, U.S. Govt. printing office, Washington D.C, 1919.
16. W. F. Langelier, *JAWWA*, 1936, **28**, 1500–1515.
17. T. S. Rao *et al.*, personal communication.
18. M. Fletcher, *Estuaries*, 1988, **2**, 226–230.
19. J. F. Wilkinson, *Bact. Rev.*, 1958, **22**, 46–73.

21

Corrosion Behaviour of a Carbon Steel Valve in a Microbial Environment

J. C. DANKO, C. D. LUNDIN, N. J. E. DOWLING and W. HESTER*

The University of Tennessee, Knoxville, TN 37996-2200, USA
*Commonwealth Edison, Chicago, IL 60690-0767, USA

ABSTRACT

A limited investigation was performed on a carbon steel valve that was removed from a nuclear power plant service water system after eight years of operation. The study included analysis of the water for total number and types of microorganisms present, chemical analysis of the valve and metallographic and scanning electron microscopy examinations. Results of the investigation indicated that microbiologically influenced corrosion was responsible for the massive pitting corrosion observed in the cast carbon steel valve.

1. Introduction

This paper presents the results of a limited investigation by the Center for Materials Processing and the Institute of Applied Microbiology of The University of Tennessee of a carbon steel valve removed from a pressurised water reactor service water system. The purpose of this investigation was to determine if microbiologically influenced corrosion (MIC) had occurred in the valve. This study consisted of analysis of the service water, a material investigation of the cast carbon steel valve, and a review of the results and conclusions.

2. Service Water Analysis

Samples of the service water were obtained from the power plant based on instructions from The Institute of Applied Microbiology. Pertinent information on the samples is given below:

2.1. Sample Sites

The following are the descriptions of the collection sites.

#1. Generator H_2 side seal oil cooler inlet tubes.
#2. Generator H_2 side seal oil cooler outlet tubes SW side.

#3. Station water cooler 2A 2GC001, End Bell.
#4. 2A station water disc. End Bell.
#5. Tubes 2GC001 station water cooler 2A.

2.2. Sample Retrieval

Scrapings of material were obtained from five different sites and placed into tissue culture bottles. Distilled water was added to each of the samples and the caps were sealed with electrical tape. The samples were placed in a cooler with ice packs and shipped overnight to The Institute of Applied Microbiology.

2.3. Microbiological Analysis

2.3.1. Acridine Orange direct counts

This is a standard procedure described by the American Society for Microbiology manual of methods for general bacteriology (Gerhardt, 1981). The data give a reasonably accurate measurement of the total numbers of bacteria involved and some distinction as to morphology, for example, whether a distinctive organism such as *Gallionella* is present. 1 mL of each sample was added to 9 mL of sterile mineral salts and vortexed. 1 mL of this dilution was then delivered to a Nucleopore™ 0.2 µm pore-size filter. The filter was pre-stained with Irgalan black to reduce autofluorescence. A solution of 0.01% Acridine Orange was then applied to the filter for 2 min and washed through with sterile tap water. The filters were then recovered from the filtration manifold, dried, and supported on a glass slide with a drop of optically pure immersion oil. Subsequent microscopy was with a ZEIS epifluorescent microscope.

2.3.2. Culture methods for heterotrophic aerobic bacteria

The media schedule for growing heterotrophic aerobic colony forming bacteria contains the following: NH_4Cl, $MgSO_4 \cdot 7H_2O$ 0.0, KH_2PO_4 0.027, $CaCl_2 \cdot 2H_2O$ 0.005, K_2HPO_4 0.871, Hepes Buffer 1.3, succinate 0.2 g, Trypticase soy 0.5, yeast extract 0.15 gL^{-1} and Hutner's Mineral Base 10 mL L^{-1}. 1.5% agar was added to make agar plates. Each sample was tested for growth on the agar plates at several dilutions, and colonies were counted after 24 h growth in an incubator at 30°C.

2.3.3. Culture methods for sulphate-reducing bacteria

The American Petroleum Institute has issued a specific medium bulletin which specifies a nutrient make-up which will isolate a broad distribution of sulphate-reducing bacteria. This medium is now issued by DIFCO Inc. and contains (gL^{-1}): Bacto Yeast extract 1.0, Ascorbic acid 0.1, sodium lactate 5.2, $MgSO_4$ 0.2, K_2HPO_4 0.01, $Fe(NH_4)SO_4$ 0.1, NaCl 10. The final pH was adjusted 7.2. The presence of these specific bacteria with the potential for corrosion by accelerating the cathodic processes is generally taken as a positive indication that MIC is occurring. Three dilutions were inoculated for each sample from the 10 mL dilution in order to have a representative assay.

2.3.4. Culture methods for iron oxidising bacteria

To test for heterotrophic iron-precipitating bacteria, we used a ferric ammonium citrate medium consisting of: $(NH_4)2SO_4$, 0.5; $NaNO_3$, 0.5; K_2HPO_4 0.5; $MgSO_4 \cdot 7H_2O$ 0.5; and ferric ammonium citrate, 10.0 gL^{-1}. The pH was adjusted to 6.6 to 6.8 and sterilised, adding 1.5% agar for plating. Several dilutions were plated for each sample and incubated at 30°C for 24 h.

2.4. Results

2.4.1. Acridine Orange direct counts — analysis

Acridine orange direct counts are analysed with the following formula:

$$\text{Cells/mL} = \frac{\text{Average Number/field} \times \text{Area of Filter} \times \text{Dilution}}{\text{Area of field} \times \text{Area filtered}}$$

The results for microscopical analysis are presented in Table 1, showing a distribution of bacteria in the order of 10^7 cells/mL up to 10^9 cells/mL. This puts the number of bacteria in the range where MIC can be expected.

The results shown in Table 2, indicate that samples #1 and #3 have slightly lower aerobic bacterial counts than samples #2, #4, and #5. Cell numbers ranged between the orders of 10^6 and 10^7 cell/sample.

Significant numbers of SRB were found in four out of the five samples. Sample #5 was the only sample that did not show any growth of SRB. Samples #2 and #4 showed the greatest numbers of SRB, and samples #1 and #3 showed moderate SRB growth. Sufficient numbers of SRB are present to assume that they could contribute to MIC. In those samples containing SRB, numbers ranged between 10^3 and 10^5 cells/sample. Results are in Table 3.

Table 1 Acridine Orange direct counts

Sample Number	Cell Numbers/Sample
#1.	3.0×10^9
#2.	1.3×10^9
#3.	3.8×10^7
#4.	1.5×10^3
#5.	3.2×10^8

Table 2. Viable counts on aerobic media

Sample Number	Cell Numbers/Sample
#1.	2.7×10^6
#2.	3.9×10^7
#3.	3.5×10^6
#4.	1.5×10^7
#5.	3.4×10^7

Table 3. Sulphate-reducing bacteria

Sample Number	Cell Numbers/Sample
#1.	1.6×10^4
#2.	5.6×10^5
#3.	1.8×10^3
#4.	2.3×10^5
#5.	SRB < 600 in 20 mL sample

Only sample #2 was found to contain iron precipitating bacteria, in the order of 10^3 cells/sample.

2.5. Chemical Analysis

2.5.1. Chloride analysis

Chloride concentration in the deposits on the walls of the pipe samples sent were determined using potentiometric titration with silver nitrate solution and a silver/silver chloride electrode system. During titration a Sycopel voltmeter was used to measure the change in potential between the two electrodes. The sample was dissolved by heating at 50°C for 1 h, and then 20 mL were added to 0.01M HNO_3. The $AgNO_3$ is added in increments and the open-circuit potential is recorded each time. The end point of the titration is that point at which the greatest change in voltage has occurred. Results are found in Table 4.

3. Materials Investigation

This section of the report covers the metallurgical examination of the valve. A variety of examinations were performed on the valve. These included: visual examination and photographic documentation, sectioning of the valve and the preparation of metallographic samples for optical light microscopy (OLM), scanning electron microscopy (SEM), energy dispersive X-ray (EDX) analysis and chemical analysis.

Table 4. Chloride analysis

	Sample No.
Edge	1 – 0.62 mg g^{-1} sample
Edge	2 – 0.55 mg g^{-1} sample
Edge	3 – 1.11 mg g^{-1} sample
Weld	1 – 1.01 mg g^{-1} sample
Weld	2 – 0.82 mg g^{-1} sample
Curve	1 – 0.93 mg g^{-1} sample
Curve	2 – 1.30 mg g^{-1} sample

Fig. 1 *Valve appearance; note rust (corrosion) on exterior surface.*

Fig. 2 *Valve interior; note extensive rust (corrosion) at entrance into the valve.*

(a)

(b)

Fig. 3 *Heavy corrosion and deep pitting penetration in valve body wall in different sections.*

3.1. Visual Examination

The valve as delivered to The University of Tennessee is shown in Fig. 1. Note the rust (corrosion) on the exterior surface of the valve and the heavy corrosion in some areas of the surface. A view of the valve interior showed extensive corrosion at the entrance into the valve at the flange section (Fig. 2).

The valve body was sectioned using a dry cut-off wheel to avoid cooling water reactions with the carbon steel material. Sections of the valve were examined and corrosion attack was observed over the entire internal surface. However, this corrosion was not uniform; some areas exhibited very heavy corrosion with extensive pitting penetration into the valve body. Sections with deep pitting corrosion into the valve wall are shown in Figs. 3(a) and 3(b).

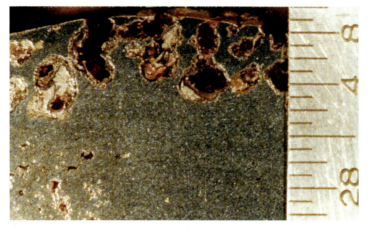

Fig. 4 Photomacrograph of sample with deep pitting corrosion.

Fig. 5 Photomacrograph of sample with extensive surface corrosion and deep pitting penetration.

3.2. Chemical Analysis of Valve

Information obtained from the utility on the material used in the valve was that it was a cast carbon steel. There were no records available on the chemistry of the material or valve fabrication. In order to confirm the chemistry of the valve material, drillings were taken and used for a complete chemistry. The results are reported in Table 5. Based on the results, the material is a C–Mn–Si steel with 0.23 wt% carbon. The valve body was fabricated by casting and the flanges were welded to the main valve section.

3.3. Microscopic Examination

Metallographic samples were removed from several sections of the valve body using the dry cut-off technique described previously. The samples were mounted in epoxy and polished. Photomacrographs of two sections are shown in Figs. 4 and 5 (see previous page).

The sample shown in Fig. 4 was removed from a section of the valve with a relatively smooth surface but with clear evidence of pitting. Note the extensive pitting and the depth of the pitting corrosion. The pitting extended approximately 20% into the wall thickness of the valve. The pits are probably connected with each other via channels. The pitting damage to the valve body is severe.

The sample in Fig. 5 was removed from a section of the valve that had a very irregular surface with features of deep craters. As seen in this figure extensive surface corrosion had occurred and pits occurred below this corrosion area. Pitting penetration of approximately 16% was slightly less than that measured for the sample in Fig. 4.

3.3.1. Macrosegregation in castings

A phenomenon that occurs in castings is macrosegregation of alloying elements and impurities. This is a common occurrence in castings and carbon steel castings are no

Table 5. Chemical analysis of valve

Element	wt%	Element	wt%
Carbon	0.23	Phosphorus	0.017
Manganese	0.84	Sulphur	0.024
Nickel	0.06	Silicon	0.41
Chromium	0.06	Copper	0.05
Molybdenum	0.02	Aluminium	0.051
Vanadium	0.003	Boron	< 0.001
Columbium	< 0.001	Arsenic	0.006
Titanium	0.001	Tin	0.004
Cobalt	0.007	Tungsten	< 0.01

exceptions. In plain carbon steel castings elements such as carbon, manganese, sulphur and phosphorus have been reported to segregate. This segregation can influence the mechanical properties, microstructure and corrosion properties.

Segregation of the various constituents in the steel may be minimised by homogenisation or annealing heat treatments. At the high temperatures of this process diffusion of the various elements will occur, thereby reducing the concentration gradients that exist. This results in more uniform properties.

Since there was no information available on the casting regarding heat treatment, samples of the castings were prepared for metallographic examination to determine the microstructural characteristics and whether macrosegregation was present. A representative microstructure of the cast valve in the etched condition is shown in Fig. 6. As seen in this figure light and dark areas are evident. In a slowly cooled carbon steel having the chemical composition of this valve, two constituents will exist at room temperature. These are ferrite, the white constituent, and pearlite, the dark constituent. Ferrite is a single phase of iron with minor alloying elements in solid solution. Pearlite on the other hand consists of two phases, ferrite and cementite (Fe_3C, a chemical compound). The amounts of each constituent and their distribution is a function of the chemical composition and the cooling rate from the high temperature of casting or heat treatment.

To determine the existence of macrosegregation, a special chemical etchant (reagent) was used. Following this etch, evidence of macrosegregation was observed as

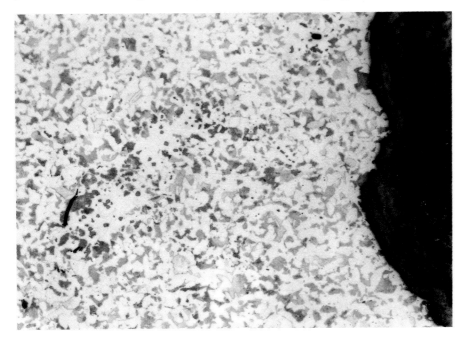

Fig. 6 Representative microstructure of cast valve body showing ferrite (white) and pearlite (dark) constituents; Nital etch, 100×.

illustrated in Fig. 7. A non-uniform microstructure is noted and the island near the centre of the photomacrograph shows pronounced segregation. This is important in understanding the behaviour of the material in corrosive environments, particularly in the presence of microorganisms. A discussion of this will be presented in the ensuing sections of this report.

3.3.2. *Corrosion of valve body*

The photomacrographs in Figs. 4 and 5 show extensive corrosion and deep corrosion pits. In this portion of the report, metallographic examination of the corrosion areas are reported and discussed.

A large corrosion pit originating at the internal surface and penetrating into the carbon steel valve is shown in Fig. 8 (unetched condition). The entire pit could not be captured at 50 magnifications, indicating the size of the pit. Note the small opening at the surface, the size of the pit cavity and the corrosion product attached to the carbon steel.

In another area of the valve, a shallow and a dip pit were observed as shown in Fig. 9. The carbon steel was etched to show the microstructure in this figure. Again, the heavy corrosion product is observed along the pit walls. To understand fully the extent of corrosion, one must visualise the third dimension of the photomicrographs, namely the depth of the pits and interconnection with other pits.

Extensive linkage and channelling was observed but could not be captured either by light microscopy or scanning electron microscopy.

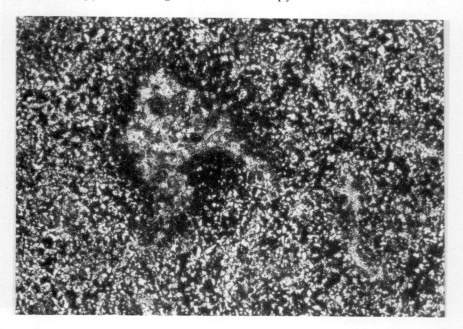

Fig. 7 *Macrosegregation in cast valve body. Note nonuniform microstructure and pronounced segregation in the island near the centre of the photomicrograph; 50×.*

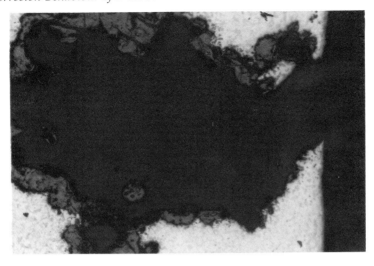

Fig. 8 Deep corrosion pit on internal surface of the valve; Unetched, 50×.

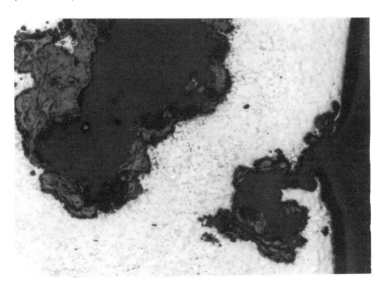

Fig. 9 Shallow internal surface and deep subsurface corrosion pit in the valve; Nital etch, 50×.

In one of the samples cavities were observed that were well within the valve body. There was evidence of corrosion products in the cavities. The shape of the cavities appeared to be more characteristic of porosity which is not uncommon in cast materials. The presence of the corrosion products at these locations suggests the possibility that the pores were connected by channels to the deep corrosion pits. Thus, the corrosion environment was in communication with the shrinkage cavities or porosity.

Examination of the samples was performed at higher magnifications (200–1000×). A most interesting observation was made regarding the corrosion attack on the pearlite constituent of the carbon steel microstructure. In Fig. 10 (200×), large corrosion areas surround what remains of the carbon steel. The preferential corrosion of the pearlite (dark areas) and the considerable ferrite/corrosion product interfaces are obvious in this figure.

At higher magnifications (1000×), the selective corrosion attack of a pearlite colony was clearly evident. As shown in Fig. 11, corrosion is occurring in the islands of pearlite while the ferrite remains unattacked. Note that some of the pearlite areas are completely consumed (uniform dark areas) while discrete regions of corrosion appear within some of the pearlite colonies.

Further evidence of the selective corrosion of the pearlite constituent is illustrated in Fig. 12. Note the final remnants of the microstructure which is completely surrounded by the corrosion product. The microstructure is being stripped of pearlite via corrosion. The remaining islands of ferrite (white) show no internal corrosion; only at the ferrite–corrosion product interface. It is important to note that the ferrite is not immune to the corrosion in the aqueous environment within the valve, but rather that it has a greater resistance to corrosion than the pearlite.

3.4. Scanning Electron Microscope (SEM) Examinations

Use of SEM provided examination of the samples at high magnifications, surface observations with great depth of focus and the ability to perform energy dispersive X-ray (EDX) analysis.

Surface examinations were performed on the corrosion sample that appeared to have a relatively smooth surface with corrosion pits and the sample with extensive

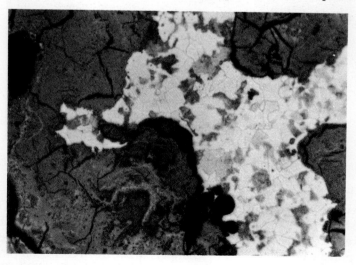

Fig. 10 *Massive corrosion around a remaining island of carbon steel valve material with selective attack on pearlite (dark constituent); Nital etch, 200×.*

Fig. 11 Selective corrosion of pearlite in the microstructure of the valve; Nital etch, 1 000×.

Fig. 12 Corrosion observed in an internal cavity (porosity) of the valve body; Nital etch, 50×.

surface corrosion with pits. These are illustrated in Figs. 13(a) and (b) respectively. Note the significant differences in these surfaces; the more regular surface of (a) and the deep surface corrosion of sample (b). Also, note the depth of corrosion pits in sample (b). These should be compared to the macrographs of these samples shown in Figs. 4 and 5 respectively.

Using the resolution and higher magnification capability of the SEM, examination of the pearlite corrosion was performed. Figure 14 at (1500×) shows corrosion product surrounding a ferrite–pearlite region of the carbon steel valve. Note the corrosion of the pearlite at the corrosion product–pearlite interface and the internal

(a) Unetched

(b) Unetched

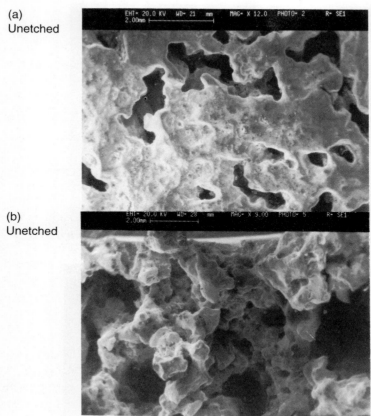

Fig. 13 SEM of internal surface of valve (a) Sample with relatively smooth surface and corrosion pits. (b) Sample with surface corrosion and corrosion pits.

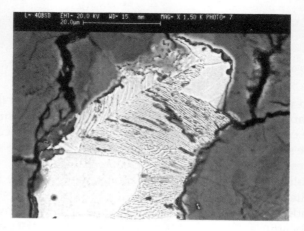

Fig. 14 SEM view of selective corrosion of pearlite and corrosion of ferrite; Nital etch.

corrosion of the pearlite colony. Also, note the corrosion of ferrite at the ferrite–corrosion product interface at the top of the micrograph. Clearly the corrosion of the pearlite appears to be more selective relative to the ferrite.

Energy dispersive X-ray (EDX) analysis was performed on the surface of the corrosion sample shown in Fig. 13(b). The results are presented in Fig. 15, as a plot of X-ray intensity vs energy. Salient features in this figure are the high peaks for sulphur (S) and evidence of chlorine (Cl), relative to that of Fe, the basic element in the carbon steel. The presence of sulphur and chlorine are generally very good indicators of microbiologically influenced corrosion (MIC).

EDX of the corrosion product shown in Fig. 14 revealed the presence of iron (Fe) as the major constituent with very minor amounts of manganese (Mn), silicon (Si) and sulphur (S). While chromium (Cr) and phosphorus (P) appear in the analysis, they are essentially at background levels. Sulphur is significant from the relationship to MIC.

Inclusions observed in the ferrite were analysed by EDX for elemental composition. The inclusions are shown in the photomicrograph of Fig. 16, while the EDX results are presented in Fig. 17. Based on the X-ray data, the inclusions are clearly manganese sulphide (MnS). Their presence may be important in the MIC behaviour of the carbon steel.

4. Discussion

The results of the analysis of the service water samples showed a distribution of bacteria in the order of 10^7–10^9 cells mL^{-1}, a range where MIC can be expected. In addition, sufficient numbers of sulphate reducing bacteria (SRB) were present to contribute to the MIC phenomenon. Based on these data, evidence of MIC was anticipated in the examination of the carbon steel valve body.

The valve body appeared to be fabricated by sand casting. Chemical analysis of the material revealed a C–Mn–Si steel chemistry. Hence, the valve body is referred to as a plain carbon cast steel. Metallographic examination of the material showed a two constituent microstructure of ferrite and pearlite, consistent with a medium carbon cast steel. Porosity and some segregation of principally manganese and sulphur were also revealed. Inclusions of manganese sulphide were observed scattered throughout the ferrite phase.

In the as-received condition, the valve had corrosion on the external surface. When the valve was sectioned, massive corrosion of the internal surface was observed. The amount of corrosion varied from one end of the valve to the other. In some areas, the surface was relatively smooth while in other parts the surface was highly irregular. Nevertheless, extensive corrosion pits were present in both areas. Many of the pits appeared to be interconnected or channelled. A heavy corrosion product was clearly evident in the pits. Analysis of the corrosion product using EDX of the scanning electron microscope revealed the presence of iron with very small amounts of sulphur and chlorine. Both of these elements usually accompany MIC.

Microstructural observations of the corroded areas showed a preferential attack

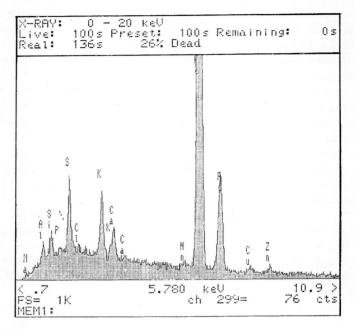

Fig. 15 *EDX analysis of surface of sample shown in Fig. 4.*

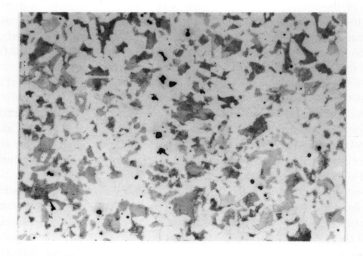

Fig. 16 *Inclusions (dark) in microstructure; Nital etch, 200×.*

on the pearlite. While the ferrite was also corroded, the corrosion process progressed through the pearlite colonies and appeared to bypass the ferrite. Some areas of the sample showed only isolated islands of ferrite remaining. Eventually the ferrite would be consumed by the aggressive corrosive environment.

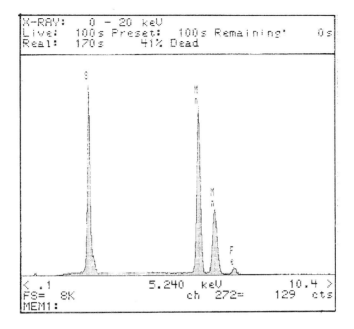

Fig. 17 EDX analysis of inclusion shown in Fig. 16.

It is indeed unfortunate that information on the valve operation was not available. For example, flow rates of the water, start-ups and shut-downs, time in the presence of stagnant water conditions, location and orientation of the valve in the service water system, etc. These factors are very important in understanding the corrosion behaviour of the valve. Stagnant conditions would encourage corrosion while high flow rates minimise MIC.

Given that sufficient bacteria including the SRB were present and the nature of the heavy pitting corrosion, MIC is likely the cause of massive corrosion of the carbon steel valve. The presence of macrosegregation would exacerbate the corrosion process. Sources of sulphur within the carbon steel are important for SRB to attack the material. Residual sulphur in the steel along with the macrosegregation would provide such a source. Also, the presence of manganese sulphide inclusions dispersed throughout the material would provide considerable sulphur. Manganese sulphide inclusions were identified by X-ray analysis and fairly large distribution observed in the microstructure.

5. Conclusion

Based on all of the results of this limited study, MIC would appear to be the major contributor to the massive pitting corrosion observed in the carbon steel valve.

6. Acknowledgements

The authors wish to recognise contributions from the individuals listed: Cary B. Jozefiak, Commonwealth Edison, J. Guezenneck, M. Mittleman and Jan Bullen from the Institute of Applied Microbiology at The University of Tennessee.

22

Sulphide-Producing, not Sulphate-Reducing Anaerobic Bacteria Presumptively Involved in Bacterial Corrosion

M. MAGOT, L. CARREAUL, J.-L. CAYOL*,
B. OLLIVIER* and J.-L. CROLET[†]

Sanofi ELF Bio Recherches, Microbiologie, F-3 1676 Labege, France
*Orstom, Laboratoire de Microbiologie, 3, Place Victor Hugo F-13331 Marseille, France
†Elf Aquitaine Production, F-64018 PAU, France

ABSTRACT

A bacteriological study of produced waters from an oil field in west Africa was started after the corrosion failure of a major pipeline. Several strictly anaerobic strains were isolated, including sulphate-reducing, methanogenic, and fermentative bacteria.

Some of the fermentative strains were shown to produce hydrogen sulphide by reduction of thiosulphate, but none of them reduced sulphate.

According to a recent model of microbially induced corrosion, such bacteria producing both hydrogen sulphide and organic acids as terminal end products of their metabolism could be involved in corrosion failures as well as sulphate-reducing bacteria. This model also stated that this corrosive activity could be dramatically enhanced when thiosulphate is used as the terminal electron acceptor. This compound was detected in the produced waters of the oil field where the failure occurred.

The main physiological and metabolic characteristics of these non-SRB, thiosulphate-reducing strains are described here, and their potential implication in bacterial corrosion is discussed.

1. Case History

In 1989, one of our subsidiaries in West Africa observed the corrosion of a 10 year old, 23 km main subsea line. This pipeline transported sour oil. The first five kilometres, corresponding to the corroded segment, were replaced. But one year later, this new line was again corroded. The penetration rate of this pitting corrosion was thus at least 1 cm per year! The line was then entirely replaced, and an extensive chemical and microbiological study was started.

A piece of the corroded metal is shown in Fig. 1. Several pits of different diameters can be seen, some of them up to 4 cm wide. From the closer view shown in Fig. 2, it is clear that the dissolution process was different within and outside the pit. A classical tunnelling, which is a signature of a locally acidic medium, can also be seen. Figure 3 shows an extensive pit nucleation process. The small nuclei can compete, and some of them become larger and larger.

294 *Proceedings of the 3rd International EFC Workshop on Microbial Corrosion*

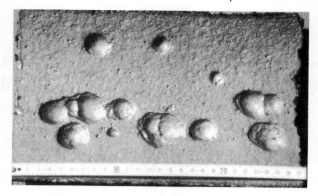

Fig. 1 *View of the corroded pipe line showing pits of different sizes.*

Fig. 2 *Closer view of the lower right part of Fig. 1.*

Fig. 3 *Closer view of the lower left part of Fig. 1.*

Sulphide-Producing, not Sulphate-Reducing Anaerobic Bacteria in Corrosion

The first microbiological data we obtained indicated that a lot of sulphate-reducing bacteria (SRB) were contaminating the pipeline water. We also noticed that thiosulphate can be detected in the water, as the result of H_2S oxidation due to oxygen entries. According to a recent model describing the acid metabolism of SRB, thiosulphate is suspected to be more dangerous for bacterial pitting corrosion than sulphate when reduced by sulphate-reducing bacteria [1]. This is illustrated in Fig. 1. When sulphate is reduced to H_2S by SRB, the resulting net metabolism induces the production of H^+ ions if the pH in the medium is higher than 6.7. At lower pHs, acidity is consumed. The consequence of the interaction of the bacterial metabolism with local chemistry is that SRBs can regulate their local pH around 6.7 in the example given here (an incomplete oxidiser using lactate as energy source). This model was recently experimentally validated [2]. In the pit, if the iron ions formed are precipitated as FeS, the production of H^+ is increased by one H^+ in the overall reaction [1], and the resulting pH decreases to 5.4 (Fig. 4). When thiosulphate is considered in the model as the terminal electron acceptor reduced by SRB, the net metabolism is characterised by a permanent production of acidity on the anode, whatever its pH (Fig. 4). In that case, the pH differential between the anode and the cathode becomes very high. The stabilisation of localised corrosion should be easier, and the dissolution rate higher.

Moreover, in some other *in vitro* experiments, thiosulphate was also shown to stimulate the growth of most SRB strains.

All these observations suggested that the reduction of thiosulphate by SRB could have played a significant role in this unusual case of bacterial pitting corrosion. But unexpectedly, several strains of non-SRB, thiosulphate-reducing anaerobic bacteria were also grown from the pipeline water samples. The question then arose of the role these bacteria could have taken in the corrosion process.

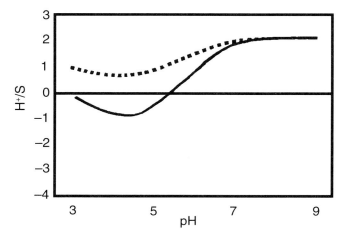

Fig. 4 Computerised modelling of the acidity production expressed as the ratio of H+ ions per atom of sulphur used by SRB, vs pH of the medium, by a sulphate-reducing bacterium using sulphate (dark line) or thiosulphate (dotted line) as electron acceptor.

2. Isolation and Preliminary Characterisation of the Bacterial Population

The bacterial counts of the main physiological and metabolic groups were determined in both the pipeline water, as planktonic bacteria, and on bioprobes, as sessile bacteria (Table 1).

The total bacterial count was shown to be 4.10^5 bacteria per mL in the water, by Acridine Orange Direct Count [3]. Most of the cultivable planktonic bacteria were SRB, with about 10^3 viable bacteria per mL. Other physiological and metabolic bacterial groups were also detected: fermentative anaerobes and strict or facultative aerobic bacteria.

A slightly different feature was exhibited with bioprobes (Table 1). Although SRB were here dominant too, fermentative anaerobes were in greater proportion in the biofilm than in the water, with about 10^5 cultivable bacteria per cm^2. The aerobic population was not studied in this experiment.

All the strains that have been grown in these experiments were characterised and tentatively identified. The sulphate-reducing population was very diverse, since at least 8 different strains were recognised. Five *Desulfovibrio desulfuricans* were identified, but these strains differed by the restriction fragment length polymorphism of their ribosomal RNA genes [4]. One strain of *Desulfovibrio longus*, and two unknown SRB strains, one of which being a new *Desulfovibrio* species, were characterised.

The aerobic bacteria were identified as being species of the *Pseudomonas*, *Micrococcus*, and *Vibrio* genera.

The fermentative anaerobes were, to our knowledge, all previously undescribed. Most of them can reduce thiosulphate into hydrogen sulphide.

We then looked for similar thiosulphate-reducing anaerobes at the wellheads from which production was transported in the pipeline. Several similar strains were isolated (Table 3). Later, different strains having the same ability to reduce thiosulphate were isolated from other wellheads in the Paris Basin or in the Cameroons (thermophilic strains), from another pipeline in the Congo (salinity 12%), and even from a pipeline in the south of France transporting a brine containing 30% sodium chloride.

3. Metabolic Characteristics of Thiosulphate-reducing Bacteria

The main morphological, physiological, and metabolic characteristics of the thiosulphate-reducing anaerobes isolated from the corroded pipeline water or from wellhead samples are summarised in Tables 2 and 3, respectively.

They are all Gram-negative, strictly anaerobic bacteria. Their physiology is in agreement with the physico–chemical parameters of their ecosystem: they are mesophilic bacteria with an optimum temperature for growth close to 30°C, and moderately halophilic, able to grow in the ecosystem of which the salinity is 4–5%.

Although some of these strains cannot use sugars as carbon and energy sources, they are all, at least under certain conditions (see below), able to use peptides as substrates. They produce organic acids as terminal end-products of their metabo-

Table 1. Bacterial counts. (ND = no data.)

	Planktonic bacteria (p/mL)	Sessile bacteria (bioprobes) (p/cm^2)
Total bacterial count (epifluorescence microscopy)	4×10^5	ND
Cultivable SRB	9.5×10^2	8×10^5
Cultivable fermentative anaerobes	4.5	8×10^4
Cultivable aerobic bacteria	4.5	ND

Table 2. Characteristics of thiosulphate-reducing strains (pipeline)

	SEBR 4226	SEBR 4205
Morphology	Gram negative rods	Gram negative rods
Respiratory type	Strictly anaerobic	Strictly anaerobic
Growth temperature	30°C	30°C
Growth salinity	2–7.5%	0.1–7.5%
Sugar fermentation	–	+
Peptide utilisation	+	ND
Fermentation products	Citrate, pyruvate, α-ketoglutarate	Citrate, pyruvate
Sulphide production from		
SO_4^{2-}	–	–
SO_3^{2-}	–	–
$S_2O_3^{2-}$	+	+
S^0	–	–
Cystein	+	+

lism. None of the strains can reduce sulphate or sulphite, but all reduce thiosulphate into hydrogen sulphide.

These strains display previously undescribed metabolic features that could be of great significance considering their potential role in bacterial corrosion. As mentioned before, peptides can be used as carbon and energy sources by these strains. This metabolic activity allow them to grow within the biofilm, where peptides can be considered as a common substrate: they originate from the spontaneous degradation of the biofilm components, e.g. dying bacteria.

Another striking characteristic is that several of these strains can use peptides only if thiosulphate is available. This is the case for strain SEBR 4268 in Table 4.

Table 3. Characteristics of thiosulphate-reducing bacteria (wellheads)

	SEBR 4224	SEBR 4207	SEBR 4198	SEBR 4211
Morphology	Gram negative rods	Gram negative vibrios	Gram negative rods	Gram negative rods
Respiratory type	Strictly anaerobic	Strictly anaerobic	Strictly anaerobic	Strictly anaerobic
Growth temperature	30°C	30°C	30°C	30°C
Growth salinity	5–15%	2–10%	0–20%	3–5%
Sugar fermentation	+	−	+	+
Peptide utilisation	+	+	+	+
Fermentation products	Acetate, (propionate)	Acetate, propionate, (lactate, isovalerate)	Acetate, propionate, pyruvate	ND
Sulphide production from: SO_4^{2-}	−	−	−	−
SO_3^{2-}	−	−	−	−
$S_2O_3^{2-}$	+	+	+	+
S^0	−	−	−	+
Cystein	−	−	+	−

Table 4. Metabolic behaviour: peptides and thiosulphate

	SEBR 4198 SEBR 4211	SEBR 4207	SEBR 4226	SEBR 4268*
Peptide utilisation				
Without thiosulphate	+	+	+	−
with thiosulphate	+	+	+	+
Inhibition by hydrogen				
without thiosulphate	−	+	+	+
with thiosulphate	−	−	+	−
Growth stimulation by thiosulphate	−	+	+	+
Total [S^{2-}] (mM)	2	15	13	30

*Thermoanaerobacter sp. isolated from a wellhead sample in the Paris Basin.

The growth of several strains is inhibited in the presence of hydrogen (strains SEBR 4207, 4226 and 4268 in Table 4). But, in some cases, the growth resumes in the presence of thiosulphate (SEBR 4207 and 4268 in Table 4) [5]. The reduction of thiosulphate thus appears to act here as a detoxification mechanism against hydrogen toxicity. The growth of several strains is also stimulated in the presence of thiosulphate: higher growth rates and higher bacterial densities were observed in laboratory experiments.

The most significant property of these bacteria regarding bacterial corrosion is probably their H_2S production. Most of the strains we have studied produce hydrogen sulphide *in vitro* in much larger amounts than SRB, which are generally inhibited above 3 to 5 mM sulphide in the growth medium. Here, we detected more than 10 mM, and up to 30 mM total sulphides in the culture media (Table 4).

4. Discussion

We described here the isolation and characterisation of strictly anaerobic bacteria which can reduce thiosulphate into hydrogen sulphide. Although thiosulphate reduction is a common metabolism in the bacterial world [6], its implication regarding bacterial corrosion has never been investigated. Nevertheless, thiosulphate is naturally more or less abundant in ecosystems where hydrogen sulphide and oxygen come into contact, as in industrial facilities such as the sour oil-transporting pipeline we studied here.

Three main characteristics we describe in this paper were not previously described, and could be responsible for the implication of these bacteria in corrosion processes: (i) these bacteria can use peptides as carbon and energy sources for their growth, sometimes only if thiosulphate is available in the medium; (ii) when the development of some strains is inhibited by hydrogen, this inhibition can be suppressed by hydrogen oxidation coupled to thiosulphate reduction; (iii) very large quantities of hydrogen sulphide can be produced together with organic acids. The occurrence of such bacteria in samples taken from a facility corroding at a very unusual high rate, the reading of their metabolism in the light of a recently described model of microbially influenced corrosion, lead to the hypothesis that thiosulphate reduction by non SRB might be responsible of some cases of MIC. New experiments are now in progress to evaluate their pitting ability more directly [7].

References

1. J. L. Crolet, S. Daumas and M. Magot, 'pH regulation by sulfate-reducing bacteria', *Corrosion '93*. Paper No. 93303, NACE, Houston TX, 1993.

2. S. Daumas, M. Magot and J. L. Crolet, 'Measurement of the net production of acidity by a sulphate-reducing bacterium: experimental checking of theoretical models of microbially influenced corrosion', *Res. Microbiol.*, 1993, **144**, 327–332.

3. J. E. Hobbie, R. J. Daley and S. Jasper, 'Use of Nuclepore filters for counting bacteria by fluorescence microscopy', *Appl. Environ. Microbiol.*, 1977, **33**, (5), 1225–1228.

4. F. Grimont and P. A. D. Grimont, 'Ribosomal ribonucleic acid gene restriction patterns as potential taxonomic tools', *Annales de l'Institut Pasteur/Microbiologie*, 1986, **137B**, 165–175.

5. M. L. Fardeau, J. L. Cayol, M. Magot and B. Ollivier, 'H_2 oxidation in the presence of thiosulphate, by a *Thermoanaerobacter* strain isolated from an oil-producing well', *FEMS Microbiol. Lett.*, 1993, **113**, 327–332.

6. A. Le Faou, B. S. Rajagopal, L. Daniels and G. Fauque, 'Thiosulfate, polythionates and elemental sulfur assimilation and reduction in the bacterial world', *FEMS Microbiol. Rev.*, 1990, **75**, 351–382.

7. J. Guezennec, M. W. Mittelmann, J. Bullen and D. C. White, 'Stabilization of localized corrosion on carbon steel by sulphate-reducing bacteria', *UK Corrosion '92*, London, publ. Inst. corrosion, Leighton Buzzard, UK.

23

Biofilm Monitoring and On-line Control: 20-Month Experience in Seawater

G. SALVAGO, G. FUMAGALLI, P. CRISTIANI[†] and G. ROCCHINI[†]

Dipartimento di Chimica Fisica Applicata, Politecnico di Milano, Piazza Leonardo da Vinci 32 - 20133 Milan, Italy
[†]ENEL Spa, Environment and Materials Research Center, Via Rubattino 54 - 20134 Milan, Italy

ABSTRACT

The behaviour of an austenitic stainless steel and an Al-brass exposed to seawater were studied. Resistance to heat transfer and electrochemical measurements under cathodic polarisation were used as parameters to monitor the presence and growth of biofilm on the internal surface of tubular specimens.

Experimental results, obtained in a 20-month-run, indicated that electrochemical measurements and monitoring of biofilm significantly depend on the presence of sulphides. Furthermore, sulphides could mask the presence of biofilm even days after their disappearance from the seawater. In the absence of sulphide pollution, the behaviour of the alloys could be divided into three stages of times that were dependent on seasons.

During biocide treatments, the electrochemical response was influenced by the nature of the antifouling compound in use and by the stage reached by the system when the operation started.

Treatments of seawater with continuous addition (1 mg L^{-1} of Cl$_2$, or 0.1 mg L^{-1} of ClO$_2$, or HCl to pH 6.3) starting at the beginning of the run, were able to keep the surface of the two materials clean from fouling. In this case, the electrochemical and heat exchange resistance measurements were in agreement.

No formation of fouling was found when the current intensity observed on the monitoring cell was used to control occasional treatments of seawater with HCl to pH 6.3.

On six-month fouled tubes, occasional additions of NaClO to 1 gL^{-1}, or ClO$_2$ to 10 mg L^{-1}, or HCl to pH 4, which lasted 0.5 h, were not able to minimise the heat exchange resistance immediately, but influenced the electrochemical measurements. The oxidising agents Cl$_2$ or ClO$_2$ led to an increase of the current on galvanically coupled iron–stainless steel samples, decreased the cathodic protection throwing power and increased the cathodic current output from the stainless steel of the monitoring cell. These are the same effects observed when biofilms are formed on materials. On the other hand, the effects of acid additions in seawater were opposite to those observed in the presence of biofilms.

Electrochemical monitoring of biological activities, based on the evaluation of the depolarisation effects they induce on cathodic processes, may be doubtful or even misleading when sulphides or electroactive oxidising substances are present.

1. Introduction

It is well-known that biofilms develop on metallic surfaces in contact with water and that this is one of the factors responsible for corrosion of metals and alloys. In the case of heat exchange equipment cooled with seawater biofouling as well as corrosion, will increase the resistance to heat transfer and lead to a decrease in efficiency. For thermal power plants, for instance, the decrease of the thermal efficiency increases the specific consumption of fuel so that costs for electric energy production increase [1, 2].

Electrochemical methods are widely used to study corrosion induced by microorganisms [3–7] and have been employed to monitor the biofilm growth [7–14]. Some monitoring techniques have been proposed in order to detect heat exchange changes [12, 14–17].

There are several techniques in use or proposed to reduce the formation of biofouling, or to restrain its adverse effects [1–18]. On-line chemical treatments are particularly important and in seawater the most widely used is chlorination [19]. Some authors consider chlorine dioxide to be the ideal biocide for use with industrial cooling plants [20]. Recently simple chemical shock treatments, based on acidification of cooling seawater, have been proposed by Salvago et al. as an antifouling system [12, 21].

Work was therefore conducted to evaluate the reliability of the monitoring of the growth and development of biofouling, as a function of exposure time, on stainless steel specimens. Electrochemical methods, including cathodic polarisation of the control electrode were used together with thermal measurements. The overall problem of control and monitoring of biofouling is, for instance, of great practical importance when large steam condenser are cooled by seawater with and without biocide treatments. Thus, a monitoring system could be used to control the operation of an antifouling program.

2. Experimental Procedure

The tests were carried out at an experimental area near to the ENEL thermal power station in Vado Ligure, on the Tyrrhenian coast of Italy.

Seawater was derived from the cooling-water supply canal. The abstraction was made downstream to the pumps and the water was discharged by bucket mill so as to interrupt the continuity of the electric path from the intake point to that of discharge. This procedure was used to avoid any stray-current flow through the sea-experimental apparatus loop.

The experimental apparatus consisted of two identical parallel branches containing tubular specimens of austenitic stainless steel (UNS S31254) and aluminium brass (UNS C68700). One line (U) was fed with untreated seawater and: controlled amounts of additives were added to the other (A).

Each branch had sections devoted to:

- optical observations of the internal surface of the specimens by endoscope;
- evaluation of the heat exchange resistance of both the stainless steel specimens and the aluminium brass specimens;
- electrochemical monitoring of the biofilm;
- examination of the behaviour of the cathodic protection of stainless steel specimens galvanically coupled with tubular samples of iron;
- examination of the trend of the cathodic protection of aluminium brass specimens by impressed current.

Particular care was taken to reduce electrical interference between the various sections and between the electrochemical cells in any one section. The order in which the cells were installed was chosen so as to avoid substances produced in preceding cells (chlorine developing on the anodes, corrosion products, etc.) affecting other sections.

Potential measurements were performed using Hg/Hg_2Cl_2, $Ag/AgCl$ or Zn electrodes as reference electrodes which were inserted through holes in the walls of the specimens. All electrode potential values were referred to the SCE scale.

To evaluate the variations of the heat-exchange coefficient stainless steel samples were inserted into blocks of lead and aluminium brass samples into blocks of brass. Each block was heated by electrical resistors, so as to create a radial flow of 5 W cm^{-2} on the surface of the stainless steel specimens and 1.5 W cm^{-2} on that of aluminium brass samples. The temperature of the heated specimen in proximity to the fouled surface was then compared to that of a reference specimen, as described elsewhere [12, 14].

The current intensity required to polarise a stainless-steel tubular to -250 mV vs, SCE, was used to monitor the biological activity as previously described [8]. A platinum wire was used as counter electrode. To reduce the effects due to uneven current distribution the length of the cylindrical electrode was limited to one diameter.

The system described in Fig. 1 was used to evaluate the extent of the penetration of cathodic protection throwing power obtained by galvanic coupling of the specimen under examination with an iron tube. A similar system was used for the aluminium brass [14]. Here, however, the potential at the rim of the tube was kept at -400 mV vs SCE by a potentiostat and a platinum counter electrode. This system was sited downstream to the preceding one. It was, in fact, the last section of the line, immediately before the water discharge, so that the considerable amounts of chlorine developing at the platinum anode would not alter the characteristics of seawater.

Tests were carried out between February 1992 and October 1993. The supply of seawater to the apparatus was sometimes interrupted between September 1992 and March 1993 because of the maintenance of the water-supply canal. On these occasions some formation of hydrogen sulphide was observed. This occurred particularly in March 1993. Apart from these occasions, the velocity of the water in the tubes was maintained at 0.5–1.0 ms^{-1}. The concentration of sulphides in the water was below 0.02 mg L^{-1}. The temperature oscillated between 14 and 22°C, according to the season, and O_2 concentration remained in the range of 8–10 mg L^{-1}.

The continuous antifouling treatments on branch A were kept up for periods of

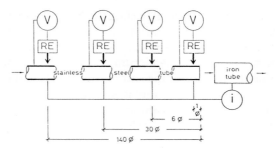

Fig. 1 *Device for evaluating the depth of cathodic protection throwing power in the case of stainless steel tubular specimens galvanically coupled with iron tubes.*

over three months, by adding either NaClO (up to 1 mg L^{-1} of active added Cl$_2$) or ClO$_2$ (up to 0.1 mg L^{-1}) or HCl (to pH 6.3). Additives were occasionally added to both branch (A) and branch (U). In one test a monitoring system was used to control the addition programme and in the others occasional treatments were made by successive additions of increasing amounts of biocide using a step mode. Each step lasted for half an hour, while the choice of addition times depended on previous experience.

3. Results and Discussion

The results obtained in the twenty months of experience show that the performance of the electrochemical monitoring systems used and the electrochemical behaviour of both aluminium brass and stainless steel depend heavily on the presence or on the absence of sulphide contamination, even when its presence is caused by occasional events.

In the absence of contamination the behaviour of both the materials can be divided into three stages each corresponding to different periods of exposure and each lasting a different length of time, depending on the season. The electrochemical behaviour of the materials and the results of the monitoring techniques were found to depend both on the type of antifouling procedures used (NaClO, ClO$_2$, HCl) and on the exposure stage in which the material was found to be, when antifouling procedures started.

3.1. First Stage

At the first stage, after a period varying from 3 to 30 days of exposure to uncontaminated flowing seawater, it was found that:

(i) there was no visible formation of fouling either on the stainless steel or on the aluminium brass;
(ii) no variation in resistance to thermal exchange was observable either on the stainless steel or on the aluminium brass;

(iii) the current intensity, measured on the cells used to monitor the biofilm, was close to zero;
(iv) the current intensity due to the galvanic coupling of iron with stainless steel increased, as found previously [10], and similarly the aluminium–brass cathodic protection current rose, again as before [14]. At the same time, throwing power of the cathodic protection fell both on the stainless steel and the aluminium brass specimens;
(v) In branch (A) with the continuous chlorination treatments (addition of NaClO to 1 mg L^{-1} of Cl_2) as well as the continuous addition of ClO_2 (0.1 mg L^{-1}) or the continuous addition of HCl (pH = 6.3), which were started at this stage, there was also no visible fouling on the specimen surfaces, with seawater flowing at 1 ms^{-1}, for periods of at least 3 months. These treatments lead to an increase of the intensity of the stainless steel–iron galvanic coupling current and of the absolute value of the cathodic current necessary to keep aluminium brass at the selected potential. At the same time, a greater extent of cathodic protection (throwing power) was observed for both materials.

3.2. Second Stage

In the second stage, which followed immediately after the first and lasted variously from 5 to 30 days, it was observed that:

(i) a characteristic, reddish layer formation could be observed on the tubes made of stainless steel, PVC and glass, while no visible fouling formed on the internal surface of the aluminium brass specimens;
(ii) the resistance to thermal exchange of the stainless steel samples (Fig. 2), increased almost linearly over the period, while no such variations were observed in the case of aluminium brass;

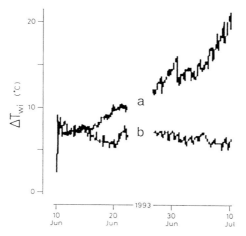

Fig. 2 *Time behaviour of heat exchange resistance relating to stainless steel (second stage): (a) cooling by natural seawater; (b) cooling by seawater with ClO_2 (0. 1 mg L^{-1}).*

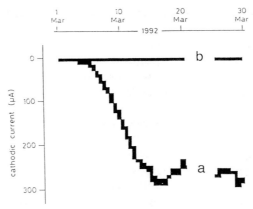

Fig. 3 Current intensity recorded on a stainless steel electrode polarised at −250 mV vs SCE (second stage): (a) fed with natural seawater; (b) fed with seawater acidified on line to pH 6.3 with HCl.

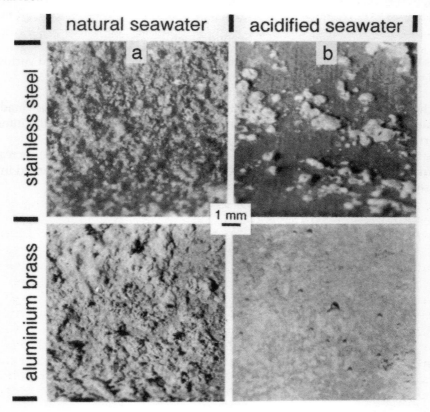

Fig. 4 Macroscopic view of the metal surfaces after: (a) three month exposure to natural seawater flowing at 1 ms^{-1}; (b) three month exposure to seawater occasionally acidified to pH 6.3 with HCl addition controlled by the monitoring system of the biofilm.

(iii) the probe used to monitor the biofilm showed a cathodic current which increased, in absolute values, over the test period, reached a maximum and tended to stabilise at appreciable values (Fig. 3);
(iv) the intensity of iron–stainless steel coupling current increased with time, while that necessary to protect cathodically aluminium brass tended to decrease and then to stabilise. The cathodic protection throwing power tended to diminish constantly in the case of stainless steel. It reached a minimum value in the case of aluminium brass when this alloy exhibited a passive behaviour [14.

In branch (A) occasional chlorination treatments up to 30 mg L^{-1} added Cl$_2$, addition of CO$_2$ to 1 mg L^{-1} or acidification with HCl to pH 6, carried out after a period of exposure to untreated seawater sufficient to attain this stage were able to bring the stainless steel and the aluminium brass back to their previous state. The current intensity observed on the monitoring cell could be used to control the additions of biocide [13] or acid. Figure 4 shows the surfaces of the two materials under examination after three month exposure to seawater in branch (U), or seawater occasionally acidified to pH 6.3 in branch (A). The acid addition was made whenever and for as long as the monitoring cell indicated that the absolute value of the cathodic current for the stainless steel specimen was higher than 0.1 mA.

3.3. Third Stage

In a third stage, immediately following the second, and which lasted till the end of the test with uncontaminated flowing seawater, it was found that in branch (U):

(i) visible fouling had also formed on the aluminium brass specimens;
(ii) heat exchange resistance increased over the test period both for stainless steel and aluminium brass;
(iii) the probe used to monitor the biofilm recorded a significant cathodic current, oscillating over the test period;
(iv) the intensity of the iron–stainless steel galvanic coupling current which oscillated over the period, tended on average to diminish. The cathodic protection throwing power tended to increase over the test period. Such progression was consistent with the formation of calcareous deposits on the stainless steel. The intensity of the cathodic current of aluminium brass specimens, polarised at – 400 mV vs SCE, also oscillated. It tended, in absolute values, to diminish on average over the test period. The cathodic protection throwing power tended to decrease, in contrast to that with the stainless steel. The analysis of the deposits, which was carried out at the end of the period of exposure on the aluminium brass, showed a great amount of diatoms, without a significant contribution to fouling by calcareous deposits. This observation is confirmed by the small content of calcium (Fig. 5);

In branch (A), half-hour treatments with biocides carried out in this stage (chlorination in the range of 10–1000 mg L^{-1} of added Cl$_2$; addition of ClO$_2$ in the range of

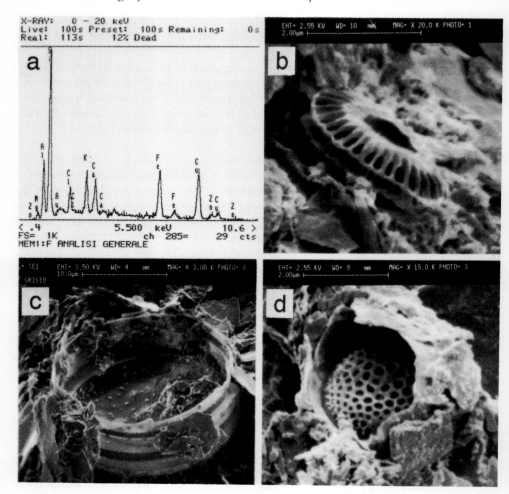

Fig. 5 Surface analysis and SEM micrograph of an aluminium brass specimen after six month cathodic polarisation at –400 mV SCE in natural seawater flowing at 0.5 ms^{-1}; (a) EDS spectrum on 12 mm^2; (b), (c), (d) typical diatoms observable in fouling.

0.1–10 mg L^{-1}; HCl reducing the seawater supply to pH 4), were found to be sufficient to attain this stage were able to eliminate the visible fouling only if they were done at the beginning of the third stage. These occasional treatments carried out in branch (U) on old fouling which had formed on the stainless steel in six months of testing, were insufficient to reduce immediately the heat exchange resistance. They were able, according to the amount added, to alter considerably the responses of the cells and systems used to monitor the fouling.

Figure 6 shows the effect, after 6-month exposure to seawater in branch (U) of the half-hour addition of biocide (NaClO, ClO$_2$, HCl) on the iron–steel galvanic coupling current, registered at the end of the addition. It can be seen that both the NaClO

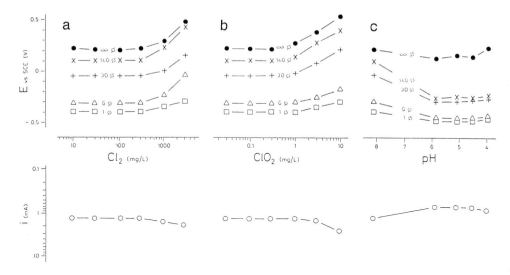

Fig. 6 Potentials measured at various distances from the stainless steel tube outlet and galvanic current intensity of the stainless steel–iron couple, according to the amount of (a) NaClO, (b) ClO_2, (c) HCl added on line after six month exposure to natural seawater.

and the ClO_2 can be active in the cathodic processes on stainless steel. Above certain concentrations such compounds significantly raised the galvanic coupling current and reduced the cathodic protection throwing power. The critical concentration above which this occurs is lower with ClO_2 than with NaClO, as it can be seen comparing the different scale values of the biocide amounts in Figs 6(a) and 6(b). These effects (higher protection current and reduced throwing power) are similar to those following the formation of biofilm. They are consistent with the ability of oxidising biocides to sustain cathodic processes at potential values comparable to those of the oxygen electrode. The opposite is the case when non-oxidising acids (HCl) were used (Fig. 6(c)). In these cases the addition of acid at increasing concentrations (lower pH values) corresponds to reduction of the current intensity due to the galvanic coupling and to an increase in cathodic protection throwing power, which is the opposite effect from those induced by the formation of biofilms.

Figure 7 shows the effect of the same additions on the responses of the probes used to monitor the biofilm by measuring the polarisation current of stainless steel. The same may be said for these measures as for those concerning the cathodic protection current and the throwing power.

3.4. Sulphide Contamination

As a consequence of the accidental contamination of seawater with sulphides, it was observed that:

(a) there were no significant effects on the heat exchange resistance;

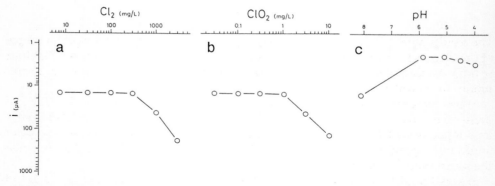

Fig. 7 *Cathodic current intensity recorded on a stainless steel specimen polarised at −250 mV vs SCE, according to the amount of (a) NaClO, (b) ClO$_2$, (c) HCl added on line after six month exposure to natural seawater.*

(b) cathodic current obtained from the biofilm monitoring-probes had zero intensity even if a biofilm was well developed. The current intensity had zero values even 3–5 days after contamination of the water supply had ceased;

(c) the iron–stainless steel galvanic coupling current was reduced drastically, and the rise in absolute values of the cathodic protection current in the case of the aluminium brass was also very marked. The effects seem to last longer (5–10 days) than in the preceding case.

Points (b) and (c) are consistent with the known effects of sulphides which, while they inhibit the cathodic reduction process of oxygen on stainless steels, are at the same time able to depolarise the process on copper-based alloys such as brasses, and also to exercise complex effects on iron [22]. Experience suggests that the effect of sulphides on the electrochemical characteristics of the materials employed is the opposite to that caused by the presence of biofilm. Such effect can also last for some days after the cause of contamination has been removed.

All the electrochemical monitoring systems used in this experiment exploit the cathodic depolarisation effects determined by the formation of biofilm on stainless steel in aerobic conditions. To have reliable results from these monitoring systems, it appears necessary:

(a) to operate in aerobic condition without any sulphide contamination;
(b) to adjust the speed of addition of the biocide so as not to exceed the critical concentration of oxidiser, as defined above.

If this is not controlled, the sulphide presence can mask the biofilm development or the biocide additions could produce the same effects that create the need for additions in the first place leading to instability phenomena.

4. Conclusions

The results obtained in twenty-month-experience have shown that the performance of the electrochemical monitoring systems used and the electrochemical behaviour of both stainless steel and aluminium brass depends heavily on the presence of sulphides in seawater.

When there is no contamination, the behaviour of these alloys can be divided into three successive stages having variable length, according to the season. This behaviour, together with the electrochemical monitoring response, depends on the type of antifouling agent used and on the stage in which the material is found to be when the initial biocide addition is made.

It was found that continuous treatments of seawater with NaClO (1 mg L^{-1} of added Cl_2), or ClO_2 (0.1 mg L^{-1}) or HCl (up to pH 6.3), if starting from the beginning of the run, impeded the formation of biofilm.

No formation of fouling was found when the current intensity observed on the monitoring cell was used to control occasional treatments of seawater with HCl to pH 6.3.

Half-hour biocide treatments of seawater such as the additions of NaClO (10–1000 mg L^{-1} of added Cl_2) or ClO_2 (0.1–10 mg L^{-1}) or HCl (up to pH 4), carried out on old fouling which had formed on the stainless steel specimens in 6-month testing, were not found to be sufficient to reduce immediately the heat exchange resistance.

The addition of oxidising biocides (Cl_2, ClO_2) was found to change considerably the responses of the electrochemical monitoring systems. Adding oxidising biocides can increase the intensities of the iron–stainless steel galvanic coupling current, can diminish the depth of cathodic protection penetration and increase the cathodic current output from the stainless steel of the monitoring cell. These effects are similar to those observed during the formation of biofilm. On the contrary, the addition of increasing concentrations of acid causes opposite effects from those induced by the development of biofouling.

The sulphides showed opposite effects on the electrochemical characteristics of stainless steel and of aluminium brass, from those induced by the development of biofouling in aerobic conditions. They seemed also capable of masking biofouling development for several days after their removal from the seawater.

It may therefore be stated that electrochemical monitoring of biological activities, based on the evaluation of the depolarisation effects they induce on cathodic processes, may be doubtful or even misleading when sulphides or electroactive oxidising substances are present.

The question of whether the determination of the heat exchange resistance, which seems to be insensitive to pollution, can integrate electrochemical measurements, in order to obtain reliable information on the trend of localised biological phenomena, and to control the performance antifouling programs, is still under study.

References

1. B. T. Hagewood, 'Improved Process for Cleaning Condenser Tubes', *Proc. Condenser Technology Conf.*, Boston, 18–20 September,1990, EPRI, Palo Alto, 1990.
2. A. M. Pritchard: 'Economic Aspects of Heat Exchanger Fouling', *Proc. Fouling of Heat Exanger Surfaces*, Munchen, 26–27 April, 1990, GVC, Munchen, 1990, p. 7.1–7.19.
3. C. A. C. Sequeira, 'Electrochemical Techniques for Studying Microbial Corrosion', *Microbial Corrosion — I* (Eds C. A. C. Sequeira and A. K. Tiller), Elsevier Applied Science, London, 1988.
4. S. C. Dexter, D. J. Duquette, D. W. Siebert and H. A. Videla, 'Use and Limitation of Electrochemical Techniques for Investigating Microbiological Corrosion', *Corrosion*, 1991, **47**(4), 308–318.
5. F. Mansfeld and B. Little, 'A Technical Review of Electrochemical Techniques Applied to Microbiologically Influenced Corrosion',*Corros. Sci.*, 1991, **32**(3), 247–272.
6. G. Salvago, G. Fumagalli, G. Taccani, P.Cristiani and G.Rocchini, 'Electrochemical and Corrosion Behaviour on Passive and Fouled Metallic Materials in Seawater', *Microbial Corrosion*, European Federation of Corrosion Publications No. 8 (Eds C. A. C. Sequeira and A. K. Tiller), p. 33–48, The Institute of Materials, London, 1992.
7. G. Salvago, G. Taccani and G. Fumagalli, 'Review of Effects of Biofilms on the Probability of Localized Corrosion of Stainless Steels in Seawater', *Microbiologically Influenced Corrosion (MIC) Testing*, ASTM STP 1232, (Eds J. R. Kearns and B. J. Little), American Society for Testing and Materials, Philadelphia, 1994, p.70-95.
8. G. Salvago, G. Taccani and G. Fumagalli, 'Electrochemical Approach to Biofilms Monitoring' in *Microbially Influenced Corrosion and Biodeterioration*, (Eds N. J. Dowling, N. W. Mittleman and J. C. Danko), p. 5.1– 5.7, The University of Tennessee, Knoxville, TN, 1990.
9. G. Licina, G. Nekoksa and G. Ward, 'An Electrochemical Method for Monitoring the Development of Biofilms in Cooling Water', *Microbially Influenced Corrosion and Biodeterioration* (Eds N. J. Dowling, N. W. Mittleman and J. C. Danko), p. 5.41–5.46, The University of Tennessee, Knoxville, TN, 1990.
10. A. Mollica, E. Traverso and G. Ventura, 'Electrochemical Monitoring of the Biofilm Growth on Active–Passive Alloy Tubes of Heat Exanger Using Seawater as Cooling Medium', in *Proc. 11th Int. Corrosion Congr.*, *4,* p. 4.341–4.349, Associazione Italiana di Metallurgia, Milano, 1990.
11. G. J. Licina and G. Nekoksa, 'On-Line Monitoring of Microbiologically Influenced Corrosion in Power Plant Environments', *Corrosion '93,* Paper No. 297, NACE, Houston, TX, 1993.
12. G. Salvago, G. Fumagalli and G. Taccani, 'Fouling Formation on Aluminium–Brass Tubes in Acidified Flowing Seawater', *Corrosion '93,* paper No. 310, NACE, Houston, TX, 1993.
13. A. Mollica and G. Ventura, 'Use of a Biofilm Electrochemical Monitoring Device for an Automatic Application of Antifouling Procedures in Seawater', *Proc. 12th Int. Corros. Congr.*, 5, p. 3807–3812, NACE, Houston, TX, 1993.
14. G. Salvago, G. Fumagalli, P. Cristiani and G.Rocchini, 'Fouling on Al Brass Tubes in Seawater and its Electrochemical Monitoring', *Proc. 8th Congress on Marine Corrosion and Fouling*, September 21–25, 1992, Taranto, Italy, *Oebalia, XIX* , suppl.1993, 331–341.
15. NACE, 'Standard Recommended Pratice On-Line Monitoring of Cooling Waters', RP0169–89, NACE, Houston, TX, 1989, *Mat. Perform.,* April, 1990, p. 42–52.
16. W. C. Micheletti, C. D. Hardy, J. F. Garey and B. A. Bennet P.E., 'A Portable Test Facility for Evaluating Condenser Microbiological Fouling Central Options', *Proc. Condenser Technology Conf.*, Boston, 18–20 September, 1990, EPRI, Palo Alto, CA, 1990.
17. W. Czolkoss, 'A New Technique for Online-Monitoring of Heat Transfer Coefficient of Single Condenser Tubes', *Proc. Condenser Technology Conf.*, Boston, 18–20 September, 1990, EPRI, Palo Alto, 1990.

18. H. M. Steinhagen, 'Mitigation of Heat Exchanger Fouling', *Proc. Condenser Technology Conf.*, Boston, 18–20 September, 1990, EPRI, Palo Alto, CA, 1990.

19. J. W. Whitehouse 'Marine Fouling Control and Chlorination', *Corrosion in Seawater System* (Ed. A. D. Mercer), Ellis Horwoood, New York, 1990, p. 53–64.

20. G. D. Simpson, R. F. Miller, G. D. Laxton and W. R. Clements, 'A Focus on Chlorine Dioxide: the Ideal Biocide', *Corrosion '93*, paper No. 472, NACE, Houston, 1993.

21. G. Salvago, G. Taccani, G. Fumagalli and L. Galelli, 'Acidification Effects on the Electrochemical Behaviour of Stainless Steel in Seawater', *Microbially Influenced Corrosion and Biodeterioration*, (Eds N. J. Dowling, N. W. Mittleman and J. C. Danko), p. 8.5–8.11, The University of Tennessee, Knoxville, TN, 1990.

22. G. Salvago, G. Taccani and E. Olzi, 'Role of Biofouling on Seawater Aggressiveness', *Proc. 2nd (1992) Int. Offshore and Polar Engineering Conf.* (Eds R. S. Puthli, J. F. Dos Santos, S. Berg, C. P. Ellinas and Y. Ueda), IV, p. 156–164, ISOPE, Golden, Colorado, 1992.

24

Challenges to the Prediction and Monitoring of Microbially Influenced Corrosion in the Oil Industry

T. S. WHITHAM

Shell Research Ltd., Environmental Research Department, Sittingbourne Research Centre, Sittingbourne, Kent, ME9 8AG, UK

ABSTRACT

The difficulties of predicting, with confidence, the occurrence of biofilm-associated corrosion in oil transport lines are discussed. The requirements for reliable prediction of microbially influenced corrosion (MIC) are outlined and mention is made of a number of techniques for gathering diagnostic information on the presence and activity of biofilm in pipelines and the associated corrosion. The limitations of these techniques are discussed and the challenges to overcoming these problems highlighted.

1. Introduction

In the oil industry, metal pipelines are used to transport water, gas and oil. The choice of material for each type of pipeline depends on a number of factors, including the purpose and location of the line, the predicted operating conditions, and cost. In general, the material of choice for overland and subsea crude oil transport lines is mild steel, mainly because this material is structurally fit for this purpose, taking into consideration the safe transport of the oil. But an additional advantage of mild steel is that it is relatively cheap compared with stainless steels or alternative materials, an important consideration for a line which may be many kilometres in length. The external surfaces of these pipelines are protected from corrosion in a number of ways, such as paint coatings or laggings of various types. The internal surfaces of some types of pipelines may have linings of plastic, stainless steel or other materials, but in most cases the internal surfaces of oil transport lines are unprotected.

A mild steel pipeline usually has a defined minimum life span which accommodates a predicted rate of metal loss from internal corrosion, based on the intended operating conditions of the line. For the pipeline manager responsible for the integrity of an oil production pipeline, a consideration which is always foremost in his/her mind is whether the risk of internal corrosion of a pipeline in operation is greater than that predicted, and allowed for, during the design stage. However, there is no standard test methodology which is routinely used as part of the design process, to determine the potential risk of internal corrosion caused by microbial biofouling.

Challenges to the Prediction and Monitoring of MIC in the Oil Industry 315

This fouling is likely to occur whenever microbially-contaminated water comes into continuous contact with internal pipeline surfaces.

In practice it is very difficult to ensure that no water enters the transport line with the crude oil. This problem can occur on some North Sea platforms where the oil is stored in subsea concrete cells prior to export to shore. The produced fluids, consisting of crude oil and produced water (formation water, possibly mixed with injection water), are dumped into the storage cells. Sea water is used as ballast in the cells and used to expel oil into the transport lines. From time to time, a mixture of produced water and sea water ballast may enter lines with oil from the storage cells. In other production operations, water can enter oil-transport lines as an emulsion with oil, only to drop out from the oil further down the line owing to low flow velocities. When water drops out to form a discrete layer in the bottom of the pipeline, which may be continuous or in isolated pockets, then the pipeline becomes susceptible to corrosion.

The ability to minimise the significant financial losses resulting from a biofilm-associated oil transport line failure, relies on the early detection of a developing microbial biofilm within the pipeline and the onset of any associated corrosion at monitoring sites. For pipelines which have been in operation for some time and are already heavily fouled, the chances of minimising biofilm associated corrosion are likely to be significantly reduced. Clearly, it would be beneficial to be able to predict reliably the likelihood of biofilm-associated corrosion within both new and established pipelines and thus have a sound scientific basis on which to undertake preventative or control measures.

2. Prediction of the Risk of Microbially Influenced Corrosion (MIC)

The ideal predictive model would furnish its user with a reliable estimate of the rate of corrosion associated with a biofilm (or an estimation of time to penetration of the pipe wall), the locations along the pipe where the attack is most likely to occur and, when corrosion will be initiated. Unfortunately, the physico–chemical and biological interactions which characterise biofilm development and activity, and the mechanisms by which these interactions influence corrosion of the underlying metal surface, are not only numerous but may also be interdependent, making a description of the entire complement of reactions and processes extremely difficult. The unpredictability of living organisms is probably the greatest problem to be faced. Indeed, for a pipeline of several kilometres in length, conditions may be such that a number of different types of biofilms are supported at different locations along the pipeline. The complexity of the interactions is summarised schematically in Fig. 1.

It is generally accepted that biofilm formation is a key step by which microorganisms become involved, both directly and indirectly, in corrosion processes [1–4]. Without a source of bacteria and suitable conditions for their survival within a pipeline, no biofilm can become established. For these reasons the first step in predicting MIC is to define the limitations for microbial activity and biofilm formation within the selected pipeline. The potential for microbial growth may be assessed by meas-

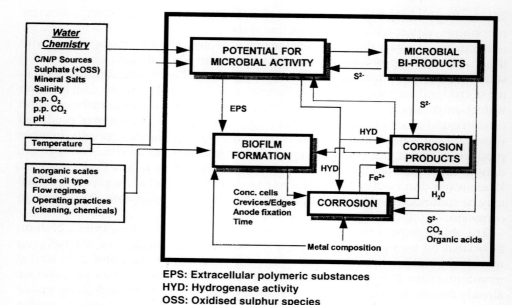

Fig. 1 Microbially influenced corrosion in oil transport lines: a complexity of interactions and influences.

uring a number of system parameters, including water availability, water chemistry, pH, redox, salinity and temperature. Other factors influence the development of a biofilm, including flow characteristics, metal composition and topology, surface scales or oil films, presence of production chemicals and other operating practices, such as physical cleaning. The activity of biofilm within a line may be estimated from nutrient and inorganics balances between the beginning and end of the line. The presence of bacteria and corrosion products (such as iron sulphide) in internal pipeline scrapings is often used to confirm (not always correctly) that biofilm is the cause of corrosion within the line [5].

So it is possible to determine the likelihood for biofilm to occur within a line. Nevertheless, this information is not always sufficient evidence for the pipeline manager to make a decision with respect to mitigation of biofouling. The control of microbial activity and the build up of biofilm is frequently tackled by the application of biocides, often in conjunction with physical cleaning using pipeline scraping devices. But, the manager equates such operations with disruption of oil production (and associated costs), costs of biocides and, increasingly, with the problems associated with the disposal of biocide-treated waters, which may not be environmentally acceptable.

Consequently, the decision to take action against biofilm rests largely on whether the biofilm present in the pipeline is likely to influence corrosion or not. If the risk is thought to be small, then operators may elect not to treat, but to monitor until corrosion rates are high enough to justify remedial action. Sometimes, this results in action being taken too late, as corrosion could be well established within inaccessible parts of the pipeline. In practice, a balance is needed between accepting the risk associated with the detection of biofilm and acting upon it alone, and waiting for proof that corrosion is occurring in association with biofilm. The development of predictive models for microbial corrosion of proven reliability will be a significant step forward in making decisions about treatment.

Uncertainty arises because the presence of biofilm does not automatically indicate that corrosion will be enhanced as a result, simply that the risk is greater. Indeed, it has been shown that biofilm may inhibit corrosion [6] or that the bacteria within the biofilm may not be capable of accelerating corrosion [7]. A number of mechanisms have been suggested by which biofilm may be involved in corrosion, some of which are chemical, others physical [8–12]. Knowing if one of these mechanisms predominates over an abiotic corrosion process would be valuable information for the selection of treatment. But often conditions are such that the mechanism of corrosion is unclear. Indeed, in a water injection line a problem which may occur periodically is the ingress of oxygen into the nominally deaerated injection water. Within the line, biofilm may occur with a risk of the localised type of corrosion associated with biofilms [5]. However, with oxygen ingress, accelerated underdeposit corrosion may also result from the formation of oxygen concentration cells [13]. The question here is whether the predominant corrosion mechanism is MIC, oxygen corrosion or a combination of both. For the pipeline manager this is an important question, i.e. can corrosion be eliminated with more effective control of oxygen ingress, or must biocides be used to control biofilm, or both?

3. Detection and Monitoring of MIC

The availability of on-line methods for the detection and monitoring of biofilm-associated corrosion within pipelines and the ability to distinguish MIC from other corrosion mechanisms will provide the data on which informed decisions about the need for, and timing of, mitigation can be based.

A difficulty with monitoring corrosion within oil production pipelines arises because the lines may be very long. Large sections of the line may be inaccessible, being submerged below the sea, in swamps or buried underground or simply remote, such as in the desert or through dense tropical forest. Consequently, monitoring devices are generally located at the parts of the lines where they can be easily reached, usually at the ends. Uncertainties arise because biofilm may be distributed heterogeneously, both quantitatively and qualitatively, along the pipeline which means that biofilm activity at the ends may not correlate with that in inaccessible regions of the line. Thus, monitoring only gives information about the site where the device is inserted or attached and cannot give information about the line as a whole. Over- or under-estimates of the 'worst-case' corrosion rate can result, and it is not possible to know which is the case.

The type of corrosion morphology commonly associated with microbial biofilm is one of localised pitting attack of the steel surface, although this morphology may also be observed with some non-biofilm-associated mechanisms of corrosion [5, 14]. Therefore, detection of such attack suggests (though not conclusively) microbial involvement and indicates a risk of rapid wall penetration via pits. Insertion of bioprobes gives an assessment of the potential for biofilm development in the line and, when removed, it is possible to observe the type of surface attack associated with the biofilm. A combination of microscopic analysis of the pitted area and a measurement of weight loss will give an indication of mean pit depth and mean pitting rate, being relevant only to the site of probe insertion.

Electrochemical monitoring, commonly employing linear polarisation resistance (LPR) or AC impedance techniques, is also limited to giving information at the site of measurement, but can give useful information on uniform corrosion phenomena and indicate changes in the system [15–17]. It has been suggested that AC impedance can be used to detect pitting, indicated by a scattering of low frequency data [18], but neither the latter or LPR are capable of monitoring or quantifying individual pitting events.

One technique which has proved useful for the detection of pitting events is electrochemical noise analysis [19–21], although it may not be possible to distinguish microbially-associated pitting from pitting caused by other corrosion phenomena. In this technique, nominally identical test electrodes, submerged within an electrolyte, are coupled to each other via a zero resistance ammeter (ZRA). Even between electrodes which are supposed to be identical, there are minor potential and corresponding current fluctuations, known as noise. When pitting occurs on one of the electrodes, a peak of current is observed over and above the noise, which is seen to decay as the pit repassivates itself, and the whole event may take several seconds. The technique can detect pitting events beneath biofilm [22]. These authors demonstrated that the frequency of pitting events continued at a high level beneath biofilm when pitting on the surface of a biofilm-free control became less as the surface was covered with corrosion products. It has been suggested that electrochemical noise analysis could be used to detect the onset of pitting events, so that remedial action could be taken before the pitting becomes self-perpetuating. When repassivation of pit sites is no longer possible, rapid penetration of the pipe wall may result.

It remains to be seen if electrochemical noise data can be used for the accurate determination of pit propagation rates, although some workers have made progress in this respect [23]. For this to be so, it is important to determine if the signals recorded come from one or several pits and if successive pitting/repassivation events occur at the same site (and if so, how often) or different sites. Thus, the source of the signals measured can be determined. The influence of biofilm on the noise signals needs to be explored for different biofilm types and the role of biofilm on the self-perpetuation (i.e. immunity from repassivation) of pits needs to be established. Without this information or an ability to determine the total pitted surface area it is not possible to reliably convert the magnitude of the signals into pit penetration rates, the key parameter of interest to the pipeline manager.

4. Conclusions and Recommendations

1. Several features of oil transport lines, including their length, location and cargo, make it difficult to say with certainty if MIC will be a significant risk to pipeline integrity.

2. The pipeline manager cannot, in many instances, be confident about long term integrity of oil transport lines since he does not have access to definitive, comprehensive information regarding the risk of MIC within these pipelines. At present it is not possible to reliably predict the occurrence, magnitude, location and timing of biofilm-associated corrosion in these lines.

3. The lack of reliable information may be remedied but a combined approach will be required. It will be necessary to develop a comprehensive model for MIC containing a database of information about the multitude of interactions occurring during biofilm development and activity in oil transport lines, and the impact of different operational practices. This is by no means a trivial task but some experience has been gained in the development of expert systems for other plant operations [24].
Additionally, it is necessary to develop techniques to sample biofilm and monitor corrosion at discrete locations within physically inaccessible regions of pipelines. It may be possible to construct a self-propelled monitoring tool which is capable of moving independently of flow along a pipeline, in either direction, and able to stop at any position. Whilst stationary, the device could take electrochemical measurements (e.g. corrosion potential, LPR, noise) at the selected site prior to the collection of water and surface scraping samples. Thus, a survey of localised corrosion activity in association with biofilm could be undertaken along the length of a pipeline considered to be at risk from MIC. The device could even be engineered to change diameter if required as a means of bypassing narrow points in the line and avoiding the serious problems associated with such a device becoming lodged in an inaccessible region of a pipeline.

4. Continued evaluation of electrochemical noise analysis is needed to explore the full potential of this method for routine monitoring of MIC, other than simply confirming the occurrence of pitting/repassivation events beneath biofilm. The role of biofilm in the perpetuation of pitting events, over and above that expected from abiotic mechanisms, is of particular interest since different types of biofilms may represent different levels of risk in terms of their ability to influence the frequency of pit initiation events and to prevent the repassivation of pit sites.

5. There is a need to develop standard, effective and reproducible tests for ranking the susceptibility of pipeline materials to MIC, in order to combat potential future biofilm associated corrosion problems at the design stage. Although some studies have addressed this requirement [25, 26] further work is needed to design tests which can readily accommodate conditions appropriate to different pipeline operating environments. Care will be needed in the design of tests to minimise deviations arising from the inherent variability and heterogeneity of living systems.

References

1. W. A. Hamilton, 'Sulphate-reducing bacteria and anaerobic corrosion', *Ann. Rev. Microbiol.*, 1985, **39**, 195–217.
2. J. L. Lynch and R. G. J. Edyvean, 'Biofouling in oilfield water systems – a review', *Biofouling*, 1988, **1**, 147–162.
3. T. Ford and R. Mitchell, 'The ecology of microbial corrosion', *Adv. Microb. Ecol.*, 1990, **11**, 231–262.
4. M. J. Franklin and D. C. White, 'Biocorrosion', *Current Opinion in Biotechnology* 1991, **2**, 450–456.
5. R. E. Tatnall and D. H. Pope, 'Identification of MIC', in *A Practical Manual on Microbiologically Influenced Corrosion* (Ed. G. Kobrin), NACE International, Houston, TX, 1993.
6. A. Pedersen and M. Hermansson, 'Inhibition of metal corrosion by bacteria', *Biofouling*, 1991, **3**, 1–11.
7. C. Gaylarde, 'Sulphate-reducing bacteria which do not induce accelerated corrosion', *Int. Biodetenor. Biodegr.*, 1992, **30**, 331–338.
8. C. A. H. Von Wolzogen Kuhr and L. S. Van der Vlugt, 'Graphication of cast iron as an electrochemical process in anaerobic soils', *Water*, 1934, **18**, 147–165.
9. G. H. Booth, L. Elford and D. S. Wakerley, 'Corrosion of mild steel by sulphate-reducing bacteria: an alternative mechanism', *Brit. Corros. J.*, 1968, **3**, 242–245.
10. R. A. King and J. D. A. Miller, 'Corrosion by the sulphate-reducing bacteria', *Nature*, 1971, **233**, 491–492.
11. J. D. A. Miller, 'Principles of microbial corrosion', *Brit. Corros. J.*, 1980, **15**, 92–94.
12. W. P. Iverson, 'Mechanism of anaerobic corrosion of steel by sulfate-reducing bacteria', *Mat. Perform.*, 1984, **3**, 28–30.
13. J. A. Hardy and J. L. Bown, 'The corrosion of mild steel by biogenic sulfide films exposed to air', *Corrosion (NACE)*, 1984, **40** (12).
14. W. Lee and W. G. Characklis, 'Corrosion of mild steel under anaerobic biofilm', *Corrosion (NACE)*, 1993, **49** (3),186–199.
15. S. C. Dexter, D. J. Duquette, O. W. Siebert and H. A. Videla, 'Use and limitations of electrochemical techniques for investigating microbiological corrosion', *Corrosion*, 1991, **47**(4), 308–318.
16. F. Mansfeld and B. Little, 'A technical review of electrochemical techniques applied to microbiologically influenced corrosion', *Corros. Sci.*, 1991, **32** (3), 247–272.
17. R. C.Newman, B. J. Webster and R. G. Kelly, 'The electrochemistry of SRB corrosion and related inorganic phenonema', *ISIJ Int.*, 1991, **31** (2), 201–209.
18. J. L. Dawson and M. G. S. Ferreira, 'Electrochemical studies of the pitting of austenitic stainless steel', *Corros. Sci.*, 1986, **26** (12), 1009–1026.
19. J. L. Dawson, K. Hladky and D. A. Eden, 'Electrochemical noise: some new developments in corrosion monitoring', *Proc. UK National Corrosion Conference*, Birmingham, UK, 1983.
20. W. P. Iverson, G. J. Olson and L. F. Heverly, 'The role of phosphorus and hydrogen sulphide in the anaerobic corrosion of iron and the possible detection of this corrosion by an electrochemical noise technique', *Proc. Int. Conf. Biologically Induced Corrosion*, NACE-8 (Ed. S. C. Dexter), NACE, Houston, TX, pp. 154–161, 1986.
21. W. P. Iverson and L. F. Heverly, 'Electrochemical noise as an indicator of anaerobic corrosion', *Corrosion Monitoring in Industnal Plants Using Nondestructive Testing and Electrochemical Methods*, ASTM STP 908 (Eds G. C. Moran and P. Labine), American Society for Testing and Materials, Philadelphia, pp. 459–471, 1986.
22. A. Saatchi, T. Pyle and A. P. Barton, 'Electrochemical noise analysis as an indicator of

microbiologically induced corrosion', *Corrosion '93, NACE International Corrosion Conference*, Houston, TX, pp. 3786–3792, 1993.

23. A. N. Rothwell and D. A. Eden, 'Electrochemical noise techniques for determining corrosion rates and mechanisms', *Corrosion '92, NACE International Corrosion Conference*, Paper No. 223, 1992.

24. G. J. Licina, C. E. Carney, D. Cubicciotti and R.W. Lutey, 'MIC Pro-predictive software for microbiologically influenced corrosion', *Corrosion '90, NACE International Corrosion Conference*, Las Vegas, USA, paper No. 536, 1990.

25. J. F. D. Stott, B. S. Skerry and R. A. King, 'Laboratory evaluation of materials for resistance to anaerobic corrosion by sulphate-reducing bacteria: philosophy and practical design', *The Use of Synthetic Environments for Corrosion Testing*, ASTM STP 970, P. E. Francis and T. S. Lee, eds., American Society for Testing and Materials, Philadelphia, PA, pp. 98–111, 1988.

26. P. J. B. Scott, J. Goldie and M. Davies, 'Ranking alloys for susceptibility to MIC — a preliminary report on high-Mo alloys', *Mat. Perform.*, 1991, **30**, 55–57.

25

Biofilm Development on Stainless Steels in a Potable Water System

L. HANJANGSIT, I. B. BEECH*, R. G. J. EDYVEAN and C. HAMMOND[†]

Department of Chemical Engineering, The University of Leeds, Leeds LS2 9JT, UK
*Department of Chemistry, Portsmouth University, Portsmouth, PO 1, UK
[†]School of Materials, The University of Leeds, Leeds LS2 9JT, UK

1. Introduction

It is widely acknowledged in the literature that metals immersed in biologically active waters undergo a sequence of biological and inorganic changes that result in an important modification of the metal/solution interface [1, 2]. Such modifications could have effects on the integrity of both the metal (i.e. leading to corrosion of pipework) and the water flowing over the metal (i.e. the dissolution of metal ions into a potable water and the development of a microbial, principally, bacterial, biofilm which might add bacteria to the water). The development of bacterial biofilms and their effects has been extensively studied on a range of materials in the laboratory, in seawater and in industrial cooling waters [3, 4, 7]. However, while there has been extensive published work on the effects of biofilms on copper in potable waters (see elsewhere in this volume) there is little published work relating to stainless steels in potable water systems.

The study described in this paper was carried out in one section of a large public building in Scotland which had been replumbed in stainless steel For study purposes, test sections were made available in both 304 and 316 grades of stainless steel tube and these were examined over a period of 19 months.

2. Materials and Methods
2.1. Stainless Steel Tube

Samples were taken from test sections of Grade 304 and 316 stainless steel tube that had been used to replumb part of the recirculating water system of a large public building in Scotland. Prior to commissioning the whole system was sterilised with peracetic acid. The test sections were so arranged that the normal water flow through them could be bypassed when required to allow samples to be removed. The furthest original sample downstream was taken each time. All test sections were horizontal. Samples for analysis were cut (with a tube cutter, not sawn) from the centre of a one metre long test piece. One 4 cm length was used in bacterial analysis and one 10 cm long piece for metal ion analysis. Other samples were taken for scanning electron microscopy.

The water system was mains water, from a peaty upland source, and supplied to the test sections as either cold unfiltered, hot unfiltered, or hot filtered (0.2 µm membrane). Both the supply and the respective return lines were sampled. In addition a number of dead-leg test spurs designed to investigate the effects of stagnant conditions were sampled.

2.2. Metal Ion Analysis

The amounts of chromium, molybdenum, iron and nickel in the biofilm were determined by ICP (Induction Coupled Plasma spectrophotometry) after the biofilm had been dissolved in an acid solution; 30 ml of acidified water (pH3) was enclosed in a 100 mm length of sample tube and the tube vigorously shaken for 2 min. The resulting sample was analysed for the metals and the results converted to mg cm^{-2} of tube surface.

2.3. Scanning Election Microscopy

Thin rings (approx. 5 mm) were cut from the tubes with a tube cutter and then cut into half or quarters. It was found that normal dehydration techniques tended to wash biofilm from the surface of the stainless steel so the preparations were rinsed and dried prior to gold coating and examination in a Cam Scan series 3 SEM.

2.4. Bacterial enumeration and identification

A 40 mm (approx.) section was cut from each tube sample on site and the biofilm removed from its internal surface using a scraping and swabbing technique with a known amount of sterile water. Cut samples were used for aerobic and anaerobic bacterial counts using the serial dilution and colony plate count methods on a total count agar for aerobic freshwater bacteria and Postgate medium B for anaerobic sulphate reducing bacteria.

3. Results and Discussion
3.1. Metal Content of the Biofilms

Leaching of metal ions from the surface of stainless steel is of concern in potable water systems. Tests elsewhere have shown that such leaching rapidly diminishes on an initial washthrough. However, little is known about the possible effects of a biofilm on such leaching. Therefore, at each sampling, the biofilm was analysed for chromium, molybdenum, iron and nickel. The detection limits of the procedure used were 5×10^{-6} mg cm^2. At the 5 month sampling all metals were below the detection limit in all samples with the exception of molybdenum at $10-15 \times 10^{-6}$ mg cm^{-2} in the biofilm on 316 stainless steel exposed to filtered hot water, the return and both filtered and unfiltered cold water, but only in one unfiltered cold water sample for the 304 grade stainless steel.

At the 10th month sampling chromium and molybdenum were absent but iron and nickel had increased reaching a maximum of 100×10^{-6} iron and 30×10^{-6} mg cm^{-2} nickel on some 304 and 316 samples.

At 19 months both chromium and molybdenum were again absent, but nickel reached 45×10^{-6} m cm^{-2} and the iron level increased to a maximum of 500×10^{-6} mg cm^{-2}.

Apart from the molybdenum levels after 5 months, which tended to be greater on the 316 grade, no pattern of metals in biofilms could be detected correlating with either water type or grade of stainless steel. The iron may well originate from the water supply rather than the stainless steel. However, molybdenum and nickel are likely to originate from the stainless steel. It is interesting to speculate as to whether the initial release of molybdenum from 316 grade, even in the very low amounts detected, has some influence on the lower bacterial numbers found on that grade of steel. However, it is considered that the effect is likely to be more complicated than this as the two grades have major differences in their surface microstructures, both on an atomic scale and with respect to the natural protective oxide film which may also influence settlement and growth of bacteria.

3.2. Comparison of Bacterial Numbers on 304 and 316 Grades Stainless Steels

Bacterial numbers on 304 and 316 grade stainless steels are shown in Table 1. In nearly all comparisons there is a significantly greater number of bacteria on 304 grade than on 316 grade steel.

Table 1. Comparison of bacterial counts on 316 and 304 stainless steel tubes (bacteria per sq. cm). Standard Deviation, SD in brackets.

Water	Grade	Months 5		Months 10		Months 19	
Unfiltered Cold water supply A	316	28	(5)	26	(4)	529	(61)
	304	52	(3)	58	(8)	467	(42)
Unfiltered Cold water supply B	316	34	(7)	31	(6)	569	(59)
	304	60	(4)	221	(23)	1153	(168)
Filtered Hot water supply	316	31	(1)	20	(1)	3	
	304	37	(1)	50	(7)	163	(17)
Unfiltered Hot water supply	316	64	(4)	42	(3)	29	(3)
	304	103	(6)	316	(35)	438	(51)
Filtered Hot water return	316	13	(1)	8	(1)	0	
	304	79	(4)	38	(6)	165	(17)
Unfiltered Hot water return	316	50	(10)	32	(3)	549	(47)
	304	29	(1)	340	(46)	1349	(150)

The effect is more marked for hot water than for cold water and this is emphasised by the two cold water comparisons. The cold water supply A is closer to the incoming main and thus is likely to be at a lower temperature than cold water supply B. It is surprising that two similar, 'inert' metal surfaces should produce a difference in biofilm bacterial numbers that lasts for at least 19 months. Laboratory experiments have not been conclusive as to the effect of surface finish, let alone small differences in metal alloy composition.

For example, an extensive study of bacterial settlement on different grades and surface finishes of stainless steel conducted by J. Benson (personal communications) failed to show significant differences. However, because most laboratory studies use nutrient rich cultures, the deposition of organics on a metal surface together with relatively large numbers of bacteria and the overall experimental conditions used may mean that any effect which morphology are immediately masked. In the case of the samples removed from plant, nutrients and bacteria present in the water will be lower for encouraging colonisation. Beech et al. [5] have demonstrated that high molybdenum steels retard the development of both bacterial number and the production of extracellular material by the bacteria.

3.3. Comparison of Bacterial Numbers in the Presence and Absence of Filtration

Table 2 shows the effect of filtration on bacterial settlement. Unfiltered water samples have significantly greater numbers of bacteria than filtered water samples.

The filtration of the water through a 0.2 μm membrane filter will significantly reduce the innoculum of bacteria arriving in the mains water supply. It is perhaps surprising that this filtration maintains such a difference in bacterial numbers for as long as 19 months but the differences continue to increase.

Table 2. The effect of 0.2 μm membrane water filtration on bacterial numbers (bacteria per sq. cm). Standard Deviation, SD, in brackets

Grade and Water Type	Filtered or Unfiltered	Months 5		10		19	
316 Hot water supply	UNF	64	(4)	42	(3)	29	(3)
	FILT	31	(1)	20	(1)	3	
316 Hot water return	UNF	50	(10)	32	(3)	549	(47)
	FILT	13	(1)	8	(1)	0	
304 Hot water supply	UNF	103	(6)	316	(35)	438	(51)
	FILT	37	(1)	50	(7)	163	(17)
304 Hot water return	UNF	29	(1)	340	(46)	1349	(150)
	FILT	79	(4)	38	(6)	165	(17)

3.4. Comparison of Stagnant and Flowing Conditions

Table 3 compares bacterial number under stagnant and flowing conditions. Comparison is difficult as there was no unfiltered hot water system. However, the conclusion that can be drawn from the results is that there is little difference between flowing and stagnant conditions for 316 grade carrying filtered water but 304 grade carrying unfiltered water shows greater bacterial colonisation in stagnant conditions.

3.5. Anaerobic Bacterial Settlement

During the sampling period anaerobic bacterial growth was so low as to be difficult to quantify. Indeed there was evidence of a decrease in anaerobic bacterial activity in some cases. Some sulphate-reducing bacterial activity was present after 19 months in the unfiltered 304 cold and hot water systems.

Anaerobic bacteria were present in large numbers in the balancing tanks which feed the systems and thus a strong innoculum is present. However, it appears to take a significant time for conditions to build up within the system which will allow anaerobic bacteria to grow.

3.6. Bacterial Species

Bacterial identification was taken as far as the API enzyme/chemical reaction system would allow. After 5 months the bacterial biofilm comprised 30% *Pseudomonas* spp, 50% *Agrobacterium* sp. and 20% *Vibrio* spp. After 10 months *Pseudomonas* spp. (70%) and *Flavobacterium* dominated corrosion.

Bacterial deposits were generally found in the bottom half of the (horizontally mounted) tubes and around the weld lines. Scanning electron microscopy confirmed this and emphasised the association of bacteria with the weld line in the 304 grade samples. A certain number of weld flaws were detected and these appear to have been exploited by the bacteria. Discrete bacterial colonies with considerable

Table 3. The effect of flow on bacterial numbers (bacteria per sq. cm). SD in brackets

Grade and Water Type	Flowing or Stagnant	Months					
		5		10		19	
316 Filtered Hot water supply	FLOW STAG	26 31	(1) (1)	25 20	(4) (1)	0 3	
316 Filtered Hot water return	FLOW STAG	17 13	(1) (1)	31 8	(4) 1	0 0	
304 Unfiltered Hot water supply	FLOW STAG	64 103	(4) (6)	35 316	(3) (35)	267 438	(31) (51)
304 Unfiltered Hot water return	FLOW STAG	87 29	(12) (1)	84 340	(8) (46)	164 1349	(15) (150)

extracellular material were observed on the 304 grade mains water pipe. This connected the mains supply of the water company with the holding tanks and filters to the building. Some shallow pitting was observed to be associated with these colonies after 10 months. Generally, visible pitting and staining was seen on 304 steel mostly associated with the weld area. No pitting was found on 316.

4. Conclusions

In this potable water system there is a significantly higher bacterial development on 304 than on 316 grade stainless steel in the first 19 months of exposure.

Filtering the water significantly reduces bacterial numbers in the biofilm. This difference tends to become more marked with time. Filtering the water also reduces bacterial build up in stagnant conditions.

The highest bacterial loads were found on grade 304 stainless steel carrying unfiltered water in stagnant conditions.

Unfiltered water, weld flaws, and 304 grade steel are the most conducive to pitting corrosion.

5. Acknowledgements

The authors would like to thank the Scottish Office, Home and Health Department for their financial support of the project and Mr. I. Campbell and Mr. A. K. Tiller for their advice.

References

1. W. G. Characklis, Fouling biofilm development. A process analysis, *Biotech. Bioeng.*, 1981, **23**, 1923–1960.
2. H. A. Videla, Metal dissolution/Redox in biofilms, in *Structure and Function of Biofilms* (Eds W. G. Characklis and P. A. Wilderer), John Wiley & Sons, 1989, pp. 301–320.
3. R. G. J. Edyvean, Biodeterioration problems of North Sea Oil and Gas production – a review, *Int. Biodeterior.*, 1987, **23**, 199–231.
4. R. E. Tatnall, Case Histories: Bacteria Induced Corrosion, *Mat. Perform.*, 1981, **20**, 41–48.
5. I. B. Beech, S. A. Campbell and F. C. Walsh, The role of surface chemistry in SRB-influenced corrosion of steel, *Obelia*, 1993, **XIX**, Suppl., 243–250.

26

Bacteria and Corrosion in Potable Water Mains

I. B. BEECH, R. G. J. EDYVEAN*, C. W. S. CHEUNG and A. TURNER[†]

School of Chemistry, Physics and Radiography, University of Portsmouth, Portsmouth, UK
*Department of Chemical Engineering, The University of Leeds, Leeds, UK
†Essex Water Company, Central Laboratory, South Hanningfield, Chelmsford, UK

ABSTRACT

This investigation was undertaken to determine the possible involvement of microorganisms in the corrosion of cast iron pipes carrying potable water. Sections of pipes (mains), varying in diameter i.e. 76.2, 101.6, 127 and 203.2 mm (3", 4", 5" and 8"), were removed from ten different locations. Cutting of the sections of the pipes lengthwise revealed the accumulation of deposits, differing in thickness and morphology and in some cases the presence of large tubercles, often filled with water. Both, multilayered deposits and tubercles, were carefully examined microbiologically and using Energy Dispersive X-ray Analysis (EDAX).

Microbiological studies, involving isolation, enumeration and identification of microorganisms, showed that aerobic bacteria (including iron oxidising species such as *Sphearotillus* and *Galionella*) as well as anaerobic, sulphate-reducing bacteria (SRB) were present within the deposits and associated with tubercles. The results of EDAX analysis demonstrated the accumulation of sulphur, phosphorus, calcium, manganese and chlorine within deposits. The distribution of these elements varied depending on the type of the layer examined. High levels of sulphur were detected in strata which also harboured SRB.

The analysis of the combined biological and physico–chemical data as well as later studies on the influence of biofilms formed by a bacterial consortium isolated from one of the pipes, on deterioration of Type 316 stainless steel, indicated that microorganisms were likely to contribute to the corrosion of some of the mains sections [11].

1. Introduction

Ample microscopic evidence is available to show that most pipe surfaces in distribution systems are colonised by microorganisms. Le Chevallier *et al.* showed high populations of bacteria in main encrustations collected from water utilities throughout the United States [1]. Diatoms, algae, and filamentous and rod-shaped bacteria were commonly encountered. In contrast, Ridway and Olson found that bacteria were sparsely distributed along pipe surfaces they examined [2]. These scarce communities, however, showed a variety of morphologically distinguishable structures including rod and chain forming cocci and filamentous and prosthecate cell types. Scanning electron microscopy demonstrated that 17% of the 10–50 µm size particles settled on the surfaces of pipes were colonised with 10–100 bacteria per particle [2].

In potable water systems most of the microorganisms grow on the inside walls of water mains as biofilms. Relatively large biofilm cells population can accumulate in the system as the immobilised cells do not wash out as fast as their suspended counterparts [3]. Biofilm environments also offer ecological advantages to microorganisms such as partial protection against disinfectants [4]. The concentration of planktonic cells in the distribution system may increase as a result of planktonic growth and contribution from biofilm erosion and sloughing leading to unacceptable drinking water quality [5].

Biofilms can drastically alter electrochemical processes occurring at the metal surface often enhancing corrosion rates [6]. In the presence of the anaerobic sulphate-reducing bacteria (SRB), the situation can further deteriorate. These bacteria are usually present in the innermost parts of a biofilm where anaerobic conditions prevail. Recently the corrosion mechanisms of this group have been reviewed by Geesey [7].

Although microbiologically influenced corrosion of cast iron has been reported in a variety of environments by a number of authors, many of these reports do not detail the specific types of microorganisms involved in microbially influenced corrosion (MIC). Of the microbes associated with corrosion of iron, different genera of SRB and aerobic iron oxidising bacteria from genera *Pseudomonas, Sphaerotilus, Vibrio* and *Galionella* are most frequently reported to be involved with pitting.

The present study has been undertaken to investigate the significance of biofilms in the corrosion of pipes in water distribution systems.

2. Materials and Methods

Two different categories of samples, surface swabs and bulk deposits, were collected from the inside of 10 pipe sections, carrying potable water, removed from various locations in Suffolk, UK. Microbiological examination of the sections was carried out within 20 min of their removal. Detection, enumeration and identification of aerobic and anaerobic microorganisms present was performed in the manner described below.

2.1. Sampling from the Metal Surface

Sampling from a defined surface area of approx. 1 cm^2 was performed by removing the bulk deposit and scraping the exposed metal with a sterile cotton swab. The cotton tip of the swab was then aseptically removed into a sterile plastic container. Sterile water (2 mL) was added to the container and the suspension was agitated vigorously by pipetting.

2.2. Detection and Enumeration of SRB

Aliquots of 1 mL of suspensions were taken up in a sterile syringe with a sterile needle and injected into glass vials fitted with a rubber septum. The vials contained 9 mL of deoxygenated Postgate medium B [8]. The enumeration of bacteria was carried out by a dilution to extinction method. Each sample was serially diluted to

10^{-4} of the original concentration. The vials were incubated for 30 days at 37°C. Blackening of the media indicated the growth of SRB. The vials in which the blackening occurred were scored as positive and the number of viable SRB was then estimated.

2.2. Detection and Enumeration of Aerobic Bacteria

The remaining one mL of the suspension was spread-plated as 0.1 mL aliquot over surfaces of solid media prepared aseptically in plastic petri dishes. Four different types of media: nutrient agar, *Pseudomonas* isolation agar, yeast extract agar and total count agar (Difco), were used to increase the probability of maximal recovery of aerobic and/or facultatively anaerobic microorganisms. The enumeration of bacteria by a viable colony count was performed after incubating plates at 37°C and 25°C for up to 48 h at each temperature (detection of mesophylic bacteria) and at 10°C for up to 1 week (detection of psychrophilic bacteria). The identification of bacteria was carried out by light microscopy and light epifluorescence microscopy (to determine Gram reaction and the presence of iron-oxidising species such as *Galionella* and *Sphaerotillus*) and by using Analytical Profile Index (API 20NE and API 20E, Bio Merieux) method.

2.4. Sampling from the Bulk Deposit

The inside of the pipes contained a very heterogeneous deposit (Fig. 1). Its morphology varied depending on the pipe section. Typically, the uppermost layer of the deposit consisted of a hard, reddish-brown 'crust'. Beneath the crust an orange-brown, solid material was accumulated, often in very large quantities, followed by the layer of a hard, black material. A soft, highly hydrated black material was present in the innermost part of the deposit, in a close proximity to the pipe surface. Some of the pipe sections contained very large tubercles filled with water.

2.5. Microbiology of the Surface 'Crust' and Surface of Tubercles

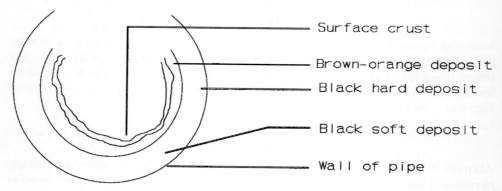

Fig. 1 Typical configuration of corrosion deposits on pipe without tubercles.

Swabs from the surface crust and from the surface of tubercles were taken and processed as described for aerobic bacteria.

2.6. Microbiology of the Soft Black Deposit

Samples of the soft black material and of the brown bulk material were collected with a pair of sterilised tweezers into preweighed, sterile plastic containers. The quantity of the collected material was determined by subsequent weighing of containers. Two ml of sterile water was added to the containers and the material was agitated by pipetting, to obtain a uniform suspension. Samples were then treated as described above. Bacterial counts were expressed either as no. cells/cm^2 or no. cells/g.

Additional sampling of water present inside the tubercles was carried out to determine the presence of SRB. Water was withdrawn from the tubercle by piercing the surface layer with the sterile needle attached to a sterile syringe. The water was then injected into vials containing SRB media, as described above.

2.7. EDAX Analysis of the Samples

Samples of corrosion products were taken from the pipework simultaneously with the extraction of the samples for microbiological analysis. Care was taken to take representative samples from different depths of the corrosion products where appropriate, i.e. from the surface of the corrosion product, at the metal surface and from any significant features through the depth of the corrosion product. Sub-samples from this material were dried, mounted on to aluminium scanning electron microscope (SEM) stubs and coated with carbon for SEM analysis by EDAX. The conditions of analysis were standardised, both in the method of analysis and in the area of the corrosion product surface scanned (all analysis being carried out at 750× magnification) to facilitate direct comparison of the samples.

3. Results and Discussion

Tables 1, 2 and 3 show the results of microbiological analysis of pipe sections. Lactate utilising, mesophyllic SRB were detected at the metal surface, under the accumulated deposit in seven out of ten examined pipe sections. SRB were also present in both brown and black portions of the bulk deposit (Table 1). Only samples 3 and 9 showed no presence of SRB regardless of the sample type. The highest number of SRB was detected at the surface of pipe No 7. The soft black deposit seemed to contain more SRB than the brown deposit. This could be attributed to a possible higher degree of anaerobiosis in the innermost part of the deposit. The difference in oxygen concentration could also explain the lack of SRB detection in 50% of the brown deposit samples. However, anaerobic 'pockets' existing within a bulk deposit could harbour the SRB.

The numbers of SRB detected in the samples were relatively low, however considering the surface area of the pipes and the quantity of the deposit these numbers

Table 1. Sulphate reducing bacteria detected at the metal surface and in bulk deposit by dilution to extinction method

Pipe Sample	Number of SRB cm^{-2} metal surface	Number of SRB g^{-1} hard deposit	Number of SRB g^{-1} soft deposit
1	10^2	1.3×10^2	10^2
2	10	16	10^2
3	ND	ND	ND
4	10	10^2	10^2
5	10	ND	10^2
6	10	ND	10^2
7	10^4	ND	10^3
8	10^2	10	10^2
9	ND	ND	ND
10	ND	2×10^2	10^2

ND = not detected.

could be significant regarding their impact on corrosion. Recent studies in the field of microbially influenced corrosion revealed that not necessarily the numbers but the metabolic activities of microorganisms (including SRB) are important in the process of biocorrosion [9, 10].

Samples taken from the various strata of bulk deposit revealed the presence of aerobic bacteria belonging to the genus *Pseudomonas* and facultative anaerobic bacteria from the genus *Vibrio* (Tables 2 and 3) in all of the samples with the exception of pipe No. 6. The composition of the consortia varied with the pipe section. *Pseudomonas* spp. seemed to be more prevalent than *Vibrio* in all of the tested samples. Only *Vibrio* spp. were detected at the metal surface most probably due to anaerobic conditions prevailing at the metal/bulk deposit interface.

Light microscopy and epifluorescence light microscopy of samples prepared from the surface crust and surface layers of tubercles demonstrated the presence of iron-oxidising bacteria. Both *Galionella* and *Sphaerotillus* were found in abundance in all of the examined pipe sections. This indicates that the true numbers of aerobic organisms present in deposits were much higher than the numbers obtained by using the viable count method. The enumeration of these spp. is very difficult due to their fastidious nature and very slow growth under laboratory conditions.

The results of microbiological analysis showed the presence of microorganisms known to be associated with corrosion of mild steel in all of the tested pipe sections [10].

The elemental analysis of the corrosion products carried out by EDAX showed the presence of sulphur in some of the samples as well as silicon, phosphorus, cal-

Table 2. Aerobic bacteria detected at the metal surface and in bulk deposit

Sample	CFU cm^{-2} Metal Surface	CFU g^{-1} Soft Black Deposit	CFU cm^{-2} Surface of Tubercle	CFU cm^{-2} Upper Surface Swab
1	ND	ND	9×10^2	10^3
2	ND	ND	ND	5×10^2
3	20	10^3	—	5×10^3
4	ND	ND	2×10^3	1.6×10^3
5	ND	—	2.6×10^2	10
6	ND	ND	ND	ND
7	1.2×10^2	—	30	10^2
8	ND	ND	—	10
9	ND	ND	20	70
10	10	10	—	80

ND: Not detected   CFU: Colony forming unit   —: No sample taken.

Table 3. Aerobic bacteria identified by API method (% of a total count)

Sample	V	PI	PII
1	60	5	35
2	5	5	90
3	0	60	40
4	20	10	70
5	0	40	60
6	0	0	0
7	0	70	30
8	0	0	0
9	0	65	35
10	10	60	30

V: *Vibrio fluvialis*   PI: *Pseudomonas cepacia*   PII: *Pseudomonas paucimobilis*

cium and a high background level, all indicative of general bacterial activity, including SRB. Figure 2 shows EDAX spectrum of the surface deposit (surface swab) recovered from sample 7. The appearance of large S peak and Fe peaks indicates the

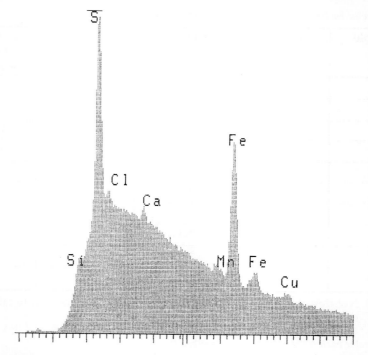

Fig. 2 EDAX spectrum of a deposit removed from the surface of pipe 7, showing the presence of sulphur, most likely accumulated as iron sulphide species

high level of iron sulphide species accumulated in this deposit. Figure 3 demonstrates the spectrum of surface deposit removed from pipe 2. Iron, most probably as oxide and/or hydroxide species, is the major element detected in this sample.

Based on the information obtained from microbiological examination, EDAX analysis and the visual inspection of samples, the risk assessment matrix has been constructed, indicating a degree of possible contribution of biological component to the overall deterioration of a given pipe (Table 4). From these data it was concluded that samples 1, 3, 4, 7 and 10 were likely to be susceptible to microbially influenced corrosion (MIC). It also appeared that samples 2, 8 and 9 were least likely to experience any MIC.

The examination of the surfaces of pipe sections, after the removal of deposits by sand blasting, revealed differing degrees of structural damage, mostly in the form of extensive pitting. Pipes 1, 3 and 4 were recommended for relining and pipe 10 had to be renewed. All macroscopic observations of the structural deterioration were in good correlation with the results of the matrix analysis with the exception of pipe 7, which did not show high degree of localised corrosion and pipe 2 which had to be renewed due to the extensive damage. It is well documented that the chemical composition of the layer in contact with the metal surface would influence the corrosion rates of this metal. Such layers could be either protective or could accelerate corrosion rates. In the case of sample 7 it is possible that the presence of the deposit such

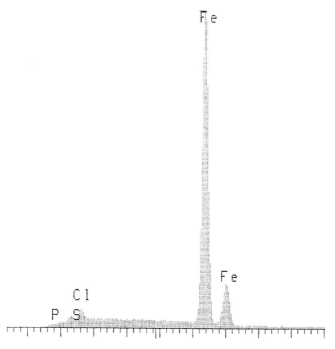

Fig. 3 *EDAX spectrum of a deposit removed from the surface of pipe 2, showing iron to be the most abundant element.*

Table 4. *Risk assessment matrix based on the visual observation, EDXA analysis and microbiological analysis of pipe sections*

Pipe No.	Most suscept. to localised corros. (visual)	SRB activity (EDAX)	Highest organic activity (EDAX)	Highest biologic activity (EDAX)	Highest count of aerobic bacteria in bulk	Highest bacterial count at metal surface	Highest count of SRB in bulk
1		X	X	X	X	X	X
2							
3				X	X	X	
4		X		X	X	X	X
5	X	X	X				
6	X		X				
7	X	X	X	X	X	X	X
8							
9							
10						(X)	(X)

as an intact film of iron sulphide (Fig. 2) would facilitate passivation of the surface, thus preventing pitting. The opposite phenomenon could be the reason for the failure of sample 2, where the porous, loosely adhering layer composed mainly of iron oxides and/or hydroxides would contribute to severe localised corrosion damage (Fig. 2).

To determine why samples 1, 3, 4, and 10 were more prone to biocorrosion comparing with other pipes, would clearly require in-depth analysis of the conditions of use of the respective mains and monitoring of the water chemistry over the long period of time. Although the results of microbiological investigation indicate the likely contribution of the biological component to the corrosion process of the mains, other factors such as physico–chemical and operational parameters should also be carefully considered when trying to establish the cause of system failure.

It should be noted that data obtained from this study has been generated through the examination of mature pipes and gives little indication as to what happens when new pipe is introduced to the system. In order to assess the potential ability of bacterial consortium to influence corrosion of steel, a separate study has been undertaken [11]. This study revealed that abundant biofilm was readily formed by the bacterial consortium isolated from one of the pipe sections (pipe 10) on surfaces of stainless steel 316. The presence of such biofilm resulted in severe pitting of the alloy.

The present investigation demonstrates that the prediction of a possible contribution of MIC to the overall life of mains piping can be attempted, even when dealing with mature pipes, and that combined microbiological and surface analysis are valuable tools in carrying out the biocorrosion risk assessment.

References

1. M. W. LeChevallier, C. D. Lowry and R. G. Lee, 'Disinfecting biofilms in a model distribution system', *J. Amer. Water Works Assoc.*, 1990, **82**(7), 87–99.
2. H. F. Ridway and B. H. Olson, 'Scanning electron microscopy evidence for bacterial colonization of a drinking water distribution system', *Appl. Environ. Microbiol.*, 1981, **41**, 274–287.
3. W. C. Characklis and K. E. Cooksey, 'Biofilms and microbial fouling', *Adv. Appl. Microbiol.*, 1983, **136**, 93–98.
4. W. C. Characklis and K. C. Marshall, *Biofilms*, John Wiley & Sons, 1990.
5. C.W. Keevil and C.W. Mackerness, 'Biocide treatment of biofilms', *Int. Biodeterior.*, 1990, **26**, 169–179.
6. W. P. Iverson, 'Microbial corrosion of metals', *Adv. Appl. Microbiol.*, 1987, **32**, 1–36.
7. G. G. Geesey, 'What is biocorrosion?', in *Biofouling and Biocorrosion in Industrial Water Systems* (Eds H. C. Flemming and G. G. Geesey), Springer-Verlag, Berlin, 1990, 155–164.
8. J. R. Postgate, *The Sulphate Reducing Bacteria*, Cambridge University Press,UK.
9. R. D. Bryant, W. Jansen, J. Bovin, E. J. Laishley and J. W. Costerton, 'Effect of hydrogenase and mixed sulphate reducing bacterial populations on the corrosion of steel', *Appl. Environ. Microbiol.*, 1991, **57**(10), 2804–2809.
10. I. B. Beech, C. W. S Cheung, M. A. Hill, R. Franco and A. R. Lino, 'Study of parameters implicated in biodeterioration of steel in the presence of different species of sulphate-reduc-

ing bacteria', *Int. Symp. on Marine Biofouling and Corrosion*, July 1993, Portsmouth, UK, *Int. Biodeterior. Biodegr.*, in press.

11. A. Steele, D. Goddard and I. B. Beech, 'The use of atomic force microscopy in study of the biodeterioration of stainless steel in the presence of bacterial biofilms', The Biodeterioration Society Spring Meeting, Stratford-upon-Avon, April 1993, *Int. Biodeterior. Biodegr.*, in press.

Study of Corrosion Layer Products from an Archaeological Iron Nail

A. A. NOVAKOVA, T. S. GENDLER* and N. D. MANYUROVA

Department of Physics, Moscow State University, Moscow, Russia
*United Institute of Physics of the Earth Russian Academy of Sciences, Moscow, Russia

ABSTRACT

The very thick corrosion layers on a nail of archaeological origin were investigated, layer by layer, by various techniques: Mossbauer spectroscopy. X-ray diffraction, thermomagnetic and chemical analysis. The following model of podzol soil corrosion layer was derived: near the nucleus, crystalline iron oxides, such as Fe_3O_4 and Fe_2O_3 predominate, and on approaching the surface of the sample the quantity of magnetite diminishes and virtually disappears, haematite is found on the surface only in fine dispersed form, while the concentrations of iron–sulphur and iron–phosphate compounds are increased. This model can be explained only by active bacteria (especially sulphate reducing) playing a role in the process of soil corrosion.

1. Introduction

This work is concerned with studies of the corrosion scale on a nail of archaeological origin (12th century), which had been buried in podsol soil (near Smolensk) for 8 centuries.

In soils objects will be exposed to moisture, oxygen, and various chemical agents, that will be characteristic of the particular soil type and microbial activity. Moreover, the humidity and temperature of the ambient medium undergo periodic changes. Therefore, the phase composition of the corrosion scales, on an iron object will differ depending on the nature and site of burial.

Investigations are also complicated by the fact that, in addition to crystalline compounds of varying degrees of dispersion, amorphous iron compounds may also be present in the corrosion layer.

2. Experimental

The very thick corrosion scale on the nail could be divided by colour into the following three layers: greyish blue, brownish orange, and grey (i.e. proceeding from the surface to the bulk). The complexity of the problem makes it necessary to investigate the rust layer by layer by means of a complex of methods that are mutually complementary, such as Mossbauer spectroscopy, X-ray diffraction, thermomagnetic analysis, and chemical analysis. The ^{57}Fe Mossbauer absorption spectra were measured with a source ^{57}Co (Rh). The isomer shifts are reported relative to α-Fe. The Mossbauer spectra were fitted by Lorentzian line shapes. X-ray diffraction using Cu–K$_\alpha$ radiation was applied to identify the chemical compounds produced on iron. Magnetic measurements (specific saturation magnetisation and its temperature dependence) were carried out with a vibro-magnetometer having a high sensitivity about 10^{-6} – 10^{-7} Am. This allowed the identification of the magnetic phases at low concentrations in polyphase samples.

3. Results and Discussion

The soils of the Smolensk region are soddy podzol soils and are known [1] to contain sulphates. After prolonged exposures of metal objects to these soils, the main component of rust is usually $FeSO_4\,nH_2O$ [2]. These data have been confirmed by our own investigations. The chemical analysis throughout the rust layer covering the iron nail showed the presence of the following elements (besides iron): S~5, P~3.6, Cl~0.1, Si~0.1, Mn~0.1, Al~0.1. It was difficult to analyse the presence and volume of some compounds such as Fe_3O_4, γ–Fe_2O_3, and FeS_2 from the X-ray diffraction patterns because these compounds have overlapping peaks. However, peaks of α–Fe_2O_3, $Fe_3(PO_4)_2\,8H_2O$, and $FeSO_4\,nH_2O$ were recognised in X-ray diffraction patterns quite accurately. Having taken into account the data and the spectral parameters reported for several salts of ferrous iron, we carried out a fitting procedure of the Mossbauer spectra.

The Mossbauer spectra (Fig. 1) of three layers, measured at 300 K, consist of a set of doublets (in the central part of the spectra) corresponding to the paramagnetic compounds of ferric and ferrous iron. An obvious increase of magnetic fraction (six line components) with depth is observed in these spectra. Sextets are nearly absent in the spectrum of the upper layer. This may be explained by the fact that the very loose upper corrosion layer, which is in direct contact with the soil, usually contains amorphous and dispersed substances as well as soil minerals. The data obtained from the fitting of the spectra (Fig. 1(a)) imply that FeS_2 and γ-FeOOH, as well as finely dispersed α-Fe_2O_3 particles, may be responsible for the central narrow doublet, but these compounds have very similar Mossbauer spectra [3]. Another doublet with broadened lines and high quadrupole splitting (about 3 mm s^{-1}) resulting from overlapping set of doublets is also present in the spectra. The corresponding Mossbauer parameters are typical of the following Fe(II) compounds: $FeSO_3\,3H_2O$, $FeSO_4\,4H_2O$, $Fe_3(PO_4)_2\,8H_2O$, $FeCl_2\,2H_2O$ and $Fe(COO)_2\,2H_2O$. The components which show magnetic hyperfine structure in the spectrum of the middle layer are assigned to magnetite and haematite. In the spectrum of the deepest layer there is, in addition to the magnetite and haematite lines, a sextet of metal iron. Therefore, the spectra clearly reflect the dynamics of corrosion, in the course of which the metal iron is oxidised to magnetite, haematite and other iron compounds.

The ferromagnetic components and transitions of nonmagnetic into magnetic com-

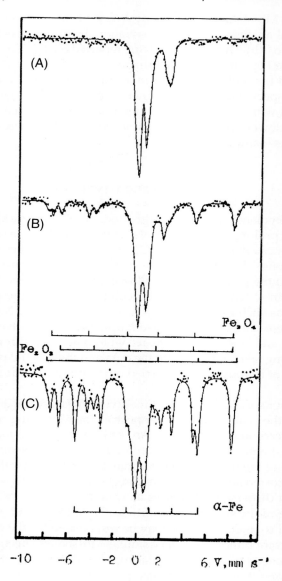

Fig. 1 Mossbauer spectra of the outer (A), middle (B), inner (C) corrosion layers measured at 300K.

pounds have been detected by thermomagnetic analysis (TMA) (Fig. 2), which allowed their nature to be determined. Between 480 and 520°C a specific behaviour (stepwise jump) is observed in the $J_s(T)$ curves of the first heating. These plateaux correspond to the phase transition of the partially oxidised pyrite to haematite. On the magnetisation curves of the second heating the plateaux disappear and the mag-

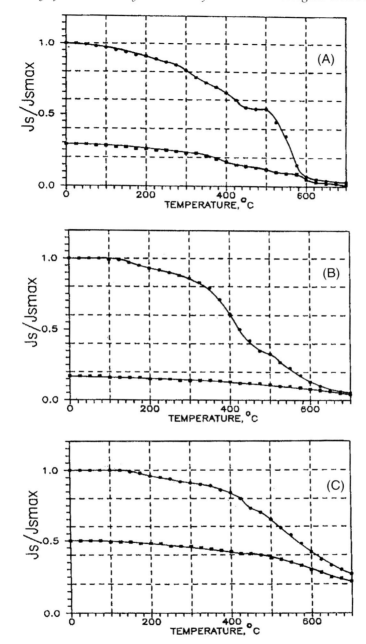

Fig. 2 *Dependence of the saturation magnetisation J on the heating temperature of the corrosion layers:*
● *first heating;* ■ *second heating.*
(A) the outer layer; (B) the middle layer, and (C) the inner layer.

netisation diminishes. This leads us to assume the presence of haematite and thermally unstable oxidised magnetite. Complete demagnetisation does not occur at $T = 700°C$ (the maximum temperature attained in experiments). Hence we can conclude that a small amount of iron with $T_c = 770°C$ is present even in the upper corrosion layer although no magnetic hyperfine splitting is observed in the Mossbauer spectrum of this layer.

The phase transitions that occur in the course of heating up to 700°C can be seen easily in the spectrum of the upper layer annealed during the TMA (Fig. 3) The peaks belonging to the compounds of Fe^{2+} disappear. Before heating, the magnetic oxides of iron occupy only about 4% of the total area of the spectrum of the initial sample (Fig. 3(a)), whereas 46% are occupied by the spectra of ionic compounds of ferrous iron, pyrite, and finely dispersed $\alpha-Fe_2O_3$. After the TMA (Fig. 3(b)), the area corresponding to the spectrum of the magnetic fraction of the upper layer increased up to 50% due to the decomposition of ferrous iron and pyrite and the partial oxidation of magnetite. Mossbauer spectra corresponding to the middle and inner layers show the same processes resulting from the heating up to 700°C.

The same iron compounds are present throughout the rust bulk, but the percentage phase content is different. This can be seen in the chart of phase volume in relation to the depth of the corrosion layer, (Fig. 4), which we have calculated from our experimental data.

4. Conclusions

The following model for a podzol soil corrosion layer was derived. Near the iron nucleus crystallised iron oxides, such as Fe_3O_4 and Fe_2O_3 predominate; the quantity of magnetite decreased to effectively zero as the surface of the layer is approached; haematite is found on the surface, in the finely dispersed form, while the concentration of iron–sulphur and iron phosphate compounds is increased.

Furthermore:, the outer layer, which is in direct contact with the soil, contains about 60% ferrous iron compounds with ~40% of this amount corresponding to FeS_2 (~16%), $FeSO_3\ 3H_2O$ (~12%), and $FeSO_4\ 4H_2O$ (~12%), i.e. Fe(II) sulphur-containing compounds.

These compounds are in a finely dispersed form (which accounts for the absence of strong X-ray reflections for some of them).

These data lead to the conclusion that concurrent reduction processes take place in the soil. The sulphur cycle, reported by Tiller [4], allows us to make some assumptions about the nature of this process. The fact is that the reduction of sulphate to hydrogen sulphide is carried out by anaerobic bacteria, the sulphate reducers. That type of corrosion (obtained by sulphate-reducing bacteria proliferation) is usually recognised by the smell of hydrogen sulphide, which was detected when scratching the rust surface of our sample. We can therefore assume that sulphate-reducing bacteria are involved in the corrosion process under examination. Our assumption has been confirmed by the presence of the poorly crystallised fine dispersed pyrite content observed in this case.

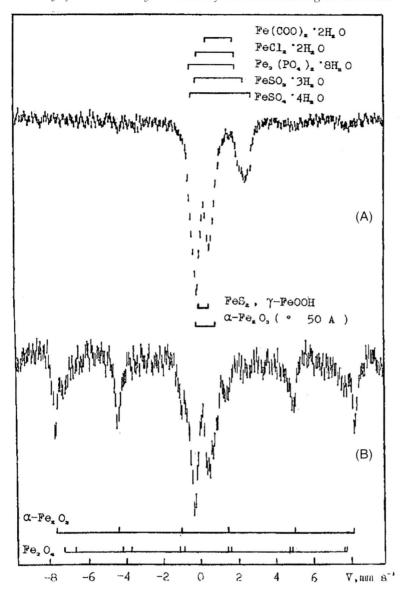

Fig. 3 Mosbauer spectra of the outer corrosion layer before (A) and after (B) the thermomagnetic analysis (heating up to 700°C).

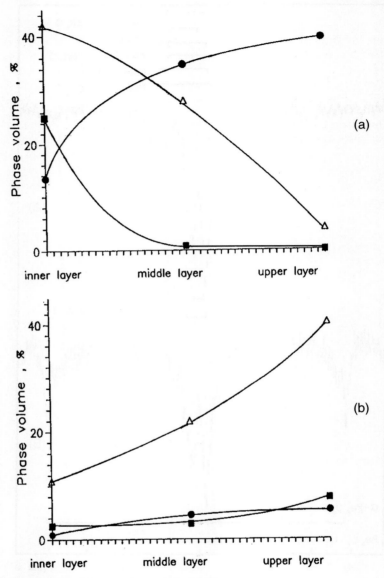

Fig. 4 *Phase analysis of the corrosion scale in relation to depth;*
(a) • *fine dispersed* Fe_2O_3
  ■ α–Fe
  △ *iron oxides.*

(b) • $FeCl_2\ 2H_2O$
  ■ $Fe_3(PO_4)_2\ 8H_2O$
  △ *iron compounds with S.*

References

1. S. V. Zonn, *Iron in Soils* (in Russian), Nauka, Moscow, 1982.
2. G. E. Zaikov, S. A. Maslov, and V. L. Rubailo, *Acid Rains and Environment* (in Russian), Nauka, Moscow, 1991.
3. A. A. Novakova, T. S. Gendler, N. D. Manyurova, and R. A. Turishcheva: 'Products of Soil Corrosion of an Iron Sample as Studied by Mossbauer Spectroscopy', *Vestn. Mosk. Univ.,* 1992, **48**, (3), 76–79.
4. A. K. Tiller, 'Aspects of Microbial Corrosion, in *Corrosion Processes* (Ed. N. Parkins), Applied Science Publishers, London, New York, 1982.

Part 4

Microbial Corrosion: Prevention and Control

Part 4

28

Electrochemical Sensor of Water and of the Amount of Microbial Proliferation in Fuel Tanks

B. M. ROSALES

Corrosion Research Center, Citefa-Conicet, Zufriategui 4380, Villa Martelli, Argentina

ABSTRACT

An instrument, described as Detector H_2O–X–MP has been developed to provide fast and easy indication of the presence of settled water and the magnitude of microbial proliferation in liquid fuel tanks. It measures the aggressiveness of the aqueous phase towards aluminium alloys, giving not only information about the microbial growth in the tank but also about the salt content in water. It is useful in preventing the risk of bottom perforation by corrosion from inside the tank and also detects water entering the tank as a result of corrosion perforation from the outside.

1. Introduction

Corrosion is a common fact in two phase systems consisting of a hydrocarbon and non-dissolved water. The risk increases with the salt content and with the amount of microbial proliferation. A well known trouble in aviation is the development of a biological sludge in wing tanks or in storage tanks when turbo fuel is in contact with a water layer.

The problems caused by settled water and microbial growth in aircraft include: clogging of the filters of turbines, malfunction of pumps and fuel level indicators, corrosion of the alloy of the wings, deterioration of the interior coatings of the tanks and reduction in the fuel quality.

Ecological damage resulting from ground–water contamination by gasoline is a consequence of leakage from underground tanks in service stations. Even when water enters with the fuel as humid air (clean water) the solubilisation of acid residues from the oil distillation increases its aggressiveness on the inside of the tank bottom. These contaminants bring about various forms of corrosion of the steel tank, which can eventually result in perforation.

Underground fuel deposits installed in aggressive soils or in contact with brackish ground-waters, can also be corroded by perforation from the outside. The subsequent entry of the brine water into the vessel can be detected by this device.

2. Experimental

The detector H_2O–X–MP (under patent) consists of a bi-electrode sensor and an electronic command instrument.

The sensor is formed from two concentric metallic electrodes, separated by an insulator material, between which an electric current flows when a given potential difference is applied. Special construction characteristics and materials are used to make this current proportional to the amount of contaminants in the water and to the microbial proliferation. The sensor is fixed in the deposit at the bottom of the tank where water accumulates when the temperature decreases.

The command instrument is installed outside the tank: in the cockpit of an aircraft, in the console of service stations. A portable model can be connected to sensors installed in ground deposits, in airports or multi tank plants.

The command instrument remotely senses the presence of water and the amount of salt content or microbial proliferation that has occurred in the tank.

According to the research work performed [1–12], for various salt concentrations and/or microbial proliferation, different current intensities will flow in the water for a given potential applied between the sensor couple of electrodes. A microprocessor provides an optional programme for the frequency of reading, and to initiate a sound alarm or start a pump which manually or automatically drains the water once this is detected, regardless of its contamination level.

The control of separate tanks can be achieved through remote electric switching of channels of the instrument in a fixed installation, or by means of a portable equipment connecting it to the sensors installed in each tank, where a permanent control is not required.

The portable model is useful for the control of numerous tanks when a fixed installation is not available. That is the case, for example, for distant tanks not requiring very frequent control. The only fixed element, which should be placed in the lower point of the tank, will be the sensor. This is connected to the command unit for each test of the tanks. Only one instrument need be used for as many sensors as there are tanks to be checked. If water elimination is wanted then a tube fixed to the bottom of the tank should be installed with the sensor together with equipment to pump the water out once it is detected.

The advantages with respect to the present housekeeping control in aviation are the following:

1. No incubation is needed, as is the case, for example, with a commercial monitor test kit, which is used at present and which requires 48 h incubation before a change in colour indicates the presence of aerobic microbial contaminants.

2. It is not necessary to empty the tank, as when visual inspection of the bottom is carried out. This avoids taking equipment out of service.

3. Since the sensor is installed at the lowest point in the tank it detects the presence of separated water. If proliferation occurs the instrument indicates its magnitude, and the effectiveness of biocide treatments as soon as these are initiated.

4. It can also be used out of the tank by immersing the sensor in drained water, even when the fungal mat has remained inside the tank during draining.

5. No increase in fuel price, as when treated fuel is specified.

6. Obvious readings are obtained. Thus, no experience is needed to obtain the correct answer (concerning the presence of water or assessing magnitude of microbial proliferation). This differs from the situation, when sump water is inspected visually for any peculiarity that could suggest microbial proliferation.

7. Sumping practices can be reduced to the necessary minimum because the presence of water can actually be checked.

8. The equipment is easy to install and operate. No design modifications are needed in aircraft fuel tanks or ground storage tanks.

9. It can be used either as an independent control or to feedback to any associated scavenging system.

10. This equipment is essential in aircraft for the diagnosis of the presence of water when capacitive fuel level indicators work erratically.

In the case of service stations, the following advantages can be mentioned:

1. The sensor detects separated water and determines the salinity level in one operation.

2. Maintenance cost in tanks can be lowered to necessary minimum when salt or microbial contaminations are routinely determined by the sensor.

3. It allows automatic detection and instantaneous solution of the problem.

4. It prevents inner corrosion of fuel tanks.

5. Water contamination of the fuel resulting from ingress of ground, river or sea water resulting from perforation by corrosion from the outside (due to high aggressivity of the soil or insufficient external protection of the tank) can be detected.

3. Operation

Once turned 'ON', the 'SCAN' button commands the autotest of the components through the sequential polarisation of all the sensors, tank by tank, in correspondence with the number of the channels. During this checking 'INT' pilot is lighted.

The 'START' button initiates the analysis of the tank whose channel number is lighted. Selection of the next tank is done by pushing the 'SCAN' button. When any discontinuity occurs in the circuit of a sensor, 'OPEN CH' lights advising the failure during the test of the corresponding tank or channel.

The response, on the left hand side of the panel, indicates the presence of 'FUEL' or 'WATER' in the bottom of the respective tank. In the first case the green pilot indicates the absence of water. In the latter, the amount of microbial proliferation in the tank, 'NO', 'MEDIUM' and 'DANGER' is visualised by means of green, yellow and red lights. These stay alight for 5 s, the polarisation time for each sensor.

With a fixed installation, the control of four tanks can be performed in less than 1 min.

One prototype has been in service for 8 years in a plane of the Argentinean Air Force.

The device is intrinsically safe, according to the international standard IEC 79-11.

References

1. B. M. Rosales, E. R. de Schiapparelli and D. Cabral, 'Corrosión por microorganismos en tanques de combustible de aviones. Estudio de aleaciones'. *Corrosión y Protección*, 1979, **10**, (3).
2. B. M. Rosales and E. R. de Schiapparelli, 'Microbiological corrosion prevention in jet fuel tanks. Paper No. 3, Rio de Janeiro, Brasil, *Proc. 7th. Int. Congr. on Metallic Corrosion*, 1978, **3**, 1424–1430.
3. B. M. Rosales and E. R. de Schiapparelli, 'A corrosion test for determining the quality of maintenance in jet fuel storage', *Mat. Perform.*, 1980, **19**, (8), 41–44.
4. E. R. de Schiapparelli and B. M. Rosales, 'Microbial corrosion in terminal storage tanks for aircraft fuel', *Mat. Perform.*, 1980, **19**, (10), 47–50.
5. E. R. de Schiapparelli and B. M. Rosales, 'Mechanism of microbiological corrosion'. Barcelona, Spain, *Proc. 5th Int. Congr. on Marine Corrosion and Fouling*, 1980, Marine Biology 1–6.
6. D. Lupi, L. Fraigi and B. M. Rosales, 'Potenciostato Automático', *Proc. II Congreso de Electrónica*, 1981, Bs. As., Argentina.
7. M. C. Leiro and B. M. Rosales, 'Factores metalúgicos que controlan la corrosión por picado de las aleaciones aeronáuticas de Al', *Revista Latinoamericana de Metalurgia y Materiales*, 1984, **4**, (1), 8–16.
8. M. C. Leiro and B. M. Rosales, 'Predominance of microbial growth on the metallurgical characteristics in the corrosion of the 2024 Al alloy through electrochemical parameters', *Proc. Corrosion '86*, 1986, NACE,Houston, TX, paper No. 124.
9. B. M. Rosales, 'Corrosion microbiologique de l'Aluminium et ses alliages dans les reservoirs de carburant des alies d'avion', Surfair VI, Cannes, France, Nov. 1986.
10. E. S. Ayllon and B. M. Rosales, 'Corrosion of AA 7075 aluminium alloy in media contaminated with *Cladosporium resinae'*, *Corrosion NACE*, 1988, **44**, (9), 638–643.
11. B. M. Rosales, E. S. Ayllon and M. C. Leiro, 'Accelerated test for determining

microbiologically influenced corrosion resistance of aluminum alloys', in *New Methods for Corrosion Testing of Aluminium Alloys*, ASTM STP 1134 (Eds V. S. Agarwala and G. M. Ugiansky), ASTM, Philadelphia, PA, 50–59, 1992.

12. E. S. Ayllon, M. C. Leiro, M. A. Esteso and B. M. Rosales, 'Análisis electroquímico de la influencia del hongo *Cladosporium resinae* en la resistencia a la corrosión de Cu, acero AISI 316L y níquel químicamente depositado', Proc. IX Congreso Iberoamericano de Electroquímica, 1990, Tenerife, España, 499–502.

29

Biocorrosion in Groundwater Engineering Systems

P. HOWSAM, A. K. TILLER* and B. TYRRELL

Silsoe College, Cranfield University, UK
*Lithgow Associates, Kingston-upon-Thames, UK

ABSTRACT

Corrosion is often out-of-sight out-of-mind. Water sources in subsurface aquifers are very much in this category, as are the engineered system for abstracting, controlling and monitoring groundwater. If on top there is added the misconception that most groundwaters are sterile, then it is easy to see why many groundwater engineers and hydrogeologists are often not aware of biocorrosion. Whilst the understanding and experience of corrosion of borehole components is widespread, there appears to be less awareness of the significance of microbial activity in corrosion processes. A lack of understanding of biocorrosion processes may lead in some cases to inappropriate borehole or tubewell design, construction and operation.

Groundwater engineering would benefit from an increased input of biocorrosion expertise. This paper serves to explain the components and factors in groundwater engineering and how, where and when biocorrosion prevails.

1. Introduction

Reference to corrosion can be found in very early groundwater engineering literature [1, 2]. The explanation of the electrochemical process by invoking physics and chemistry has stood the test of time very well. This understanding has permitted the avoidance of many corrosion problems by the use of appropriate materials or protection measures. Yet despite this history of knowledge and experience, corrosion problems are still far too prevalent in engineered systems. This applies no less to groundwater engineering.

Considerable progress has been made in the past decade in understanding many of the factors influencing microbially induced corrosion (biocorrosion). The volume of research literature now available reflects the substantial international effort that is being devoted by some industries to the problem [3]. This does not appear to be the case in the water industry as a whole.

It is likely that a primary reason for this state of affairs is the lack of understanding of the role played by microorganisms in many processes which occur in engineering environments, including corrosion [3, 4].

2. Groundwater Engineering
2.1. Groundwater and Aquifers

Groundwater is a major source of water for domestic, industrial and agricultural purposes world-wide. In Britain groundwater accounts for 35% of all supplies whereas in some arid regions it accounts for nearly 100%.

Groundwater is stored in a wide variety of formations [5]. The quality of the groundwater in these water-bearing formations (aquifers) depends on the geochemistry of the formation, how long the water has resided in the aquifer, how the aquifer is or was recharged and the quality of the recharge water. Most aquifers occur in sedimentary formations such as unconsolidated alluvium, consolidated sandstones and limestones. In some cases the groundwater is that which was present at the time the aquifer was formed. These connate waters are very old and highly mineralised (saline). In other cases, this connate water has during geological time been flushed out and replaced or diluted by fresher water. Where modern rainfall recharge occurs the groundwater quality is very good — low in minerals and acceptable for domestic, industrial and agricultural use.

In a layered aquifer composed of different formations the groundwater quality may also vary with depth.

Some aquifers are confined by overlying aquicludes. Under these confined conditions groundwater is under pressure and when the confining layer is penetrated by a borehole, the groundwater will rise towards or beyond the ground surface. Such deep confined groundwaters are anaerobic and again the intrusion of drilling a borehole can upset the natural equilibrium, in this case with respect to hydrochemistry. In deep boreholes the groundwater quality profile can change significantly from its natural state as it enters the borehole to a new equilibrium state as it leaves the borehole (e.g. parameters such as pH, E_h, alkalinity, iron). This profile will of course also be influenced by any biofouling and biocorrosion processes occurring in the borehole column. For instance, hydrogen sulphide can be produced naturally in some aquifers and it is not easy to distinguish background values from values occurring as a result of biocorrosion in the system.

The type of aquifer also dictates the way groundwater moves and is stored. This can be via interconnected pores (as in a sandstone) or fissures (as in a limestone).

Springs are where groundwater emerges at the earth's surface. In most cases however man has to tap into the aquifer in order to abstract groundwater There are many ways that groundwater can be abstracted, from shallow hand-dug vertical wells and horizontal adits to deep vertical drilled boreholes. The latter are the most common and it these types of engineered systems that will be discussed further.

2.2. Groundwater Engineering — Design

Borehole design will relate to the need to match the purpose of the borehole with the prevailing ground conditions. Typically a borehole design will include the following components: casing, screen, grout seal, headworks, pump and riser (Fig. 1). The variables are: depth, diameter and choice of materials.

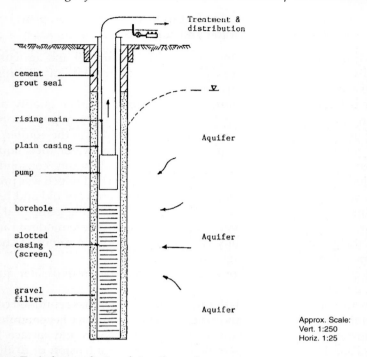

Fig. 1 Typical groundwater abstraction system

A casing/screen string is used to line and support the drilled hole. The screen or perforated casing is the section where water is drawn in to the borehole from the aquifer. The blank casing is for non-yielding sections. The grout seal is used to isolate different horizons and to prevent vertical movement of water leading to mixing and cross-contamination, in the borehole.

Depth and diameter will depend on the hydrogeological conditions and on the amount of water supply required. Borehole yield will depend on both the transmissivity and the storage capacity of the aquifer.

Choice of materials is one of the most important factors with regard to biocorrosion. Traditionally steel is the most common borehole construction material. It has the toughness to cope with the rough handling often encountered in borehole construction and has the compressive/tensile strength required for most depths and ground conditions. Mild steel is the most common form but stainless and other special alloy steels are used where it is felt that the cost can be justified.

Conditions for corrosion are common in many aquifers. The way groundwater engineers have often coped with corrosion is by increasing the thickness of the steel.

Coatings have been used but generally the risk of damage during site handling and installation is always high, and in aggressive downhole environments even a few isolated breaks in protection can lead to intense localised corrosion and perforation of a lining. Most groundwater engineers are cautious about using coatings as the sole means of protection against corrosion.

Plastics were introduced in the 1970s. This solved the corrosion problem but the limited strength of the early plastics limited the depths and diameters where it could be used. The use and development of glass-reinforced plastics (GRP) and glass-reinforced-epoxy (GRE) have significantly reduced the depth and diameter limitations.

2.3. Groundwater Engineering — Construction

There are a variety of methods for drilling holes in the ground. Major developments in water well drilling have been derived from experience and developments in the petroleum industry.

As for design, the construction method will relate to the need to match the purpose of the borehole with the prevailing ground conditions.

The verticality and diameter of most constructed boreholes is in reality somewhat different from that shown in design drawings. This has implications for casing/screen installation, sealing annular areas and installing pumps.

Many problems in water boreholes relate less to problematic ground conditions and more to poor design, poor quality control during construction, inappropriate operation and poor maintenance.

2.4. Groundwater Engineering — Pumps and Associated Equipment

Having constructed a borehole to gain access to the aquifer, in most circumstances a pump is required to lift the groundwater to the surface. In some deep confined aquifers the groundwater is confined under sufficient pressure such that it will rise up to or above the ground surface. In these cases of artesian or free-flowing boreholes pumps may not be required.

Borehole pumps are usually either: electric powered submersibles where both the drive motor and the pump are a single assembly set at the required depth in the borehole. or: vertical turbine pumps where a electric or diesel powered, usually variable speed, drive motor is fixed at the surface and the submerged pump is driven via a vertical shaft.

Groundwater is conveyed to the surface via a rising main or pump column pipe. The rising main is often galvanised iron or mild steel, flush or flanged jointed, and may have a protective coating of bitumen paint.

Pump impellers are often bronze, sometimes stainless steel and pump bodies can be cast iron or stainless steel.

2.5. Groundwater Engineering — Operation and Maintenance

There is a tendency for groundwater engineers to try to abstract the maximum yield from a borehole. However, in the long run this can often be a false economy since it can lead to hydraulic and mechanical stress which eventually affects borehole performance and may even lead to failure.

Operation in terms of operating schedule as opposed to operating yield can also affect performance. Continuous pumping at moderate rates is much preferred to intermittent pumping at high rates. High velocities can generate sand pumping/hole collapse, and pump wear/failure. Water quality changes can occur under stagnant non-pumping conditions and under turbulent/aerating conditions generated by high rates of pumping.

3. Biocorrosion

Having described the nature of groundwater abstraction systems, it is perhaps useful here briefly to summarise the nature of biocorrosion. The subject of biocorrosion can be introduced to the non-expert by first discussing basic corrosion principles, then introducing the concept of biofilms and then seeing how these two phenomena interact.

3.1. Corrosion

Corrosion is a electrochemical process which is enhanced:

- in strong electrolytes
- in low pH waters
- by the presence of dissolved gases (O_2 CO_2, H_2S)
- by high velocity flow
- where high cathode–anode area ratios exist
- where there is a large difference in electrical potential between two different metals
- by the presence of stray electrical currents
- where temperature differences occur (cold-cathode hot-anode).

Differences in electrical potential can exist between different areas of the surface of the same metal. Such differences in potential will exist between a plain metal surface and that which has been affected by

- heat (e.g. at welded joints: becomes anode)
- work hardening (e.g. at machined edges/slots: becomes anode)
- cutting (e.g. where threads have been cut: becomes anode) – coating (e.g. where coating is missing/damaged: becomes anode)

Corrosion can also be used to describe the chemical process of metal dissolution by strong acids and erosion (cavitation) corrosion can occur under localised conditions of high flow velocities and turbulence.

3.2. Biofouling

Biofouling describes the activity of sessile microorganisms which attach themselves to, and develop biofilms on, surfaces. The biofilm typically consists of bacterial cells, extracellular slime and inorganic material trapped or precipitated within the biofilm.

Biofilms are common at fluid/solid interfaces in systems where there is flow and also in stagnant conditions. Biofilms are often stratified, consisting of an upper aerobic layer and a lower anaerobic layer in contact with the component surface. This creation of micro-environments is typical of biofilms and is a factor in enhancing corrosion processes [6].

3.3. Biocorrosion

Biocorrosion is the inducement or enhancement of electrochemical process of corrosion by the activity of sessile bacteria in forming biofilms. In many cases the activity results in conditions in localised micro-environments which are different from the macro-environments.

The common processes are:

- the setting up of concentration cells on the surface of the metal where a biofilm develops. Basically in an environment where oxygen is present, microbial activity within the biofilm will reduce oxygen, thus forming a differential aeration cell. The centre of the biofilm becomes anodic compared to the edge of the biofilm where oxygen is present, which becomes cathodic. This process results in localised pitting.

- the removal of cathodic hydrogen by sulphate reducing bacteria promotes continuation of the corrosion process by depolarising the cathodic area of the metal surface.

- the formation of corrosive metabolic by-products from the bacterial activity within the biofilm. Examples include the production of sulphuric acid by sulphur-oxidising bacteria, and the production of organic acids, hydrochloric acid, hydrogen sulphide, ferrous sulphide, ferric chloride and elemental sulphur, by anaerobic sulphate reducing bacteria.

- the metabolic activity within an iron biofilm which leads to a depletion of oxygen from the lower biofilm layers, where redox conditions can be created suitable for the growth of anaerobic sulphate reducing bacteria which are able to enhance/induce corrosion processes.

- the depassivation of the metal surface produced by the microbial reduction of insoluble ferric deposits to soluble ferrous compounds.

4. The Nature and Occurrence of Biocorrosion in Groundwater Engineering Systems

Biocorrosion in boreholes is not well documented for two reasons:

(i) there is a common misconception that groundwater and aquifers are microbiologically pure and therefore physical and chemical reasons are sought for corrosion;

(ii) most borehole components are down-hole and as such are not readily observable.

Observation of components is achieved by their retrieval from the borehole. Examples are shown in Plates 1 to 4, on pp.363–366. This particularly applies to the pump and rising main but more rarely to the casing/screen string. The latter can usually only be observed internally by use of a down-hole CCTV camera. Recent developments in CCTV cameras now allow very good quality observations to be made.

4.1. Casing

As in other pipe systems biocorrosion in boreholes tends to concentrate at joints or where protective coatings have been damaged.

The welded or machined zones associated with flanged and screwed joints provide ideal sites for biofilm development, as well as generating differences in potential between worked and unworked metal. Properly placed grout seals will provide a measure of support and protection to the external casing surfaces, but this is difficult to achieve. Access of contaminated nutrient-rich surface waters down an incompletely sealed borehole annulus will enhance biocorrosion, as will direct connection with horizons bearing aggressive groundwaters. Poor quality jointing will also permit ingress of such waters inside the casing and into the pump.

4.2. Screen

The same applies to screens as to joints and to perforated casings as to blank casings. Enlargement of screen slots or perforation due to biocorrosion is often subsequently exacerbated by a combination of sand pumping and high velocities inducing a sand blasting action to enlarge perforations even further.

The screen, as the point of entry from the natural aquifer to the engineered system, is a primary zone where hydrochemical and biochemical changes occur commonly as a result of pressure changes (e.g. with respect to carbon dioxide equilibrium) and aeration (e.g. with respect to iron biofouling).

4.3. Pump

The pump is the focus of groundwater flow out of the system and as such will be influenced by natural groundwater quality as well as the quality of other waters allowed to enter the system.

The velocity distribution into and within the pump is important. There is some field and operational evidence of increased biofouling linked with higher velocities in pumps and pumping systems. There is no available evidence for borehole pumps of biocorrosion/velocity inter-relations, but the indications are that biofilms can withstand, indeed can be enhanced by, high velocities and turbulence.

4.4. Rising Mains

Biofouling and biocorrosion processes which are initiated in the borehole will continue and will be extended within the rising main. They are subject externally to the groundwater quality within the borehole column and internally to that being pumped to the surface.

4.5. Headworks and Distribution

Problems of biocorrosion of the headworks and the distribution are not always traced back to the raw water source and processes occurring in the borehole.

5. Monitoring

Where awareness of biocorrosion exists within groundwater engineering methods for monitoring and diagnosis tend to be a mixture of standard borehole inspection techniques, water quality analysis and adaptions of oil industry techniques.

5.1. Borehole Inspection

Any components which can be withdrawn from the borehole i.e. pump and rising main can be relatively easily inspected. Other components, i.e. screen and casing are inspected by means of CCTV cameras.

5.2. Water Quality Analysis

General parameters such as pH, E_h, alkalinity, TDS are commonly analysed for groundwater samples. The problem with groundwater quality analysis lies in the method and location of sampling and the time and location of analysis. Probes are available for downhole measurements, as are downhole samplers, but most analyses are conducted on pumped samples collected at ground level. Unstable parameters should be, but are not always, measured on site immediately the sample is taken. For other parameters, samples are taken to a laboratory for analysis.

In the event it is often difficult, because of the hydrochemical and biochemical processes occurring within the aquifer, the abstraction system and then at the surface, to identify how a water quality parameter measurement relates to processes and conditions within the system. As such water quality information has often proved to be unhelpful in predicting the potential for, or existence of, biofouling and biocorrosion.

Measurement of general bacterial activity in groundwaters is, apart from public health tests for coliform bacteria, not widely carried out. Simple field test kits for measuring bacterial activity in groundwater systems do exist. Known as BARTs, kits are available for sulphate reducing bacteria, iron related bacteria, slime-forming bacteria, amongst others. Very little data have been accumulated on nutrient concentrations in groundwaters but it can be assumed that adequate nutrients are available in most systems to permit biofilm growth.

5.3. Biofouling/Biocorrosion Monitors

Various devices have been developed by groundwater engineers and scientist to assess corrosion/biocorrosion [4, 7, 8]. These include downhole devices such as corrosion coupon and biofilm slide collectors. At the wellhead devices such as the Howsam-Tyrrel moncell and the Smith flowcell have been used [4, 8].

6. Summary

A degree of awareness and understanding of corrosion is common amongst groundwater engineers. However, because of the nature of corrosion and the nature of the borehole environment a less than 100% understanding can lead to unexpected consequences. This is particularly the case with respect to biocorrosion since many engineers do not fully perceive the natural or engineered parts of groundwater systems as microbially active. Groundwater engineers and biocorrosion experts need to meet and talk to each other. Many established biocorrosion monitoring and diagnosis techniques could usefully be applied to groundwater abstraction systems as could some principles of prevention and control used in other sectors.

7. References

1. M. A. Borch, S. A. Smith and L. N. Noble, 'Evaluation and restoration of water supply wells', American Water Works Association Research Foundation (AWWARF), 1993, Denver, CO, USA.
2. F. G. Driscoll, *Groundwater and Wells*, 1986, Johnson Screen Division UOP, St. Paul, USA.
3. A. K. Tiller, 'Biocorrosion in Civil Engineering', in *Microbiology in Civil Engineering* (FEMS Symposium No. 59, edited by P. Howsam, 1990), pp.24–38. E. & F. N. Spon, London.
4. P. Howsam and S. F. Tyrrel, 'Diagnosis and monitoring of biofouling in enclosed flow systems — experience in groundwater systems', *Biofouling*, 1989, **1**, 343–351.
5. M. Price, *Introducing Groundwater*, George Allen & Unwin, London, 1989.
6. R. G. McLaughlan, M. J. Knight and R. M. Stuetz, *Fouling and corrosion of groundwater wells*. Research Publication 1/93, National Centre for Groundwater Management, UTS, Sydney, 1993.
7. P. Howsam (Ed.), *Water Wells — Monitoring Maintenance and Rehabilitation*, E & F N Spon, London, 1990.
8. P. Howsam, B. D. Misstear and C. R. Jones, 'Monitoring, maintenance and rehabilitation of water supply boreholes', Construction Industry Research & Information Association (CIRIA), London.

Plate 1A. *Vertical turbine borehole pump drive shaft (316 stainless steel) recovered from borehole, after 9 months installation, in pristine condition.*

Plate 1 B. *Vertical turbine borehole pump drive shaft (416 stainless steel), recovered from borehole after only 6 months installation, having been subject to severe biocorrosion.*

Plate 1C. *Vertical turbine borehole pump drive shaft and coupling (both 416 stainless steel), recovered after only 6 months, having been subject to severe biocorrosion.*

Plate 2A. *Coal tar epoxy coated mild steel pump column, showing large perforation arising from concentrated localised corrosion, probably at point of coating damage.*

Plate 2B. *Flanged end of coal tar epoxy coated mild steel rising main, showing large perforation arising from biocorrosion concentrated around weld zone of flange section, where coating application is often inadequate.*

Plate 3A. Steel groundwater supply pipeline showing internal pitting arising from biocorrosion.

Plate 3B. J55 grade mild steel rising main showing general encrustation, blistering and intense localised biocorrosion resulting in a significant perforation.

Plate 4A. Stainless steel submersible pump coupled with galvanised iron less than two years in the borehole.

Plate 4B. Steel well head works showing signs of the biocorrosion processes occurring within the borehole.

Laboratory Evaluation of the Effectiveness of Cathodic Protection in the Presence of Sulphate Reducing Bacteria

K. KASAHARA, K. OKAMURA and F. KAJIYAMA

Tokyo Gas Co., Ltd., Fundamental Technology Research Laboratory, 1-16-25 Shibaura, Tokyo 105, Japan

ABSTRACT

Off-potential requirements were studied in the laboratory to achieve complete cathodic protection on steel pipes buried in microbiologically aggressive soils. Burial corrosion tests with simulated cathodic protection were carried out potentiostatically at the off-potentials between –0.65 and –1.3 V (Cu/CuSO$_4$). Emphasis was placed on the identification of the role and influence of sulphate reducing bacteria (SRB, *Desulfovibrio desulfuricans*) in the corrosion and protection processes in the soil.

Weight loss measurements and macroscopic observations revealed that, sufficiently practical protection in SRB active soils is obtainable at off-potentials more negative than –0.85 V and/or by applying a cathodic current of not less than 0.03 A m^{-2}. Bacterial counts clearly indicated that SRB could proliferate even in the potential range where cathodic protection was being sufficiently achieved.

1. Introduction

Cathodic protection has received a worldwide recognition as the most effective and economical technique to protect buried coated steel pipes against corrosion [1, 2]. This technique first became practical in 1933 when Kuhn proposed a –0.85 V (Cu/CuSO$_4$) protective potential criterion, that is, the corrosion potential of the pipe to be protected has to be reached to a range more negative than –0.85 V, the threshold value below which an acceptable corrosion resistance is achieved.

However, possible microbiologically influenced corrosion (MIC) of steel pipes buried in anaerobic clayey soils containing SRB, which was first reported in 1934 by Kuhr and van der Vlugt [3], raised a need for an alternative protective potential criterion. Based on thermodynamic considerations, a –0.95 V criterion was proposed by Hovarth and Novak in 1964 for cathodic protection in the presence of SRB [4]; and was afterwards experimentally verified by Fischer [5].

Although SRB have almost exclusively been the focus of many investigations involving MIC, the roles in MIC of other species of bacteria, i.e. acid producing bacteria and metal depositing bacteria, have increasingly been emphasised. The present authors have carried out a number of site-surveys on buried pipes to clarify the causes of a wide variety of corrosion failures by local action cells or localised cells, and demonstrated that iron-oxidising bacteria (IOB, *Thiobacillus ferrooxidans*), sulphur-oxidising bacteria (SOB, *Thiobacillus thiooxidans*) and iron bacteria (IB, *Gallionella*) could have played a major role in significant corrosion of buried pipes [6–8]. The most significant finding was that the contribution of SRB was of little importance in the process of soil corrosion.

Under such circumstances, the need has naturally been strengthened for the revision of cathodic protection criteria. In the present study, the effectiveness of cathodic protection was reinvestigated in the laboratory to protect steel pipes against corrosion in soils with high activity of bacteria as SRB. Potentiostatic tests were conducted with an emphasis on examining the action and behavior of SRB under conditions applied with cathodic currents.

2. Experimental Procedures

In the present laboratory study, a simulated cathodic protection cell as shown in Fig. 1 was used; in this three electrodes: a counter; working and reference electrodes, respectively, were buried in a PVC soil box of 100 mm wide × 150 mm long × 100 mm deep (internal measurement).

The working electrode of a carbon steel plate to simulate a coated steel pipe having a 7.1 cm^2 holiday in the coating was obtained by covering the steel plate surface

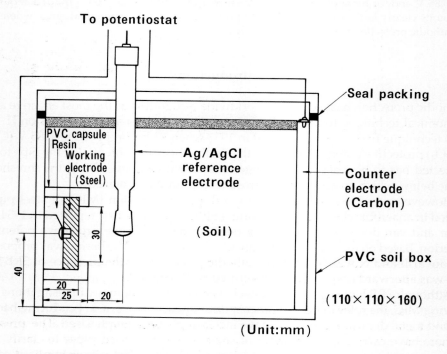

***Fig. 1** Schematic of burial soil corrosion testing cell with simulated cathodic protection.*

with a PVC capsule of 5 mm in thickness. The surface of the working electrode was slightly abraded with 2/0 emery paper, followed by degreasing with acetone. A graphite plate of 100 mm wide × 100 mm long × 10 mm thick was used to simulate a counter electrode in an impressed current type of cathodic protection system. This counter electrode was placed at a distance of 120 mm from the surface of the working electrode. A commercially available double-junction type saturated Ag/AgCl electrode was used as a reference electrode and was placed between the working and the counter electrodes at a distance of 20 mm from the surface of the former.

The PVC soil box with these three electrodes was then filled fully with the soil to create a simulated cathodic protection test cell, and wired to a potentiostat as indicated in the Figure. These test cells were then set in an incubator at 298 K for 90 days. The soil used in the test was a clay soil which was sampled on-site where high activity of SRB had been recognised. Simulated cathodic protection was applied potentiostatically by applying cathodic currents via the counter electrode to the working electrode so that off-potentials of the working electrode should fall in the range from −0.533 to −1.183 V vs Ag/AgCl (potential values, however, are hereinafter given with reference to $Cu/CuSO_4$ at 298 K). During the testing period, currents flowing between working and counter electrodes were monitored continuously by means of a potentiostat.

Spontaneous burial corrosion testing was also carried out in the same manner so as to highlight the effectiveness of cathodic protection.

Soils used in the tests were investigated with respect to pH, FeS content and bacteria counts (IOB, SOB, IB and SRB) before and after the exposure. The soil samples were collected from the immediate vicinity of each working electrode surface (< 1 mm). Bacterial counts were made in accordance with the most probable number (MPN) method. Postgate's medium B [9] was used for SRB enumeration. IOB, SOB and IB enumerations were made according to the recommendation appearing in Ref. [10]. The rest of the environmental factors was determined in accordance with the Japanese Standardized Manual for Soil Analysis. Upon conclusion of the tests, each working electrode was removed from each cell, cleaned and weighed. The weight loss measurements were then converted into a uniform corrosion rate (mm/year). Maximum corrosion rate (mm/year) was obtained by means of an optical microscope equipped with a depth gauge.

3. Results and Discussion
3.1. Cathodic Protection Tests in the Presence of SRB

Figure 2 shows the effect of applied potential on the uniform corrosion rate in a SRB active clayey soil. Free corrosion potentials, E_{corr}, before test ranged from −0.76 to −0.82 V.

The data grouped in the category of 'free corrosion' represent those for working electrodes without cathodic protection, that is, those subjected to free corrosion. Measurements were taken at 90 day on termination of exposure testing. The maximum corrosion rates were identical with respective uniform corrosion rates because the type of attack was classed as typical uniform corrosion.

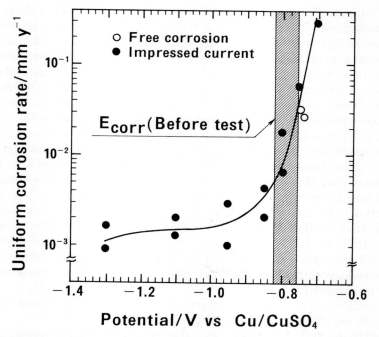

Fig. 2 *Effect of applied potential on the uniform corrosion rate in a clayey soil containing active SRB.*

The rates of uniform corrosion at free corrosion potentials were around 0.03 mm/year; which fell well within the range of average uniform corrosion rates experienced in the Kanto-area (in and around the Tokyo metropolitan area) [8]. Such levels of uniform corrosion rates suggested that the role of SRB in the soil corrosion process should be of little importance.

It can be seen in the Figure that uniform corrosion rates decreased rather drastically as off-potentials were lowered from the free corrosion potentials, and in the potential range more negative than –0.85 V, uniform corrosion was depressed significantly to rates as low as 0.004 mm/year or less, thus leading to a preliminary conclusion that complete, or practically sufficient, cathodic protection in a clayey soil containing active SRB may be achieved at off-potentials more negative than – 0.85 V.

Figure 3 shows the relationship between the uniform corrosion rate and the cathodic current density averaged over the testing periods. It can be concluded from this result that the minimum required cathodic current density to achieve complete cathodic protection would be 0.03 A m^{-2}. It may be worth noting that this current density is lower than the level critical for the common soils in the Kanto-area, 0.1 A m^{-2} [11].

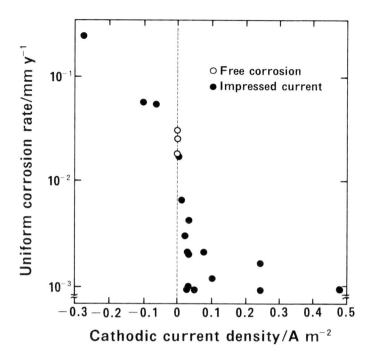

Fig. 3 Relationship between uniform corrosion rate and cathodic current density averaged over the testing period.

3.2. Cathodic Protection and Microbial Action

Although the contributions of bacterial species to soil corrosion have not yet been fully clarified, it may easily be imagined that environmental changes taking place during cathodic protection, such as pH, aeration, etc., would significantly influence the proliferation and activities of bacteria species involved in the soil corrosion process.

Table 1 compares bacterial counts in the soil close to the working electrode surface (< 1 mm) before and after 90-day laboratory burial tests. As one-by-one counting of IB is not easy because these bacteria usually form agglomerated strings, a 3-level semi-quantitative indicator system, +, ++, and +++, was employed here to clarify the IB population. Particularly noteworthy in the table is the SRB proliferation even in the potential range where cathodic protection was being adequately achieved, that is, more negative than –0.85 V. It is also noteworthy that at the off-potential of –1.30 V, SRB counts were still detected, though the activities were slightly reduced. In agreement with such SRB counts, thin and black films of iron sulphide were detected without exception on every working electrode surface.

Figure 4 shows the effect of applied potential on the pH, FeS content and SRB counts in the soil close to the working electrode surface (< 1 mm). Readings for free corrosion are also plotted in the Figure. It can be seen in the Figure that the potential dependencies of both SRB counts and FeS content were the same. Thus they increased, starting at around −0.75 V of free corrosion, with enhancing cathodic protection, rising to a maximum level at around −1.00 V and thereafter decreasing gradually with more negative potentials. Such a similarity in the potential dependencies of SRB counts and FeS content would lend a strong support to a mechanism wherein FeS is formed as a result of metabolism of SRB and could play an important role in the process of protection. It is to be noted that FeS could still be detected at −1.30 V.

It can be seen also from Fig. 4 that the change of pH with applied potential was shaped like a long and narrow "S", and such that it increased rather rapidly, starting at around the value of pH 7 of free corrosion, with increasing cathodic potential, rising to around 8.0 at −0.85 V, and thereafter increasing very slowly up to 1.0 V. It may be worth noting that the inflection point which occurred at around −0.85 V exactly corresponded to the potential where effective practical protection against corrosion could be attained. Below the off-potentials of −1.0 V or less, the pH began to increase again and was almost linearly related to the decreasing off-potential; indicating that the mechanism whereby protection was brought about fitted well with the ordinary cathodic protection theory.

Such changes in SRB counts, FeS and pH observed in the potential range between −0.85 and −1.0 V may be correlated with the cathodic reactions as eqns (1) and (2):

$$H_2O + 1/2 O_2 + 2e^- \rightarrow 2OH^- \tag{1}$$

$$2H_2O + 2e^- \rightarrow 2OH^- + 1/2 H_2 \tag{2}$$

The following sequence could be assumed to operate:

(a) The poor dissolved oxygen supply in an anaerobic clayey soil could not always support the cathodic reaction of eqn (1), and hence resulted in a sharp reduction in cathodic current density. At this point, the cathodic reaction was replaced by eqn (2).

(b) The consumption of dissolved oxygen by eqn (1) would make the environment more anaerobic, thereby enhancing the proliferation of SRB.

Table 1. Bacterial counts before and after burial tests

Time/Days	Location	Applied potential/V vs Cu/CuSO$_4$	IB	IOB	SOB	SRB
Before test	Bulk	—	+	0	9×10	1×10^5
90	Interface	Free corrosion	+	0	2×10^2	5×10^5
90	Interface	−0.85	+	0	0	3×10^5
90	Interface	−1.3	+	0	0	2×10^3

Interface: Electrode/soil interface.

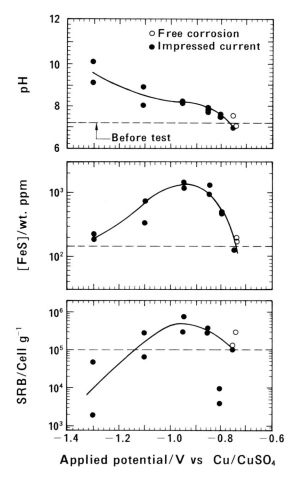

Fig. 4 Effect of applied potential on the pH, FeS formation and SRB counts in the soil close to the working electrode surface.

(c) These cathodic reactions were bound to generate alkalinity which could accumulate in the soil close to the working electrode surface. The amount of alkalinity, however, was bound to remain at levels not sufficiently significant to bring about the decline in the SRB activity because of the reduced cathodic current density in anaerobic clayey soil, and hence the resultant pH would still remain within the optimum range for the proliferation of SRB.

(d) The enhanced proliferation of SRB could then stimulate the cathodic reaction of eqn (2) through hydrogen consumption as follows (the process usually referred to as the hydrogenous activity):

$$SO_4^{2-} + \overset{SRB}{\rightarrow} S^{2-} + 4H_2O \qquad (3)$$

$$Fe^{2+} + S^{2-} \rightarrow FeS \qquad (4)$$

(e) Thus FeS was formed even under cathodically protected conditions. SEM observations revealed that FeS films thus formed were rather dense by nature, though thin, and were thought to give protection no matter how low the cathodic current density might be.

4. Conclusions

Based on the results presented herein, the following conclusions can be drawn concerning the effectiveness of cathodic protection in the presence of SRB:

(1) Complete cathodic protection in an anaerobic clayey soil containing active SRB can be achieved at the off-potentials more negative than –0.85 V and/or by applying the cathodic current of not less than 0.03 A m^{-2}.

(2) Cathodic protection in a clayey soil containing active SRB could mainly be brought about by a synergistic effect of thin films of FeS and impressed cathodic current.

5. References

1. NACE, NACE Standard RP0169-92, NACE, Houston, 1992.
2. K. Kasahara and H. Adachi, *Proc. 8th Int. Congress on Metallic Corrosion*, International Corrosion Council, Mainz, Paper No.49, 1981.
3. C. A. H. von Wolzogen Kühr and L. S. van der Vlugt, *Water*, 1934, **186**, 147.
4. J. Hovarth and M. Novak, *Corros. Sci.*, 1964, **4**, 159.
5. K. P. Fischer, *Mat. Perform.*, 1981, **20** (10), 41.
6. K. Kasahara and F. Kajiyama, *Proc. Int. Congress on MIC and Biodeterioration*, NACE, Houston, TX, p.2, 1990.
7. F. Kajiyama, K. Okamura, Y. Koyama and K. Kasahara, *Proc. Int. Symp. on MIC Testing*, ASTM, Philadelphia, PA, 1992.
8. K. Kasahara, F. Kajiyama and K. Okamura, *Zairyo-to-Kankyo*, 1991, **40**(12), 806 (in Japanese).
9. J. R. Postgate, *The Sulfate-Reducing Bacteria*, Cambridge University Press, Cambridge, p.26, 1979.
10. R and D Planning; *Handbook of Classification of Microorganisms*, R and D Planning, Tokyo, 1986 (in Japanese).
11. K. Kasahara, *Corros. Engng*, 1981, **30**, 524 (in Japanese).
12. K. Okamura, Y. Koyama, F. Kajiyama and K. Kasahara, *Proc. 12th Int. Corros. Congr.*, NACE, Houston, TX, **4**, p.2293, 1993.

31

Studies on the Response of Iron Oxidising and Slime Forming Bacteria to Chlorination in a Laboratory Model Cooling Tower

K. K. SATPATHY, T. S. RAO, V. P. VENUGOPALAN, K. V. K. NAIR
and P. K. MATHUR

Water and Steam Chemistry Laboratory, Applied Chemistry Division (BARC), Indira Gandhi Centre for Atomic Research Campus, Kalpakkam - 603 102, Tamil Nadu, India

ABSTRACT

Experiments were carried out to study the effects of various chlorination regimes on iron oxidising and slime-forming bacteria in a laboratory scale model cooling tower. Results indicated the dependence of bacterial regeneration on dose and frequency of chlorination. Following chlorination, the planktonic TVC (total viable counts) was reduced by two orders of magnitude. The reduction in iron bacteria count was less pronounced than that of TVC. No significant reduction in wet film thickness of biofilm following chlorination was observed. Changes in bacterial number in the bulk water were not truly reflective of those in the film, indicating the need for monitoring bacterial count in the film and not in the bulk water, as is routinely done, for effective fouling control. At the end of the experiment, it was observed that there were qualitative differences in bacterial population between pre- and post-chlorination periods. Post chlorination bacterial community consisted mostly of slime-formers. Corrosion rates of mild steel suspended in bacterial cultures isolated from cooling tower pit were not significantly different from those in the control system. The results of this study are discussed in the context of microbial corrosion in recirculating cooling system.

1. Introduction

Water has been used as an industrial cooling medium since the dawn of industrialisation. This water could be river water, seawater or brackish water, depending upon the availability and proximity of the source to the industry. Irrespective of the source, water contains dissolved ions, dissolved gases, organics, inorganics and micronutrients in varying concentrations and a host of micro- and macro-organisms. The continuous interactions between the cooling water and materials of construction of cooling system result in problems such as scaling, corrosion and biological growth (biofouling) which in turn cause problems for smooth operation of the plants. In fresh water cooling systems, biofouling is mainly due to the growth of microorganisms such as bacteria and protozoa. Bacteria are the most problematic type of microorganisms encountered in cooling systems [1–3]. They can grow anywhere in a system and can cause serious problems due to their extremely rapid growth rate and associated slime accumulation [4]. In addition to fouling, activities of certain bacteria (sulphate reducing and iron oxidising bacteria) cause corrosion in the cooling system [5–7]. Many chemical additives such as chlorine, bromine, bromine–chloride, ozone etc. are employed either to eliminate or control the growth of microorganisms in cooling water systems. Dosage, duration of contact and frequency are the features which ultimately determine the degree of control [1–3].

Among these, chlorine has been widely used due to its favourable economics as well as easy availability [8, 9].

The Fast Breeder Test Reactor (FBTR) at Kalpakkam (12°33' N, 80°11' E)) had been experiencing in the past problems such as flow blockage of pipes and valves, pipe punctures and unusually high corrosion rates in the service water systems. Analyses of the cooling water revealed the presence of iron oxidising and sulphate reducing bacteria in relatively high numbers (5×10^5 and 20 cfu. mL^{-1}, respectively). In the service water system, a chlorination regime was maintained with a residual of 1 mg L^{-1} for 15 min, once a day. Because of these problems with the FBTR, a study was initiated to understand the process of chlorine decay and bacterial regeneration and its implication following a chlorination event. A laboratory model open recirculating cooling tower was used for the experiments. Experiments were designed to study the effects of various chlorination regimes (shock, intermittent and continuous dosing) on the growth rates and regeneration of iron oxidising bacteria (IOB) and slime forming bacteria. The effect of bacterial population on the corrosion rate in a static environment was also studied using bacteria cultured in nutrient broth.

2. Experimental

The study was carried out using a model cooling tower designed and fabricated in the laboratory. It consists of a fibreglass basin with a total hold-up volume of 60 L, a Perspex tower ($35 \times 35 \times 140$ cm) and plastic trough at the top (Fig. 1). Water was recirculated using a monoblock pump at a flow rate of 12 L min^{-1}. Water falling from the trough was allowed to trickle through a series of perforated baffles, before being collected in the tower basin. Care was taken to minimise the loss due to spray. Since the aim of the study was only to observe the pattern of chlorine decay and bacterial mortality in a circulating system, no attempt was made to heat the circulating water to simulate actual condenser coolant temperatures. The cooling water was chlorinated to the desired strength using NaOCl stock solution. Complete water quality (temperature, pH, conductivity, total alkalinity, total hardness, sulphate, silicate, chloride, DO, nitrate, nitrite, phosphate and chlorophyll) analyses were carried out at the beginning and at the end of the experiment, following standard methods [10]. Total viable counts (TVC), counts of iron oxidising bacteria (IOB) and residual chlorine were monitored every 4 h in the cooling water. TVC was made by pour plate method [11], IOB were counted [11] on enrichment media supplied by M/S Hi-Media India and residual chlorine [12] by iodometric method at ≥ 1 mg L^{-1} and by DPD comparator at < 1 mg L^{-1} was used. In addition, glass microscopic slides with pre-developed (48 h old) biofilms were suspended in a side stream bioreactor [13] cou-

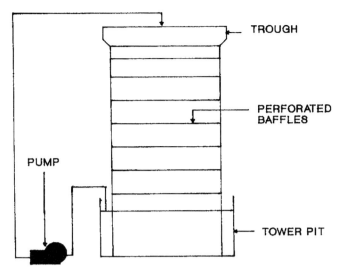

Fig. 1 *Schematic diagram of the model cooling tower.*

pled to the cooling tower. Biofilm thickness, colony morphology, TVC and iron bacteria count in the film were also monitored at similar intervals. For corrosion studies a total of 9 conical flasks containing the following were used: (1) sterile distilled water, (2) sterile cooling tower water, (3) sterile broth, (4) Gram positive filamentous bacteria isolated from FBTR service water system and grown in liquid culture), (5) Gram negative short rod shaped bacterial culture, (6) Gram negative long rod shaped bacterial culture, (7). Gram positive coccus culture, (8) Gram positive bacillus culture and (9) a mixture of all the species of bacteria that had regenerated (after 48 h) in the cooling tower basin after a bout of chlorination (5 mg L^{-1}). The medium used was the same as used by Rao *et al.* [14]. Mild steel coupons were used for corrosion studies following the ASTM procedures [15].

3. Results and Discussion

Chlorination experiments were conducted in three different modes, viz. (a) shock dose at 5 and 10 mg L^{-1} concentrations, (b) intermittent (every 8 h) chlorination to give a residual of 1 mg L^{-1}, and (c) continuous chlorination to give a residual of 1 mg L^{-1}. The pattern of chlorine decay for shock doses is given in Fig. 2 and those for intermittent and continuous are given in Fig. 3. Bacterial mortality in terms of TVC in the shock dose mode is given in Fig. 4. TVC and IOB were monitored for intermittent and continuous chlorination regimes and mortality rates of bacteria in water and biofilm for TVC and IOB are given in Figs 5 and 6 respectively. Water quality data are presented in Table 1. Table 2 gives data on corrosion rates of mild steel coupons suspended in different bacterial broths.

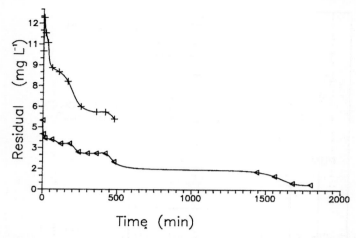

Fig. 2 *Pattern of chlorine decay during shock dose chlorination (5 and 10 mg L^{-1}).*

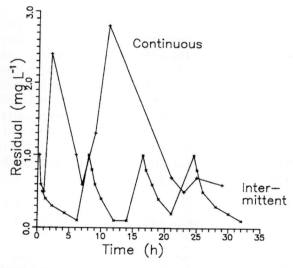

Fig. 3 *Pattern of chlorine decay during intermittent and continuous chlorination regimes (1 mg L^{-1}).*

3.1. Shock doses

At a dose 10 mg L^{-1} (Fig. 2) a relatively high residual was observed during the initial period. Subsequently the residuals showed a progressive decrease. The mean TVC (Fig. 4) came down from 1.8×10^6 cfu mL^{-1} in the beginning to 7.5×10^4 cfu mL^{-1}, 30 min after addition of chlorine. In the case of 5 mg L^{-1} (Fig. 2) shock dose, the total residual chlorine (TRC) gradually decreased from an initial value of 4 mg L^{-1} (at 5

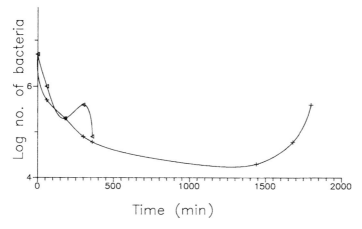

Fig. 4 Variations in bacterial mortality (TVC) during shock dose chlorination.

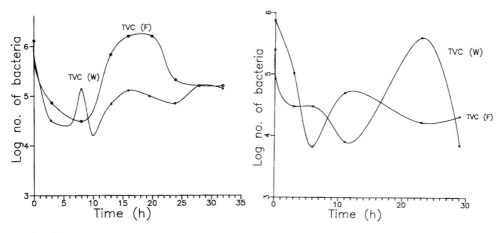

Fig. 5(a) Variations in bacterial mortality (TVC) in biofilm (F) and water (W) during intermittent chlorination; *(b)* Variations in bacterial mortality (TVC) in biofilm (F) and water (W) during continuous chlorination.

min) to 0.7 mg L^{-1} after 1800 min (30 h). The mean TVC (Fig. 4) at the start of the experiment was 1.3×10^6 cfu . m^{-1} and gradually decreased following the addition of chlorine. The decrease in number was observed up to 1440 min (24 h) after which there was an increase in TVC indicating a fresh onset of regeneration. However, this could not be confirmed in the case of 10 mg L^{-1} dose as the bacterial mortality data were not available beyond 360 min. A rapid chlorine consumption at the beginning of chlorination and subsequent slow but continuous decrease in residuals observed in the case of 5 as well as 10 mg L^{-1} doses indicated the possibility of two types of reaction rates being involved during shock dose chlorination.

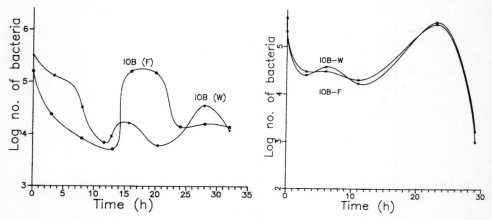

Fig. 6(a) *Variations in iron oxidising bacteria in biofilm (F) and water (W) during intermittent chlorination; (b) Variations in iron oxidising bacteria in biofilm (F) and water (W) during continuous chlorination.*

Table 1. *Water quality of the Palar water used for the study*

		Range
Temperature	27.0	27.5 °C
pH	8.4	8.6
Conductivity	386	534 µS.cm^{-1}
Total Alkalinity	94	100 mg L^{-1} CaCO$_3$
Total Hardness	88	92 mg L^{-1} CaCO$_3$
Sulphate	13.3	16.1 mg L^{-1}
Silicate	7.1	7.8 mg L^{-1} as Si
Chloride	31	40 mg L^{-1}
Nitrate	37	663 µg L^{-1} as N
Nitrite	BDL	2.4 µg L^{-1} as N
Phosphate	7	100 µg L^{-1} as P
Chlorophyll	10	20 µg L^{-1}
Dissolved oxygen	5.5	6.5 mg L^{-1}

3.2. Intermittent Chlorination

Chlorine residuals (Fig. 3) varied from 0.1 to 1 mg L^{-1}. A decrease of two and one order of magnitude was observed for TVC (Fig. 5(a)) and IOB (Fig. 6(a)) respectively, 8 h after chlorination. There was no further decrease in the microbial count, but the bacterial flora were found to have changed their complexion at the end of the experiment. TVC reached the original levels 64 h after the cessation of chlorination, indicating the completion of the regeneration process.

In the biofilm there was no variation in the wet film thickness, possibly due to inability of the low chlorine dose to slough off the existing film. Changes in bacterial number in the bulk water did not truly reflect the bacterial numbers in the film. The

Table 2. Corrosion rate (mpy) of mild steel coupons

Sample No.	Mean corrosion rate (mpy)*
1- Sterile distilled water	7.4
2 - Sterile cooling tower water	8.6
3 - Sterile broth	4.7
4 - Gram + FBTR Strain	5.3
5 - Gram - Rodshaped bacteria	3.9
6 - Gran + cocci	2.7
7 - *Bacillus* sp.	2.7
8 - Mix of Bacteria	3.9

*Thousandths of an inch per year, 1 mpy ≡ 0.025 mm/year.

data thus reinforce the argument that bacterial number in the biofilm, and not in the bulk water, as is routinely done, needs to be monitored for effective biofouling control in cooling water systems [16]. An initial reduction of two orders of magnitude in TVC (Fig. 5(a)) in the biofilm was observed (10^6–10^4 cfu mL^{-1}). However, the original number was restored by a different bacterial strain, 4 h after the fourth bout of chlorination, indicating the selection of chlorine tolerant strains. The iron bacteria (Fig. 6(a)) population in the film followed a similar trend to that of TVC. After the end of the experiment, it was observed that there were qualitative differences between the pre- and post-chlorination bacterial populations. Slime forming bacteria (especially, *Pseudomonas* sp.) were the dominant ones during the post chlorination period. Since slime formers are known to be promoters of corrosion, this fact must be taken into consideration when chlorine alone is used for chemical control of biological slime.

3.3. Continuous Chlorination

During this investigation (1 mg L^{-1} continuous residual) it was observed that unlike in intermittent chlorination, IOB (Fig. 6(b)) in the water as well as those in the biofilm exhibited nearly similar patterns of variation. However, the heterotrophic bacteria, including slime formers, showed dissimilar pattern (Fig. 5(b)) of behaviour; i.e. planktonic bacterial counts were found not to reflect the numbers in the biofilm. The study showed that response of different bacterial groups (i.e. those attached and those freely suspended in water) would vary, depending on the dose and frequency of biocide addition.

The corrosion studies on mild steel carried out using the bacteria isolated from water showed interesting results (Table 2). It was observed that after 7 days, the corrosion rates in the sterile water (both distilled water and cooling tower water) were higher than those of the coupons suspended in the inoculated flasks. It is probable that addition of chemicals (as nutrients) might have provided some sort of protection to the coupons. This is corroborated by the decrease in corrosion rate in the sterile broth, as compared to sterile distilled water. Further, it is also probable that the bacterial film that forms on the coupons provide protection to the metal in the initial periods, but may not do so after the film increases in thickness and complexity [16]. However, further studies using scanning electron microscopy and other surface analysis techniques would be required to confirm this possibility. It is suggested that it may be more representative to use culture media which are as close to the cooling water in composition as possible, rather than doing the experiment using nutrient broth [17, 18]. This may be achieved by suitably reducing the concentration of added nutrients. The necessity for carrying out the experiments for longer durations of time, of course, cannot be overemphasised. The mean corrosion rate of the coupons in the sterile broth was 4.7 mpy which was less than that in the sterile water. Amongst the different bacterial strains the one isolated from FBTR service water exhibited the maximum corrosion rate of 5.3 mpy.

Bacterial strains isolated from the cooling tower basin before and after chlorination were used in corrosion studies with mild steel. Corrosion rates after 7 days of exposure to the bacteria were not so different from those in the control system, indicating that the bacterial films probably did not reach the optimum conditions to promote corrosion.

4. Conclusion

The experiments have proved the usefulness of the model cooling tower in studying the decay of chlorine and bacterial mortality in an open recirculating system, under controlled laboratory conditions. The study also indicated that the efficacy of chlorine as a biocide may not improve just by increasing its concentration. It was seen that bacterial mortality did not show any significant increase when the dose was increased from 5 to 10 mg L^{-1}. It appears prudent to increase the frequency of the biocide addition rather than the dose. Moreover, qualitative differences in bacterial population between pre- and post-chlorination period was observed. It was also found that the bacterial numbers in the bulk water were not truly reflective of those in the biofilm, hence the present data reinforce the argument that bacterial number in the biofilm and not in the bulk water needs to be monitored.

5. Acknowledgement

The Authors are grateful to Sri J. B. Gnanamoorthy, Head Metallurgy Division, IGCAR Kalpakkam for presenting the paper in the Workshop. Authors are thankful to Dr P. N. Moorthy, Head Applied Chemistry Division BARC, Bombay for his keen interest and support.

References

1. J. W. McCoy, *Microbiology of Cooling Towers*, Chemical Publishing Co., New York, USA, 1980.
2. E. Troscinski and R. G. Watson, *Chem. Engng*, 1970, **77**, 125.

3. J. W. McCoy, *The Chemical Treatment of Cooling Water*, Chemical publishing Co., New York, USA, 1974.
4. C. Ramakrishna and J. D. Desai, 'Biological Fouling in Industrial cooling Water Systems and its Control', *J. Sci. Ind. Res.*
5. W. A. Hamilton, 'Sulphate reducing bacteria and anaerobic corrosion', *Ann. Rev. Microbiol.,* 1985, **39**, 195–217.
6. O. H. Tuovinen and J. C. Hsu, 'Aerobic and anaerobic micro-organisms in tubercles of the Columbus, Ohio, water distribution system', *Appl. Environ. Microbiol.,* 1982, **44**, 761–764.
7. B. J. Little, P. Wagner and F. Mansfeld, 'Microbiologically influenced corrosion of metals and alloys', *Int. Mat. Rev.*, 1991, **36**, 253–272.
8. K .K. Satpathy, Biofouling control measures in power plant cooling systems, a brief overview, in *Proc. Specialists Meeting on Marine Biodeterioration with Reference to Power Plant Cooling Systems*, Kalpakkam, 153–166,1990.
9. K. K. Satpathy, T. S. Rao, R. Rajmoahn, K. V. K. Nair and P. K. Mathur, 'Cooling water quality and its bearing on condenser tube failures of Rajasthan Atomic Power Station (unit II)', *Nat. Symp. on Electrochemistry in Nuclear Technology*, Kalpakkam, C-9, 1994.
10. APHA, Standard Methods for the Examination of Water and Waste Water, American Water Works Association, Washington, 1984.
11. M. J. Pelczar (Jr.), E. C. S. Chan and N. R. Krieg, *Microbiology*, Mc Graw Hill Book Co., New York, 99–147, 1986.
12. G. C. White, Chemistry of chlorination, in *Handbook of Chlorination*, Van Nostrand Reinhold Company, New York, 1972.
13. K. Pedersen, 'Method for studying microbial biofilms in flowing water systems', *Appl. Environ. Microbiol.,* 1982, **43**, 6–13.
14. T. S. Rao, M. S. Eswaran, V. P. Venugopalan, K. V. K. Nair and P. K. Mathur, 'Fouling and Corrosion in an Open Recirculating Cooling System', *Biofouling*, 1993, **6**, 245– 259.
15. Annual Book of ASTM Standards. Section 3. Metals Test Methods, 1986.
16. D. Thierry, 'Field Observations of microbiologically induced corrosion in cooling water systems', *Mat. Perform.*, 1987, **26**, 35–41.
17. G. J. Licina and D. Cubicciotti, 'Microbially Induced Corrosion in Nuclear Power Plant Materials', *J. Metals*, 1982, December, 23–27.
18. S. W. Borenstein and G. J. Licina, 'Avoid MIC - related problems in nuclear cooling systems', *Power*, 1990, June, 13–19.

32

Monitoring and Inspecting Biofouled Surfaces

S. W. BORENSTEIN and G. J. LICINA

Structural Integrity Associates, 3150 Almaden Expressway, Suite 123, San Jose, CA 95118, USA

ABSTRACT

Electric power generation plants typically use untreated surface water to transfer heat from various components to the ultimate heat sink for the plant. These systems, especially at nuclear plants, spend most of their time in a stand-by or stagnant condition in case an emergency condition arises. The failure of these systems could result in a failure of the systems they support. There is a variety of degradation mechanisms that can affect flow, heat transfer rates or structural integrity. Many times the systems are degraded by fouling and corrosion. This paper will discuss the failure modes, particularly microbiologically influenced corrosion (MIC), as well as the tools and methods used for inspection and monitoring of these degradation mechanisms. In addition, the design and operating parameters that minimise corrosion and fouling problems, are discussed.

1. Introduction

Bacteria have the capability to adhere to surfaces, grow and produce biofilms. Biofilms create a variety of problems including degradation of heat transfer capabilities or corrosion beneath the biofilm. Biofilms are typically comprised of consortium of sessile organisms and their extracellular polymers, also termed exopolymers, [1, 2]. The extracellular polymers are secretions that form a fibrous matrix that envelops the organisms. Biofouling is the fouling on surfaces with macro- or microscopic organisms [2]. Microbiologically influenced corrosion (MIC) involves the initiation or acceleration of corrosion by microorganisms or their metabolites [1]. Biofouling often leads to corrosion beneath the deposits.

The study of MIC is relatively recent. More than one corrosion mechanism may be involved. Problems in detecting, monitoring and inspecting for corrosion become even more difficult for MIC. Monitoring and inspection methods for MIC need to be able to detect diverse corrosion mechanisms and evaluate the influence of the microorganisms.

2. Background

Several types of corrosion mechanisms are associated with power plants and other large industrial plants. These include general corrosion, pitting, underdeposit or crevice corrosion, and MIC. Less commonly occurring problems include corrosion-assisted cracking and velocity-assisted corrosion.

Classifying corrosion can be difficult. General corrosion is a fairly uniform loss of metal; usually over a large area. General corrosion is the form of corrosion used for corrosion rate measurements. Localised corrosion is selective attack, such as pitting. For example, a large portion of the surface area of a pipe may be unaffected, and intense corrosion occurs at weldments in the form of pitting. This may lead to rapid penetration and costly problems. Monitoring and inspecting for localised corrosion can be extremely difficult.

Crevice corrosion is another form of localised corrosion. Crevice corrosion typically occurs under deposits or at crevices between mating surfaces, such as gasketed flanged connections. Often it occurs on film-protected (passive) metal surfaces, such as austenitic stainless steels. Some instances of MIC may be considered a form of crevice corrosion when the exopolymer film formed by the microorganisms creates a 'living crevice', as shown in Fig. 1. The role of organisms in underdeposit corrosion may relate influence and formation of differential aeration cells under a biological deposit and pit initiation [1].

3. Discussion

One of the difficulties with characterising MIC attack involves understanding the various aspects of MIC. There are three important facets of MIC. What organisms are involved and is the **microbiological** species or microbial ecology relevant? How much or to what extent are the microorganisms involved and how do they **influence** surfaces? What are the **corrosion** mechanisms involved? The following discussion highlights a review of the technical literature concerned with monitoring and inspection techniques for the study of each of these aspects of MIC, particularly as they relate to biofouled surfaces.

3.1. Monitoring Techniques

Microbiological analysis has long been associated with monitoring MIC of metals, particularly with the oil and gas industries. Sulphate reducing bacteria (SRB) and their action upon metals resulting in corrosion was the basis of a theory by von Wolzogen Kuhr and van der Vlugt. They theorised that corrosion of iron buried in

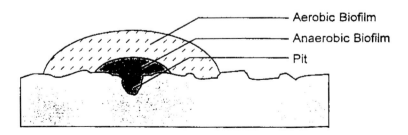

Fig. 1 Biofilm over pitting corrosion.

an anaerobic soil was the result of activity by SRB [5]. Since then, Little and Wagner [6] note that, in addition to SRB the iron-oxidising, sulphur-oxidising-bacteria and many other organisms have been shown to influence corrosion of metals [7–41]. Borenstein and Licina [42] discuss microbiological analysis methods such as:

- cultures
- microscopy
- lipid analysis
- gene probes.

When inspections are costly and are scheduled years apart, it is extremely useful to have corrosion monitoring methods available and functioning properly.

It must be made clear that mere presence of microorganisms in a fluid or on a surface does not indicate how, or how much those organisms may have influenced corrosion if they influenced it at all. Surfaces exposed to environments rich in microbiological activity would **always** be expected to have such organisms present on surfaces; possibly in large numbers. Those organisms may have been the primary factor in the corrosion; they may have had nothing whatsoever to do with it. Examination of surfaces for chemical species produced by microorganisms which interact with the underlying metal to produce corrosion, or for mineralogical 'fingerprints', can be useful in determining the microbiological influence. Similarly, since corrosion is an electrochemical process, methods that can measure electrochemical activity and correlate it with biological activity provide a useful means for assessing the microbiological influence.

3.1.1. Chemical analysis for specific metabolites
The chemical analysis of deposits removed from a corrosion site or even deposits in the water may provide insight into whether microorganisms contributed to the corrosion. Several analytical methods or combinations of methods are commonly used. Elemental analysis of the deposit is often done by energy dispersive X-ray analysis, or atomic absorption techniques. In addition, wet chemistry or other techniques may be used to differentiate between valence states. Some of the chemical species of relevance to MIC monitoring include:

- ammonia
- chlorides
- sulphides
- sulphates
- sulphur
- nitrates
- nitrites
- iron
- manganese
- iron or manganese + chlorides
- phosphorus.

An organic content of 20% or greater in the deposits indicates likely microbiological activity [45]. Analysing for organic carbon is the most helpful although analysing for total carbon or loss on ignition (LOI) is also useful. Other sources of carbon may contribute to the measured value and results in LOI being a less direct indicator than other methods.

Microbiological analysis can determine what organisms are present in a given sample. As White et al., note, 'Problems in demonstrating mechanisms of MIC have been complicated by the fact that classical methods of microbiology that were so successful in the study of infectious diseases (isolation and characterisation of the pathogenic species) have proved of little use in understanding biofilm dynamics' [46]. Laboratory analysis by microbiologists involves the use of methods for defining the biomass, community structure, nutritional status, and metabolic activities of mixed microbial communities. Interpretation of the data is sometimes difficult and complex. It is important to understand that there is no direct correlation between the numbers of bacteria detected and the corrosion that can be predicted by their presence [1, 9, 47].

On-line monitoring methods for MIC need to permit the operator to evaluate trends and make system changes prior to damage from a biofilm [3]. Licina and Nekoksa discussed on-line biofilm monitors and suggested they must be:

- simple to install
- simple to use
- simple to interpret
- sufficiently rugged for field use
- sensitive to biofouling changes of significance
- accurate
- economical.

A number of corrosion monitoring techniques are available. Some are suitable for field use, some for laboratory use, and some for both.

3.1.2. Coupons

The most commonly used technique for corrosion monitoring is that employing weight loss coupons. Many coupons can be exposed simultaneously to an actual process stream or to a side stream. Coupons are commonly used for measuring uniform corrosion but are considered poor for pitting corrosion. Coupons are sometimes used to evaluate MIC because they provide a useful tool for assessing various control treatments simultaneously.

Another problem with corrosion monitoring methods using coupons is that coupons do not detect some forms of corrosion. For example, weldments and areas adjacent to weldments are susceptible to pitting corrosion and MIC, particularly for austenitic stainless steels. Fabricated spool pieces are sometimes used as a monitoring method for this situation either in-line or as sidestream loops.

Traditional corrosion monitoring probes using the electrical resistance technique (such as CORROSOMETER* or using the linear polarisation technique (such as CORRATER* do not lend themselves very well to the monitoring of MIC, most likely because the corrosion has such a small surface area for an initiation site.

3.1.3. Corrosion potential

The corrosion potential, E_{corr} is easy to measure but can be difficult to interpret, particularly as it relates to MIC. Mansfeld and Little examined such techniques and noted that changes in both, E_{corr} and polarisation resistance, R_p, can occur at the same time so both parameters should be measured simultaneously [48]. They observed that negative shifts of E_{corr} have been explained by different authors as resulting from a reduction of the cathodic reaction rate or as a result of an increase in the anodic reaction rate. Both theories lead to the same resulting observation. Without additional data, no definitive conclusions can be drawn.

The case of stainless steel in seawater and fresh water is the topic of many reports on the time dependence of E_{corr} [48–56]. Most reports show an ennoblement of E_{corr} during exposure. This could be a thermodynamic effect, a kinetic effect, or both. Thermodynamic changes are based on free energy changes (e.g. changes in oxygen concentration), while kinetic effects are based on changes in the types or rates of reactions (e.g. a decrease in pH). Mansfeld and Little note that for MIC, the situation is even more complex because naturally occurring microorganisms within a marine biofilm can either increase or decrease the local oxygen concentration and the pH [48]. Dexter and Gao have argued that the biofilm catalyses the reaction as the reason for the cathodic activity in Fig. 2 [56]. As Scotto *et al.* showed, the effect of the biofilm is to shift the corrosion potential in the noble direction [50]. This could affect initiation of localised corrosion. Both Dexter and Scotto discuss possible changes in the Tafel slope and/or exchange current density for the oxygen reduction. These changes would be consistent with changes on the metal surface due to colonisation by the organisms and the production by the organisms of enzymes that accelerate the electrochemical reaction on the surface of the metal.

3.1.4. Redox potential

The reduction–oxidation (redox) potential or solution potential indicates the oxidation power of an electrolyte. Mansfeld and Little found that although redox potential has been used in the study of MIC to monitor changes in a solution as a result of bacterial metabolism, redox potential can produce widely varying degrees of corrosion for different metals exposed to the same solutions, [48]. In other words, redox potential provides a useful parameter for tracking the oxidising power of a solution. However, it is not a definitive diagnostic parameter for MIC. For the study of MIC they suggest that measurements of E_{corr} and polarisation resistance are probably more useful.

---

*CORROSOMETER and CORRATER are registered trademarks of Rohrback Cosasco Systems, Inc. Santa Fe Springs, CA, USA.

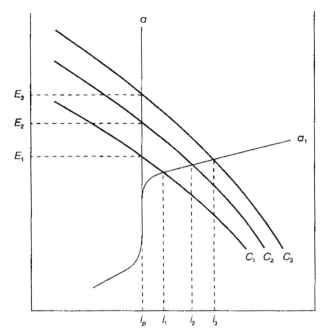

Fig. 2 The influence of marine aerobic microbial film on stainless steel corrosion behaviour. Stainless steel corrosion diagram in natural seawater. C– cathodic reduction curves varying with test progress; a – anodic polarisation of a passive stainless steel; and a_1– anodic polarisation of an active stainless steel (Ref. [50]).

3.1.5. Linear polarisation resistance

The linear polarisation resistance technique is common for measuring uniform corrosion either in the field or in the laboratory. Bockris and others have shown that when anodic and cathodic polarisation is within 10 mV of the corrosion potential, the applied current density is approximately linear with potential [57]. The slope of the linear curve $(E–I)$ is the polarisation resistance, R_p [57]. Stern and Geary derived an equation showing that the corrosion rate is inversely proportional to R_p at potentials close to E_{corr} [58]. Again, the problems related to using this technique in monitoring MIC lies with interpretation of the data. The linear polarisation technique is good for determining uniform or general corrosion but poor for assessing pitting. As noted previously, MIC is most often localised.

3.1.6. Dual-cell technique

The technique known as the dual-cell technique is the first and only definitive demonstration of MIC. It is purely a laboratory technique. This technique, developed by Little *et al.*, to study MIC, allows for continuous monitoring of the changes in the electrochemical response of a metal surface due to the presence of a biofilm [30, 59]. The technique uses two identical cells; one inoculated with bacteria and one as sterile control. The cells are separated by a semipermeable membrane (which permits ions to pass but not microbes). The method does not allow the researcher to calculate the corrosion rate but rather, by measuring the galvanic current, demonstrates whether the microbial side is the anode or cathode and records the time-dependence of the corrosion process. The difficulties are that the process is fragile and easily disturbed. It provides a powerful demonstration of the existence and degree of microbial influences on corrosion.

3.1.7. Electrochemical biofilm monitoring probe

Several experimenters have shown that cathodic polarisation can encourage biofilm formation [43, 44]. Licina and Nekoksa describe an approach for on-line monitoring of biofilm activity in plant environments that uses that effect [3]. The BIoGEORGE* probe monitors changes in electrochemical reactions produced by biofilms on stainless steel electrodes.

Their probe consists of a series of stainless steel discs, separated by an insulating epoxy resin (Fig. 3). Alternate discs are connected electrically but are isolated from their near neighbour discs and the body of the probe. The stainless steel discs and epoxy insulator form a smooth right, circular cylinder. The smooth, cylindrical arrangement produces a relatively short path for a biofilm to 'bridge' between electrodes. For a short period of time each day, a potential of 50–200 mV is applied from an external power source between the two sets of discs. The rest of the time, the electrodes are shortened through an external shunt. This gentle polarisation cycle

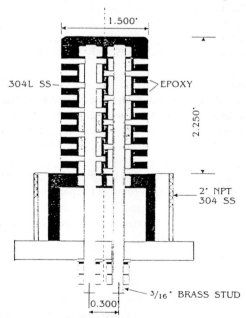

Fig. 3 *Electrochemical biofilm monitoring probe.*

---

*Patent No. 5,246,560. Electric Power Research Institute, 1991. Trade name of Structural Integrity Associates.

modifies the local environments on the discs, similar to that resulting from inclusions or locally anodic areas such as welds.

Laboratory tests have shown that the applied current on the probe (i.e. the current during the polarisation) increases abruptly as biofilm begins to form on the electrode surfaces. The onset of biofilm formation is marked by an abrupt increase in a trend plot of the applied current. A generated current, a current which persists even without the presence of any applied potential, also begins to appear at the same time. The magnitude of both currents increases as conditions suitable for biofilm growth are maintained, as shown in Fig. 4. Gentle rinsing of the probe to remove biofilms or admission of light to the probe surfaces causes the applied current to drop to the initial values and makes the generated current disappear [3]. Post-test examination of probes removed from tests reveals the presence of SRB and acid producing bacteria (APB) but **no corrosion**. If the observed currents were due to corrosion, very significant general corrosion or pitting (one or more large pits or numerous smaller ones) would be anticipated.

By monitoring the current required to achieve the pre-set potential over a period of days or weeks, the influence of biofilms on operative half-reactions may be readily detected. These effects may include catalysis of oxygen reduction in aerated environments [64–66] or alternative cathodic reactions (e.g. the result of SRB or acid producing bacteria) [67]. The intermittent polarisation schedule also generates conditions that produce differences in both the types and numbers of microorganisms present on each electrode. The polarisation schedule can simulate electrochemical conditions similar to those resulting from local anodic sites (e.g. inclusions or weldments). The 'gentle' cathodic polarisation can also encourage microbial colonisation, similar to that observed on cathodically protected structures in biologically active environments, [43, 68].

The probe is amenable to a wide variety of configurations. In the standard design probe, the electrode stack is connected to a threaded pipe plug of nominal 2 in. (50 mm) size. A much smaller probe, approximately 1.1 in. (28 mm) dia., compatible with a fitting that permits insertion or removal at full temperature and pressure, was

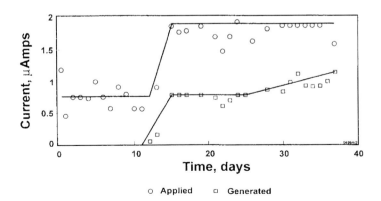

Fig. 4 Magnitude of generated corrosion current and applied current increases as biofilm develops.

installed in the fire protection system at the Tennessee Valley Authority's Browns Ferry Nuclear Plant.

3.1.8. Electrochemical impedance spectroscopy

The electrochemical impedance spectroscopy (EIS) technique, also referred to as a.c. impedance, uses a small amplitude a.c. signal applied to the test electrode. The researcher measures the response as a function of the frequency of the signal. The technique is very powerful, however, it is complex and the interpretation of the data is difficult, particularly as related to the study of MIC. It is a laboratory technique but is currently beginning to move into field applications.

3.1.9. Electrochemical noise

The electrochemical noise technique measures fluctuations of the potential and the current as a function of time. It is very good for on-line applications and has good possibilities for real-time monitoring of MIC. It has also been used in field applications including service water system applications where MIC has been observed [60–63].

3.1.10. Other monitoring methods

The large signal polarisation technique uses potential scans from several hundred mV to several volts. It is a laboratory technique that has a mixed review as to its value in studying MIC and corrosion mechanisms. This is probably due to the complexity and variety of conditions and materials examined as well as the complexity of the influence of the microorganisms.

3.2. Inspection Techniques

During the 1980s, electric power generation utilities realised the importance of the performance and structural integrity of cooling water systems, such as the service water system, to overall plant safety and economy. This increased awareness of the impact of fouling and corrosion of cooling water systems led to increased attention to inspection, monitoring, and maintenance activities. U.S. Nuclear Regulatory Commission Generic Letter 89-13 required U.S. nuclear plants to demonstrate that their service water systems were in good repair and would provide adequate performance under all conditions, including accident conditions. These activities required plants to expand the maintenance, testing, and inspection activities devoted to service water systems significantly.

3.2.1. Performance monitoring

The monitoring of systems may include performance monitoring. This would commonly include recording temperatures, pressures and flow rates. This technique is for measuring performance but may also give an indirect indication of fouling, including biofouling. It is not a corrosion measurement at all. The installation and interpretation of data analysis may be complex and expensive, but has often been found to be extremely cost effective as a power plant tool.

3.2.2. Non-destructive examination
The examination of systems using non-destructive techniques during in-service inspection is a common form of inspection. For example, pipes can sometimes be inspected during operation using ultrasonics or radiography, [60–62]. Wendell discussed using radiography on piping systems without removing insulation [61]. Both methods may be suitable in certain instances to detect wall thinning. Inspection methods include:

- Visual
- Ultrasonics
- Eddy Current
- Radiography
- Advanced Techniques.

Laser profilometry, such as LOTIS™ developed by Quest Integrated, Inc, of Kent, Washington, uses a laser source, optics, and a photodetector in the front section of the probe for internal inspection of tubing [4]. Doyle discusses how this is useful for inspecting systems, such as the rapid scanning of the full inside length of a marine boiler tube, generating a quantitative topographic map of the tube surface, and tabulating a summary of the results Fig. 5. Internal corrosion and pitting are the primary mechanisms of degradation the system was designed to examine.

FATS (Focused Array Transducer System) developed by Infometrics of Silver Spring, MD, is a UT system that improves the performance of ultrasonic examinations. Conventional UT techniques can be improved by using a focused transducer, made up of phased arrays of transducers and sophisticated electronics. In addition, the technique can be partnered with TestPro, also developed by Infometrics. TestPro is a data acquisition system with enhancement features.

4. Conclusions

Cooling water systems are extensive. In a large nuclear plant, for example, the service water system may contain miles of piping and several hundred heat exchangers. Obviously, 100% inspection is out of the question. Equipment inspection activities can be augmented by monitoring to track biofouling and corrosion. Selection of equipment to be inspected or monitoring locations is critical to the development of a surveillance program.
A variety of techniques exist for monitoring MIC and biofilm formation. Many of these methods are indirect (i.e. they measure pressure drop or loss in heat transfer performance) or simply enumerate organisms at the collection point. Other approaches are specific to corrosion but give little or no information as to the source of the attack. The selection of monitoring methods must focus on the parameters of greatest interest.

Selection of inspection tools must focus on resolution capabilities, speed of inspection, and compatibility with plant operations. Visual inspection, radiography,

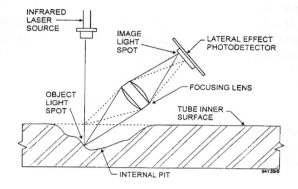
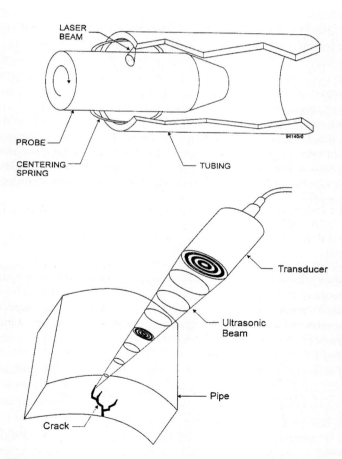

Fig. 5(a) Operation of laser scanner; (b) Optics housing of LOTIS™ generates a helical map of tube ID; (c) Focused ultrasonic beam using multiple transducers.

ultrasonic thickness measurements, and eddy current testing have been shown suitable for some applications. Improved techniques, such as laser profilometry and the focused array transducer system, may provide greater resolution and be directly applicable to the unique problems associated with MIC.

5. Acknowledgements

We are sincerely grateful to George Nekoksa of Corrosion Failure Analysis and Control, San Ramon, CA and Patricia Wagner of Naval Research Laboratory, Stennis Space Center, MS for their assistance and for their technical review.

References

1. S. W. Borenstein, *Microbiologically Influenced Corrosion Handbook,* Woodhead Publishing, Ltd., Cambridge, UK, 1994.
2. B. J. Little, 'Succession in Microfouling', in *Proc. Office of Naval Research Symp. on Marine Biodeterioration,* 1985, U.S. Naval Inst. Press, Bethesda, MD.
3. G. J. Licina and G. Nekoksa, 'An Electrochemical Method for On-line Monitoring of Biofilm Activity', *CORROSION '93,* 1993, paper No. 108, NACE, Houston, TX.
4. J. Doyle, 'Laser-Based Surface Profilometry: An Emerging Technology for Nondestructive Evaluation', *Mat. Eval.,* 1991, **49** (7).
5. C. Von Wolzogen Kuhr and I. S. Van Der Vlugt, *The Graphitisation of Cast Iron as an Electrochemical Process in Anaerobic Soils,* Water, **18**, (1934), The Hague, 147–165.
6. B. J. Little, And P. Wagner, 'Quantifying Sulfate Reducing Bacteria in Microbiologically Influenced Corrosion', *NSF-CONICET Workshop on Biocorrosion and Biofouling,* Mar del Plata, Argentina, Oct. 1992.
7. P. F. Sanders and W. A. Hamilton, *Biologically Induced Corrosion* (Ed. S. C. Dexter), NACE, Houston, TX, 1986, p.47.
8. G. Eidsa and E. Risberg, *Biologically Induced Corrosion* (Ed. S. C. Dexter), NACE, Houston, TX, 1986, p. 109.
9. R. J. Soracco, D. H. Pope, J. M. Eggars and T. N. Effinger, 'Microbiologically Influenced Corrosion Investigations in Electric Power Generating Stations','*CORROSION '88,* paper No. 83, NACE, Houston, TX, 1988.
10. D. H. Pope, *Microbial Corrosion in Fossil-Fired Power Plants — A Study of Microbiologically Influenced Corrosion and a Practical Guide for its Treatment and Prevention, EPR NP-4582,* Palo Alto, CA, Electric Power Research Institute, 1987.
11. D. H. Pope, '*A Study of MIC in Nuclear Power Plants and a Practical Guide for Countermeasures,*' Final Report NP-4582, Palo Alto, CA, Electric Power Research Institute, 1986.
12. W. C. Ghiorse, *Ann. Rev. Microbiol.,*1984, **38**, 515.
13. R. L. Starkey, *J. American Water Works Association 37,* 1945, **10**, 963.
14. R. E. Tatnall, *Mat. Perform.,* 1980, **19**, 8, 88.
15. G. Kobrin, *Mat. Perform.,* 1976, **38**, 15.
16. G. Kobrin, *in Biologically induced Corrosion* (Ed. S. C. Dexter), NACE, Houston, TX, 1986, p. 33.
17. M. Eashwar, G. Subramanian, P. Chandrasekaran and K. Balakrishnan, 'Probing Microbiologically Influenced Corrosion of Steel During Putrefaction of Seawater,' *CORROSION '90,* paper no. 120, NACE, Houston, TX, 1990.

18. R. E. Tatnall, 'Case Histories: Bacteria-Induced Corrosion,' *CORROSION '81*, paper no. 130, NACE, Houston, TX, 1981.
19. C. O. Obuekwe, D. W. S. Westlake, J. A. Plambeck and F. D. Cook, *Corrosion,* 1981, **37**, 8, 461.
20. B. Little, P. Wagner, R. Ray and M. McNeil, *Marine Technol. Soc. J.*, 1990, **24**, 10.
21. J. A. Nickels, R. J. Bobbie, R. F. Matz, G. A. Smith, D. C. White and N. L. Richards, *Appl. Environ. Microbiol.,* 1981, **41**, 1262.
22. J. D. A. Miller and A. K. Tiller, *Microbial Aspects of Metallurgy*, Elsevier, New York, 1970, p. 61.
23. K Kasahara and F. Kajiyama, in *Biologically Induced Corrosion* (Ed. S. C. Dexter), NACE, Houston, TX, 1986, p. 171.
24. W. P. Iverson, *Microbial Iron Metabolism*, Academic Press, New York, 1974.
25. J. R. Postgate, *The Sulphate Reducing Bacteria,* Cambridge University Press, Cambridge, UK, 1979, p. 26.
26. M. Bibb, in *Biologically Induced Corrosion,* (Ed. S. C. Dexter), NACE, Houston, TX, 1986, p. 96.
27. P. J. D. Scott and M. Davies, *Mat. Perform.*, 1989, **28**, 5, 57.
28. B. Little, P. Wagner, J. Jacobus and L. Janus, *Estuaries,* 1989,**12**, 3,138.
29. G. J. Licina, (ed.), *Sourcebook for Microbiologically Influenced Corrosion in Nuclear Power Plants EPRI NP-5580,* Palo Alto, CA: Electric Power Research Institute, 1988.
30. B. Little, P. Wagner, S. M. Gerchakov, M. Walch and R. Mitchell, *Corrosion,* 1986, **42**, 9, 533.
31. B. Little, P. Wagner and D. Duquette, *Corrosion,* 1988, **44**, 5, 270.
32. M. Stranger-Johannessen, *Biodeterioration, VI,* CAB Int., Slough, UK, 1986.
33. R. C. Salvarezza and H. A. Videla, *Acta Gientifca, Venezalana* 1984, **35**, 244.
34. D. H. Pope, T. P. Zintel, A. K. Kuruvilla and O. W. Siebert, 'Organic Acid Corrosion of Carbon Steel: A Mechanism of Microbiologically Influenced Corrosion,' *CORROSION '88*, paper no. 79, NACE, Houston, TX, 1988.
35. O. C. Dias and M. C. Bromel, *Mat. Perform.*, 1990, **29**, 4, 53.
36. D. G. Honneysett, W. D. Van Den Bergh and P. F. O'Brien, *Mat. Perform.*, 1985, **24**, 34.
37. R. Tatnall, A. Piluso, J. Stoecker, R. Schultz and G. Kobrin, *Mat. Perform.*, 1981, **19**, 41.
38. V. K. Gouda, M. I. Banat, T. W. Riad and S. I. Mansour, 'Microbial-Induced Corrosion of Monel 400 in Seawater,' *CORROSION '90*, paper no. 107, NACE, Houston, TX, 1990.
39. D. E. Nivens, P. D. Nichols, J. M. Henson, G. G. Geesey and D. C. White, *Corrosion,* 1986, **42**, 4, 204.
40. D. H. Pope, D. J. Duquette, A. H. Johannes and P. C. Wayner, *Mat. Perform.*, 1984, **23**, 4, 4.
41. M. Walch and R. Mitchel, 'Role of Microorganisms in Hydrogen Embrittlement of Metals,' *CORROSION '83*, paper No. 249, NACE, Houston, TX, 1983.
42. S. W. Borenstein and G. J. Licina, 'An Overview of Monitoring Techniques for the Study of Microbiologically Influenced Corrosion', *CORROSION '94*, paper No. 611, NACE, Houston, TX, 1994.
43. G.Nekoksa, 1989, 'Cathodic Protection to Control Microbiologically Influenced Corrosion', in *Microbial Corrosion: 1988 Workshop Proceedings* (Ed. G. J. Licina), EPRI ER-6345, Electric Power Research Institute, Palo Alto, CA.
44. J. Guezennec and M. Therene, 'Study of the Influence of Cathodic Protection on the Growth of SRB and Corrosion in Marine Sediments by Electrochemical Techniques', *1st EFC Workshop on Microbiological Corrosion*, Sintra, Portugal,1988, Elsevier, p.93.
45. G. J. Licina, 1988, *Detection and Control of Microbiologically Influenced Corrosion, EPRI, NP-6815-D,* Electric Power research Institute, Palo Alto, CA, 6-6.
46. D. C. White, D. E. Nivens, P. D. Nichols, A. T. Mikell, B. D. Kerger, J. M. Hensen, G. G. Geesey and C. K. Clarke, in *Biologically Influenced Corrosion* (Ed. S. C. Dexter), NACE, Houston, TX, 1986.

47. B. J. Little and P. Wagner, 'Standard Practices in the United States for Quantifying and Qualifying Sulfate Reducing Bacteria in MIC' *Proc. Symp. Redefining International Standards and Practices for the Oil and Gas Industry*, London, UK, 1992.
48. F. Mansfeld and B. J. Little, 1990, 'The Application of Electrochemical Techniques for the Study of MIC — A Critical Review,' *CORROSION '90*, paper No. 108, NACE, Houston, TX.
49. A. Mollica, A. Trevis, E. Traverso, G. Ventura, V. Scotto, G. Alabisio, G. Marcenaro, U. Montini, G. Decarolis and R. Dellepiane, "Interaction Between Biofouling and Oxygen Reduction Rate on Stainless Steel in Seawater', *Proc. 6th Int. Cong. Marine Corrosion and Fouling*, Athens, Greece, 1984, p. 269.
50. V. Scotto, R. Decintio and G. Marcenaro, *Corros. Sci.*, 1985, **25**, 185.
51. R. Johnsen and E. Bardal, *Corrosion*, 1985, **41**, 296.
52. R. Johnsen and E. Bardal,"The Effect of a Microbiological Slime Layer on Stainless Steel in Natural Seawater,' *CORROSION '86*, paper No. 227, NACE, Houston, TX, 1986.
53. H. A. Videla, M. F. L. Demele and G. Brankevich,"Microfouling of Several Metal Surfaces in Polluted Seawater and Its Relation with Corrosion', *CORROSION '87*, paper No. 365, NACE, Houston, TX, 1987.
54. A. Mollica, A. Trevis, E. Traverso, G. Ventura, G. Decarolis and R. Dellapiane, *Corrosion*, 1989, **45**, 48.
55. A. Mollica, A. Trevis, E. Traverso and V. Scotto, *Int. Biodeterior.*, 1988, **24**, 221.
56. S. C. Dexter and G. Y. Gao, *Corrosion*, 1988, **44**, 717.
57. J.O'M. Bockris, *Modern Aspects of Electrochemistry*, Butterworths, NY, 1954.
58. M. Stern and A. L. Geary, *J. Electrochem. Soc.*, 1937, **104**, 33.
59. S. M. Gerchakov, B. J. Little and P. Wagner, *Corrosion*, 1986, **42**, 689.
60. G. J. Licina, (Ed.), *Detection and Control for Microbiologically Influenced Corrosion*, EPRI NP-6515-D, Palo Alto, CA: Electric Power Research Institute, 1990.
61. J. Wendell, Applications of Radiography', *1987 Seminar on Nuclear Plant Layup and Service Water System Maintenance*, Electric Power Research Institute.
62. A. F. Deardorff *et al.*, "Evaluation of Structural Stability and leakage from Pits Produced by MIC in Stainless Steel Service Lines', *CORROSION '87*, paper No. 514, NACE, Houston, TX, 1987.
63. A. Brennenstuhl, 1993, 'Service Water Electrochemical Monitoring Development at Ontario Hydro', *12th Int. Corros. Congr.*, NACE, Houston, TX.
64. V. Scotto, 'Electrochemical Studies of Biocorrosion of Stainless Steel in Seawater', *1988 Proc. EPRI Microbial Corrosion Workshop* (Ed. G. J. Licina), EPRI ER-6345, Electric Power Research Institute, Palo Alto, CA 1989.
65. S. C. Dexter and B. Y. Gao 'Effect of Seawater Biofilms on Corrosion Potential and Oxygen Reduction of Stainless Steel', *CORROSION '91*, paper No. 114, NACE, Houston, TX, 1991.
66. G. Gundersen, B. Johansen, P. O. Gartland, P. O. Tunold, R. Vintermyr and G. Hagen, "The effect of Sodium Hypochlorite on the Electrochemical Properties of Stainless Steel,' *CORROSION '87*, paper No. 337, NACE, Houston, TX, 1987.
67. G.J. Licina, "Electrochemical Aspects of Microbiologically Influenced Corrosion', in *Microbial Corrosion: 1988 Workshop Proceedings* (Ed. G. J. Licina), Electric Power Research Institute, Palo Alto, CA,1989.
68. J. Guezennec, *et al.*, "Cathodic Protection in Marine Sediments and the Aerated Seawater Column', in *Microbially Influenced Corrosion and Biodeterioration* (Eds N. J. Dowling, M. W. Mittleman and J. C. Danko), University of Tennessee, 1991.

33

Corrosion Behaviour of Steel in Coal Mining Water in the Presence of *Thiobacillus thiooxidans* and *Thiobacillus ferrooxidans*

S. M. BELOGLAZOV and A. N. CHOROSHAVIN

University of Kaliningrad, Department of Chemistry, 14 Alexander Nevsky str., Kaliningrad 236041, Russia

ABSTRACT

The mine water in the Kizel coalmine has been studied both in the mine and in the laboratory. Measurements were made of pH, redox-potential, bacterial activity and metal weight losses. The action of *Thiobacillus thiooxidans* and *Thiobacillus ferrooxidans* resulted in the development of acidity and high concentrations of ferrous and ferric ions. The effects of N-containing organic substances on corrosion, bacterial activity and hydrogen uptake by the metal were determined.

1. Introduction

It is known that thiobacilli exist in the coal mining waters of the Kizel mine near Perm (Russia, Urals), Donbass (Ukraine) as well as in other coal basins. These bacilli produce sulphuric acid and ferric sulphate, $Fe_2(SO_4)_3$ from pyrites in the earth (aerobic conditions). In the former Soviet Union 5×10^5 $m^3 d^{-1}$ mine water polluted with H_2SO_4, and $0.005–1.22$ gL^{-1} $FeSO_4$ and $0.06–1.85$ gL^{-1} $Fe_2(SO_4)_3$ were pumped and neutralised in settling basins [1]. Steel troughs and drainage pipes failed after 15–30 days and 1.5–3 months respectively of use in coal mines with acid-containing pit water (in neutral pit waters the lifetimes were 1–2 years). Helbronner *et al.* [2] were the first to establish the microbiological oxidation of pyrites in the aerobic conditions of coal mining. Wachsman and Joffe [3] isolated *Th. thiooxidans* and Colmer *et al.* [4] separated from acid-containing pit-water a new autotrophic microorganism —*Th. ferrooxidans*. Further studies showed that the oxidation rate of margasites and pyrites in sterilised water was increased on introduction of *Th. thiooxidans* and *Th. ferrooxidans* [5]. The water in a fresh coal bed has a pH of 7 when Thiobacilli are absent but the acid content will rapidly increase as large amounts of Thiobacilli are detected [6]. Eighty percent of the sulphuric acid in coal mining in Scotland is reported to be associated with bacterial metabolism [7].

Temple *et al.* [6] point out that oxidation of pyrites passes through three stages:

$$FeS_2 + {}^3/_2 O_2 + H_2O \rightarrow FeSO_4 + H_2SO_4 \tag{1}$$

$$2FeSO_4 + {}^1/_2 O_2 + H_2SO_4 \rightarrow Fe_2(SO_4)_3 + H_2 \tag{2}$$

$$2FeS_2 + 2Fe_2(SO_4)_3 \rightarrow 6FeSO_4 + 4S \qquad (3)$$

Russian investigators [8] assumed the participation of *Th. ferrooxidans* in (1) and (3). Lalikova [9,10] made a fundamental study of the action of *Th. ferrooxidans* in the copper–nickel ore deposits in Kazakhstan and in the Kolsk peninsula. Bryner *et al.* [11] isolated *Th. ferrooxidans* from copper mining in the Bingem canyon in the USA and from copper ore deposits in Mexico.

The microbial cells of *Th. ferrooxidans* have a very complicated shell and many intercellular elements. They can live only in acid media; an environmental pH of 2.0–4.0 is optimal.

There are few papers concerning the participation of thiobacilli in corrosion processes on a metal surface. Beckwith [13] working in California was one of the first to study the acid corrosion of steel pipes arising from contact with sulphur being oxidised to sulphuric acid by means of thiobacilli. Chromium–nickel and chromium–molybdenum steels are sensitive to attack by thiobacilli contaminated environments [14].

2. Experimental

Hydrogeological characteristics and chemical analyses were determined during winter and summer for mine water from the No. 9 coal mine, 'Delyanka' and 'Uladimirskajal/2' in the Kizel field. Thus, determinations were made of concentrations of dissolved salt, Fe^{2+} and Fe^{3+}, pH and redox potential. The mine water of coal mine No. 9 had concentrations of Fe^{3+}= 12–992 ; Fe^{2+}= 24–300 and SO_4^{2-}= 1230–3680 mg L^{-1}; pH = 0.15–2.0; rH_2 = 19.4–34.5; T = 279...284K.

The enormous scatter in the data is readily explained by:

(i) differences in the source of the mine water inside the mine, i.e. through the galleries and levels, and

(ii) the seasonal changes in the activity of the microorganisms, the mine water being collected all the year round from the settling basins.

The mine water of coal mine Uladimirskaja 1/2 at the beginning of working had the following characteristics: Fe^{2+} concentration = 3.5 mg L^{-1}; Fe^{3+} = 0; SO_4^{2-} = 0; pH 6.15; rH_2 = 34.5; T = 277.5 K. But after 7 months working these values were, respectively: 14; 17; 3.6; 29; (temperature not reported), i.e. acidification had occurred as a result of ventilation of drifts, drainage and crushing of pyrites.

At the beginning of the experiment *Th. thiooxidans* was isolated from mine water by the use of Wachsmann and Joffe medium [3] with pH 4.2–4.5 and *Th. ferrooxidans* using the Silverman and Lundgren medium [15] at pH = 3.0–3.5. The pure strains of bacteria were obtained by repeated subculturing in Wachsmann medium.

For the corrosion tests mild steel (C 0.14–0.22%) specimens were used of dimensions $40 \times 20 \times 1.5$mm. These were polished with fine-grain electrocorundum paper, degreased and sterilised in ethanol. Weighed specimens were placed in tubes containing 50 mL of corrosive medium. The specimens were exposed for 4 months to natural mine water from coal mine No. 9, 'Delanka' with and without the presence of organic corrosion inhibitors, as described in Table 1. The specimens were removed at the end of the test, the black corrosion film removed by washing and the specimens were then dried and weighed.

The quantity of hydrogen absorbed by the steel during the corrosion process was measured using the vacuum heating technique.

3. Results and Discussion

Some results from the studies are given in Table 2 on p.404. The corrosion inhibiting efficiency, as represented by the mass loss, K, given in gm^{-2}/year of the compounds listed in Table 1, in natural mine water increased in the order IX (amino acridine derivative) < V–VIII (amides of diarylglycollic acids) < I–IV (aminothiophenol derivatives). Of the last group, the o-amino-β-cyano ethyl thiophenol (I) appears to be the most effective followed by the compounds II, III and IV.

Despite the reduction in corrosion rate (K) brought about by the thiophenols these compounds had little effect, or even stimulated, the hydrogen absorption by the steel, i.e. when compared to the values for specimens exposed to the natural (additive-free) mine water. We have drawn attention to this fact previously [16]. With respect to bactericidal action, the compounds I and II of the thiophenol group were more effective than III and IV—probably because of the presence of the –CN group in I and II. The amides of the diarylglycollic acids (V–VIII), with very little effect on corrosion rates, showed variable bactericidal action. Thus, the most effective for both thiooxidans and ferrooxidans types of bacteria was the dimethoxy derivative V which even at 0.05 mmol L^{-1} was better than the corresponding dimethyl derivative VII which acted only on *Th. ferrooxidans*.

Nevertheless, the dimethyl was better than the dimethoxy derivative as an inhibitor of hydrogen absorption.

In view of the bactericidal properties shown by the aminoacridine derivative(IX) towards *Th. thiooxidans* a series of 15 other derivatives were tested (in the form of hydrochloride salts). These were of the basic structure of IX but with halogen, methyl and methoxy substituents at various positions in the acridine ring system. Only the diethylaminoethyl ester of p-(2-methoxy-6-chlor-9-acridyl) aminobenzoic acid (X) was an effective bactericide for both types of bacteria, the effect being obtained at a concentration of 0.1 mmol L^{-1}.

The effects of some acridine derivatives on the pH and Fe^{3+} content of some natural mine waters were studied. In the absence of any additive the pH fell over a 10 day period from 3.3 to 2.3 while Fe^{3+} contents increased from 5.3 to 16.4 mgL^{-1}. In presence of 0.1 mmol L^{-1} of the acridine derivatives X, XI and XII the pH over this period stabilised at 3.1–3.2 and the Fe^{3+} concentration was zero.

The results of the present investigation provide evidence of the effects of the molecular structure and concentration of organic substances on inhibition of corrosion

Table 1. Structures of some of the organic compounds tested as inhibitors

Aminothiophenol derivatives	Amides of diarylglycollic acids
I: cyclohexane with S–CH$_2$–CH$_2$–CN and NH$_2$ substituents	V: bis(4-methoxyphenyl)glycollic acid amide with –CONH(CH$_2$)N(C$_2$H$_5$)$_2$
II: cyclohexane with S–CH$_2$–CH$_2$–CN and NH–COCH$_3$ substituents	VI: bis(4-fluorophenyl)glycollic acid amide with –CONH(CH$_2$)N(C$_2$H$_5$)$_2$
III: cyclohexane with S–CH$_2$–COOC$_3$H$_7$ and NH$_2$ substituents	VII: bis(4-methylphenyl)glycollic acid amide with –CONH(CH$_2$)N(C$_2$H$_5$)$_2$
IV: cyclohexane with S–CH$_2$–C(CH$_3$)=CH$_2$	VIII: diphenylglycollic acid amide with –CONH(CH$_2$)N(C$_2$H$_5$)$_2$

and bacterial activity. Effects on hydrogen absorption, such as shown by the thiophenols, are more complex and outside the scope of this paper.

Table 1 (continued). Structures of some of the organic compounds tested as inhibitors

Aminoacridine derivatives

IX

X

The differences in the resistance of the two kinds of thiobacillus towards the various organic compounds can be accounted for by the difference in penetrability of cell membranes [17]. *Th. ferrooxidans* may have cell membranes of low permeability which would be more resistant to the action of organic molecules present in the corrosive environment.

4. Conclusions

Mine waters in the Kizel coal mining area develop acidity and high concentrations of ferrous and ferric ions as a result of the action of *Th.thiooxidans* and *Th.ferrooxidans*. The conditions that arise lead to electrochemical corrosion of steel and hydrogen absorption by the metal.

Various nitrogen-containing organic substances were found to act as inhibitors of corrosion, as inhibitors of hydrogen uptake, and as bactericides. However, of 42 substances examined not one possessed all three of these properties although several possessed two of the properties.

References

1. A. N. Choroshavin, Thesis, Perm Univ., 1973.
2. A. Helbronner and W. Rudolfs, *Compt. Rend.*, 1922, **174**, 1378.
3. S. A. Wachsman and I. S. Joffe, *J. Bacteriol.*, 1922, **7**, 239.
4. A. R. Colmer, M. E. Temple, and M. Hinde, *Science*, 1947, **106**, 253.
5. A. R. Colmer, M. E. Temple, and M. Hinde, *J. Bacteriol.*, **1949**, 59, 317.

Table 2. Corrosion inhibiting and bactericidal effects of some organic compounds in natural mine water

Tested compound (see Table 1)	Bactericidal efficiency			Corrosion inhibiting efficiency		
	mmol L^{-1}	thio-oxidans	ferro oxidans	mmol L^{-1}	K gm^{-2}/y	H_2 ppm
I	0.1	–	–			
	0.05	–	–	10	174.0	33.5
II	0.1	–	–			
	0.05	–	–	10	181.2	47.4
III	0.1	–	–			
	0.05	–	–	20	204.0	31.1
IV	0.1	+	–			
	0.05	+	+	10	226.8	43.5
V	0.1	–	–			
	0.05	–	–	20	230.4	37.8
VI	0.1	–	–			
	0.05	+	+	20	290.4	40.0
VII	0.1	+	–			
	0.05	+	–	10	432.0	6.0
VIII	0.1	+	–			
	0.05	+	+	5	415.2	32
IX	0.1	–	+			
	0.05	–	+	5	495.6	9.4
natural mine water					392.4	29.2
					388.8	43.5
					490.5	29.7

6. K. Temple, and E. Delchamps, *J. Appl. Microbiol.*, 1953, **1**, 255.
7. D. Ashmeed, *Colliery Guard.*, 1955, **190**, 694.
8. J. I. Karavajko, S. I. Kuznetsov, and A. I. Golomzik, *Microorganisms in Metal Recovery from Ores*, Moscow, 1972.
9. N. N. Lalikova, *Mem. Inst. Microbiology of the Soviet Union*, 1961, **9**, 134.
10. N. N. Lalikova, *Microbiology* (in Russian), 1961, **30**, (1) 135.
11. L. Brynner, J. Beck, O. Davis, and O. Wilson, *Ind. Eng. Chem.*, 1954, **46**, 2587.
12. A. A. Avakjan and J. I. Karavajko, *Microbiology*, 1970, **39**, 855.
13. T. O. Beckwith, *J. Am. Water Works Assoc.*, 1941, **33**, 147.
14. Z. A. Reinfel'd, *Microbiology*, 1939, **8**, 1.
15. M. Silverman and O. J. Lundgroon, *J. Bacteriol.*, 1959, **73**, 3.
16. S. M. Beloglazov, *Hydrogenation of Steel During Electrochemical Processes,* Univ. Leningrad Press, 1975, 411p.

Index

Accumulation
    of biomass in marine MIC 206
    of microorganisms at surfaces 18
Admiralty brass
    corrosion and biofouling of, in fresh water 262
Aerobic bacteria
    in mains water 330
    in oilfield water 296
Aggregation see Accumulation
Alginic acid (and alginates)
    conformational changes in 38
    in gel formation 9
    light scattering measurements of 30
    in model biopolymer formulation 6, 67, 86
    solution behaviour of 30
    surface tension properties of 21
Alloying elements
    in steels, effect on MIC 119
Aluminium brass, in sea water
    cathodic protection of 303
    corrosion of 243, 302
    microfouling of 244
    nature of deposits on 255
    sulphide ions, effect of, on 311
Amides (of diarylglycollic acids)
    as corrosion inhibitors and biocides 400
Aminoacridine derivatives
    as corrosion inhibitors and bactericides 400
Aminothiophenols
    as corrosion inhibitors and bactericides 400
Anodic polarisation tests
    of steels in culture medium 122, 193
Antifouling treatments
    using chlorine dioxide 304

    hydrochloric acid 304
    hypochlorite 304, 376
Aquifers
    as water sources 355
Archaeological iron nail
    corrosion products from 338
Artificial Sea water see Sea water
Austenitic stainless steels
    corrosion and biofouling of, in fresh water 262
    in sea water 302
    localised corrosion of, in river water 107
    MIC tests with 121

Bacteria
    effect of filtration on 325
    enumeration of, as affected by CP 369
    numbers of, on stainless steels 324
Bacterial adhesion
    experimental procedure 125
    to steels 120
Bacteria species see separate entries
Biocides
    amides of diarylglycollic acids as 400
    amino acridine derivatives as 400
    assay of 182
    diffusion of, into biofilm 183
    factors affecting use of 317
    factors affecting sensitivity of 181
Biocorrosion
    in groundwater engineering 360
Biofilm see also Biofouling, Exopolymers
    effect of acidification on 257
    on aluminium brass, in sea water 243
    artificial(synthetic) 64, 89
    bacterial action on 181
    in cast iron water mains 329

Biofilm *(continued)*
    collection and treatment 206
    contents of 85, 158
    from copper pipes 64
    deterioration of metals and 158
    monitoring of, in sea water 301
    monitoring probe 390
    monitors for 387
    occurrence in pipe work 86, 158
    in oilfield pipelines 315
    physicochemical properties of 85
    in potable water 322
    preparation of, produced by microorganisms 91
    in sea water, characterised by dissolved oxygen 216
    in sea water, developed on gold electrodes 213
    on stainless steel 112, 203, 322
    on steels 122
Biofouling *see also* biofilm
    definition of 359
    determination of long term rates of 262, 271
    in ground water engineering 357
Biomembranes
    in MIC of copper 8
Biopolymers *see also* Biofilms, Biomembranes, Exopolymers
    contact angle, measurements of 36
    from corroding copper pipes 68
    physical behaviour of 17
    synthesis of 67, 89
Bioprobes
    in bacterial counts 296
Brass
    corrosion and biofouling of in fresh water 252
BRITE/EURAM Project 15

Carbon source
    importance of, in production of exopolymer 51, 61

Carbon steel
    MIC tests with, in bacterial culture 121
    corrosion of, and biofilms on, in fresh water 252
    MIC of valve in water 276
    corrosion tests in coal-mining water 399
Cast iron
    pipes, MIC of 328
Cathodic protection (CP)
    of aluminium brass 303
    and microbial action 370
    need for revision of criteria of 368
    in presence of sulphate reducing bacteria 367
    of stainless steel in sea water 303
Cation selectivity
    of biopolymers 8, 83, 87, 102
Cellulose
    as monopolysaccharide 17
Chemostat *see* Continuous culture
Chloride ions
    and biopolymers 101
    transport of in exopolymers 88
Chlorination *see also* Antifouling treatments
    effect of, on slime forming bacteria 374
    measurements of residual chlorine in 374
    shock vs intermittent vs continuous in lab tests 381
Chromium
    effect of, on growth of *D. vulgaris* 122
    effect of, on hydrogenase activity 171
    ions in biofilm, analysis of 323
Coal-mining waters
    MIC of steel in 398
Composites
    fibre reinforced polymer, MIC of 143
    environmental degradation of 143
Contact angle measurements
    of metal surfaces and biopolymers 36
    tabulated data for surface parameters from 40
Continuous culture 50

Cooling tower
    laboratory model for chlorination studies 375
Cooling water
    MIC in 261, 276, 302
Copper
    background to MIC of 3
    corrosion reactions on biofilm covered 101
    effect of, on growth of *D. vulgaris* 122
    in test loops 159
    marine corrosion of 135
    surface and near-surface chemistry of 4
    types of corrosion of 49
Copper oxide
    film thickness by ellipsometry 32
    surface structure of 32
Corrosion products
    on archaeological iron nail 338
    microbiology of 331
    sampling from 330
Corrosion rates
    of aluminium brass from polarisation resistance 245
    of carbon steel from weight loss 263
    of carbon steel in relation to season 266
    from weight loss (in CP tests) 369
Critical micelle concentrations
    at solid surfaces 19
Cyclic voltammetry
    in studies of MIC of copper 92
    results from bare copper 92
    results from copper coated with biopolymer 94

Desulphovibrio desulphuricans
    growth of 189
    in oilfield water 296
Desulphovibrio longus
    in oilfield water 296
Desulphovibrio species
    unclassified strains, in oilfield waters 296

Desulphovibrio vulgaris
    attachment to steels 110
    culturing of 120
Diatoms
    on stainless steel in reservoir water 273
Donnan phenomena 9

Electrochemical biofilm monitoring probe 390
Electrochemical effects
    as affected by sulphide contamination 304
    of marine biofilms on stainless steels 208
Electrochemical impedance spectroscopy
    in MIC studies, discussion of 392
Electrochemical measurements *see* Potential-, Polarisation-, Potentiodynamic-, Potentiostatic-
Electrochemical noise analysis
    for detection of pitting 318
    possibilities for MIC monitoring 392
Electrochemical sensor
    of water in fuel 347
Electrohydrodynamical impedance diagrams
    for stainless steel in sea water 212, 218
Electrolyte concentration
    importance of, in membrane measurements 74
Ellipsometric measurements 22
Energy dispersive X-ray (EDX)
    of corroded carbon steel valve 284
    of corroded cast iron pipe 331
Enteromorpha spp.
    use of, in production of biotic H_2S 234
Environmental scanning electron microscope (ESEM)
    in study of marine MIC of copper–nickel foils 136
    in study of MIC of composites 145
Enzymes *see also* Hydrogenase
    review of participation in MIC 170
Epifluorescence microscopy
    to determine iron-oxidising bacteria 330

European coastal waters
   effect of biofilms on stainless steels in 198
Exopolymers
   composition of 17
   'living crevice' formed by 385
   preparation of 22, 86
   production of 18, 51
Exopolysaccharides
   in binding of copper ions 13
   as measure of biofilm mass 199
   production of 9
   solution studies of 28
Extracellular polymers *see also* Exopolymers
   produced by *D. vulgaris* 126
   produced by *D. Oceanospirillum* 135

Fast atomic bombardment-mass spectroscopy 23
Ferritic stainless steels
   MIC tests with 121
Fibre reinforced polymer composites
   MIC of 143
Flavobacterium
   in biofilm on stainless steel in potable water 326
   present in prepared exopolymer 22
Flactobacillus major
   in water from copper pipes 52
Flow *see also* Test loops
   effect of, on bacterial number on stainless steel 326
Fresh water
   corrosion and biofouling of stainless steel in 262
Fuel
   sensor for water in 349

Gallionella
   on cast iron water mains 332
Galvanic coupling tests
   in sea water 204
Gel composition, location etc.
   importance of, in membrane property measurements 71
Glutamate medium
   and *oceanospirillum* 138
Glycolipids *see also* Lipids
   as components of biofilms 17
Gibbs free energy of aggregation of 21, 24
   interfacial tension of 26
   molecular structures of 24
   surface tension properties of 21, 24
Glycoproteins
   as components of biopolymers 17
Gold electrodes
   marine biofilms developed on 213
Groundwater engineering
   biocorrosion in 358
   biofouling in 359
   construction in 357
   corrosion in 354, 358
   equipment in 357
   materials in 34
   operation and maintenance of 358

Heat exchange coefficient
   determination of, in sea water loop 303
Heat transfer
   effect of sulphide contamination in sea water on 310
Heteropolysaccharide 17
Histochemical
   staining procedures 8
Hydrogen
   absorption by steel in coal mining waters 400
   embrittlement 233
   inhibitors of 400
   permeation of, through metals 233
Hydrogen peroxide
   possible role of, in MIC
Hydrogen sulphide
   abiotic, production of 234
   biotic, production of, from algae 234
   biotic vs non-biotic 234

effect on corrosion 233
effect of corrosion products on corrosion by 238
levels in SRB active conditions 238
occurrence in loop tests with sea water 303, 309
production from thiosulphate reducing bacteria 297
Hydrogenase
  activity, quantified by gas chromatography 171
  role of in MIC of steel 171
Hydrological characterisation
  of coastal test sites in MIC studies 202

Inhibitors
  tests with, in coal mining waters 400
Ion transfer in exopolymers 88
Iron *see also* Steel
  effect of, on growth of *D. vulgaris* 123
  on hydrogenase activity 171
  ions in biofilm, analysis of 323
Iron oxides and hydroxides
  effect of microflora on transformations of 152
  layers contributing to localised corrosion 334
Iron oxidising bacteria
  on carbon steel in reservoir water 268
Iron sulphide
  and passivation 334
  synergism with cathodic protection 373

Kinetic model
  of biocide action 181

Laser profilometry
  for internal inspection of tubing (reference to) 392
Light scattering measurements 21, 28, 30

Lipids *see also* Glycolipids
  in membrane systems 14
Localised corrosion *see also* Pitting corrosion
  intergranular, of chromium-nickel under biofilm 140
Low alloy steels
  MIC tests with 121

Macrophytes
  occurrence in open reservoirs 263
Mains water *see* Potable water
Marine corrosion
  and MIC of copper 135
  and MIC of stainless steels 198
MAST-2 project 208
Membrane potentials 11
  of biopolymers 77
  calculation of 12
  measurements of 65, 68
  of synthetic polymers 77
Membrane properties of polymers 8, 64
Metabolites
  chemical analysis for 385
Metallic ions
  effect of, on growth of *D. vulgaris* 122
Model
  kinetic, for bactericidal action 181
  mathematical, of oxygen reduction, effect of biofilms 226
  membranes 8
  of MIC of copper, cyclic voltammetry for 6
Molecular structures
  of some organic inhibitors of MIC 403
Molybdenum
  effect of, on growth of *D. vulgaris* 122
  effect of, on hydrogenase activity 172
  ions in biofilm, analysis of 323
Monopolysaccharides
  types of 17
Monitoring
  of biocorrosion in groundwater engineering 368

Monitoring *(continued)*
    of biofilm growth
        in sea water 301
        on line 387
        with probe 390
        recommendations for 319
        techniques for in MIC, discussion of 385
Mossbauer spectroscopy
    in study of iron oxides in soil corrosion 152
    in study of corrosion products on iron nail 339

Nickel
    in biofilm, analysis of 323
    effect of, on growth of *D. vulgaris* 122
    effect of, on hydrogenase activity 171

Oilfield waters
    MIC in 293, 314
Optical density measurements
    used in studies of bacterial growth 122
Oxygen mass transport
    for biofilm determination 223

Peaty water
    in tests with stainless steel 322
Peracetic acid
    as sterilising agent 322
Performance monitoring
    possibilities for indicating MIC 392
Permeability *see also* Cation selectivity
    of exopolymeric substance 87
Pitting corrosion
    of stainless steels 107
Planktonic phase
    polymers from 9
Platinum electrode
    voltammograms on, coated with biofilm 101
Podsol
    corrosion in 338

Polarisation curves *see also* Anodic-, Potentiodynamic-, Potentiostatic-
    for aluminium brass, in marine MIC 245
    for copper–nickel, in marine MIC 136
Polarisation resistance
    for corrosion rates of aluminium brass 245
    in monitoring of MIC 389
Polarisation tests
    compared with loop tests 159
    to monitor biological activity 303
Polarising potential
    effect of, on biofilms 159
Polysaccharides *see also* Biofilm, Exopolysaccharides
    as components of biopolymers 17
    discrimination of 9
    as extracellular polymers in biofilms 8
Potable water *see also* Fresh water
    biofilm on stainless steel in 322
    MIC on cast iron in 328
Potential measurements
    in monitoring MIC (discussion of) of steel in bacterial culture 192
    of stainless steels in sea water 202, 219
Potentiodynamic curves
    for copper in NaCl 6
    for mild steel in various media 200
Potentiostatic tests
    in CP studies 368
    of copper in potable water 159
Power plant
    inspection of water systems in (discussion of) 384
    MIC in water systems of nuclear 261, 276, 375
    MIC in water systems of thermal 302
Protein
    in excretion of exopolymers 18
Pseudomonas aeruginosa
    attack of graphite fibres by 143
    in production of biofilm polymer 9, 18
    in preparation of exopolymers 22
Pseudomonas fluorescens
    in MIC of composites 144

Pseudomonas spp.
    in biofilm on stainless steels in potable water 326
    on cast iron water mains 332
    on carbon steel in reservoir water 268
    in oilfield water 296

Redox potential
    in coal mining waters 399
    in MIC studies (discussion of) 388
Risk assessment matrix
    for biological contribution to pipe corrosion 334
River water
    MIC of stainless steel in 108
    as source of cooling atomic reactors 261
    water quality change with seasons 263
Rotating disc experiments
    in characterising biofilms 216, 224

Sea water
    aggressivity, natural vs artificial towards aluminium brass 254
    artificial
        in loop tests with aluminium brass 244, 247
        in production of biotic H_2S 234
    detection and characterisation of biofilms in 211
    MIC tests with aluminium brass in 247, 302
    MIC tests with copper in 135
    MIC tests with stainless steel in 198, 302
Seasons
    effect of, on growth of biofilms
        in fresh water 261
        in sea water 223
        and pH and water quality of reservoir water 263
Scanning electron microscope (SEM)
    for examining biofilms 122, 136, 237, 323
    for examining corroded carbon steel valve 279

deposits on aluminium brass 249
Sensor
    electrochemical, for water in fuel 349
Slime forming bacteria
    response to chlorination 375
Soil corrosion
    and archaeological iron nail 339
    and cathodic protection 369
    and iron oxide transformations 152
Sphaerotillus
    on cast iron water mains 332
Stainless steels *see also* Austenitic-, Ferritic-, Super austenitic-, Super duplex-
    bacterial loads on Type 304 vs Type 316 324
    biofilms on, in potable water 322
    corrosion and biofouling of, in fresh water 262
    effect of marine biofilms on corrosion of 193, 223
    galvanic coupling tests with 201, 204
    localised corrosion of, in river water 107
Steels *see* Carbon-, Low alloy-, Stainless-
    BS4360 50D, effect of H_2S on 234
Sulphate reducing bacteria *see also Desulphovibrium* spp.
    on carbon steel in reservoir water 268
    on cast iron pipes 331
    and cathodic protection 367
    Indonesia strain of 170
    in mains water 329
    mechanism of action of, in MIC 169
    in nuclear reactor water systems 273
    in pipeline water 289
    Portsmouth strain of 170
Sulphide producing bacteria
    in MIC of pipeline 293
Superaustenitic stainless steels
    effects of marine biofilms on 200
Superduplex stainless steels
    effect of marine biofilms on 200
Surface activity
    of exopolymers 18
Surface analysis techniques *see also* ESEM, SEM, TEM, XANES

Surface studies
    of cast iron 331
    of copper 18, 135
    of steel 122
Surface tension measurements 21, 23

Test loops
    for MIC of aluminium brass 247
    stainless steel 108
    results from, compared with polarisation tests 159
Thermogravimetric analysis
    of corrosion products on archaeological iron 339
Thiobacillus ferrooxidans
    in coal mining waters 399
    in MIC of buried pipes 367
    in MIC of composites 144
Thiobacillus thiooxidans
    in coal mining waters 399
    in MIC of buried pipes 368
Thiosulphate reducing anaerobes 296
Titanium
    corrosion and biofouling of, in fresh water 262
Transference number measurements
    in biofilms 65, 80
Transmission electron microscopy
    in marine MIC of copper–nickel 136

Uronic acids
    in exopolysaccharide layers 14
UV fluorescence microscopy
    in study of biofilms 158

Vibrio genera
    in oilfield water 296
Vibrio spp.
    in biofilm on stainless steel in potable water 326
    on cast iron water mains 332

Water *see* Coal mining-, Cooling-, European coastal-, Ground-, Mains-, Oilfield-, Peaty-, Potable-, River-, Sea-

Water layering
    with copper, water and EPS 44
Weight loss measurements
    for corrosion monitoring 387
    of steel, effect of SRB on 191
    of steel in fresh water 263
Weld line
    bacteria and, on stainless steel 326
Wettability
    of interfaces in contact with microorganisms 19
Work of adhesion
    in wettability of surfaces 19

Xanthan
    folding of chains of 21
    as heteropolysaccharides 17
    light scattering measurements of 28
    in model biopolymer formulation 6, 67, 86
    multilayer films of 22
    surface tension properties of 21
Xanthomonas comprestis
    in preparation of exopolymers 22
    in production of exopolymers 18
X-ray absorption near edge structure (XANES)
    in marine MIC of copper 135
X-ray diffraction analysis
    of corrosion products on carbon steel 268
    of corrosion products on iron nail 339
    of corrosion products on deposits on aluminium brass 249
X-ray reflectivity measurements 21
    in surface structure of copper and its oxides 32

Zimm plots
    in light scattering of xanthans 28
Zisman parameter
    for characterising energy of surface 20